Gun Digest Book Of HANDGUN RELOADING

By
DEAN A. GRENNELL
And
WILEY M. CLAPP

With
KEN HOWELL
And
MIKE VENTURINO

DBI BOOKS, INC.

ABOUT OUR COVER

Hornady has been, if you'll pardon the pun, one of the most "progressive" outfits in the handloading industry. And their products grace the covers of the Gun Digest Book of Handgun Reloading.

Seen on the front and back covers is one of the newest members of Hornady's growing family of reloading gear. It's called the Pro-Jector progressive reloading press. The Pro-Jector is a 5-station, auto-indexing press that features auto-priming, auto-primer shut-off, one set of dies, angled O-frame design, a brass-kicker, cartridge catcher, offset operating handle and Power-Pack™ Linkage. And, if all that's not enough, how about a completed ammo output of a minimum of 300 rounds an hour? Sounds good to us too!

Lastly, you'll see a nice selection of Hornady reloading accessories that includes case lube, powder scale, bullets and handbook. Good equipment. Photo by John Hanusin.

Publisher
Sheldon Factor

Editorial Director
Jack Lewis

Production Director
Sonya Kaiser

Art Director
Denise Comiskey

Associate Artists
Gary Duck
Paul Graff

Copy Editor
Shelby Pooler

Production Coordinator
Pepper Federici

Photo Services
Kelley Grant
C'est DAGuerre Labs

Lithographic Service
Gallant Graphics

Produced by
GALLANT/CHARGER PUBLICATIONS

Copyright MCMLXXXVII by DBI Books, 4092 Commercial Ave., Northbrook, IL 60062. All rights reserved. Printed in the United States of America. No part of this book may be reproduced, stored in a retrieval system or transmitted in any form or by any means, electronic, mechanical, photocopying, recording, or otherwise, without the prior written permission of the publisher.

The views and opinions expressed herein are not necessarily those of the publisher and no responsibility for such views will be assumed.

Since the authors, editors and publisher have no control over the components, assembly of the ammunition, arms it is to be fired in, the degree of knowledge involved or how the resulting ammunition may be used, no responsibility, either implied or expressed, is assumed for the use of any of the loading data in this book.

Arms and Armour Press, London, G.B., exclusive licensees and distributors in Britain and Europe, New Zealand, Nigeria, So. Africa and Zimbabwe, India and Pakistan; Singapore, Hong Kong and Japan. Capricorn Link (Aust.) Pty. Ltd. exclusive distributors in Australia.

ISBN-0-87349-014-2

Library of Congress Catalog Card Number 87-71766

CONTENTS

ACKNOWLEDGEMENTS 5

Chapter One: BASICS OF HANDGUN AMMUNITION RELOADING 6
A Brief Discussion of Pertinent Steps in
The Reloading Process — *Dean Grennell*

Chapter Two: GOOD ENOUGH ISN'T 26
Selecting Tools. Advice, Both General And
Specific, On Choosing Your Tools — *Ken Howell*

Chapter Three: SPEEDY RELOADING 36
Examining The Progressive Presses; Tools That
Ease The Burden Of Production — *Mike Venturino*

Chapter Four: ORGANIZE YOUR RELOADING 48
Making The Reloading Process Easier, Safer And
Quicker: Tools And Techniques — *Wiley Clapp*

Chapter Five: THE SHORT, VIOLENT LIFE OF A LOAD 66
Handgun Ballistics: What Happens From Hammer
Fall To Target Impact And All Between — *Ken Howell*

Chapter Six: TESTING THE HANDLOAD 82
Good Equipment And Common Sense Make The Test
Results More Meaningful — *Mike Venturino*

Chapter Seven: COMMERCIAL BULLETS FOR HANDLOADERS 90
Surveying The Expanded Field Of Pistol Bullets
From Commercial Suppliers — *Mike Venturino*

Chapter Eight: HANDGUN BULLETMAKING 100
Producing Your Own Bullets For Versatility
And Economy By Casting And Swaging — *Dean Grennell*

Chapter Nine: POWDERS FOR HANDGUN RELOADERS 118
Checking Out The Wide Variety, with Comments
On Best Bets For Various Needs — *Mike Venturino*

Chapter Ten: THE BUCK STARTS HERE **126**
    The Ways And Means Of The Indispensable
    Primer: How They're Made & What They Do — *Ken Howell*

Chapter Eleven: HANDLOADER'S GOLD **136**
    Brass: The Cartridge Case, From Manufacturer
    To Handgunner,...and Even Wildcatter — *Ken Howell*

Chapter Twelve: DOUBLE DATA **144**
    Up-To-Date Testing Of Load Data In A Pair
    Of Today's Handguns For Each Caliber — *Wiley Clapp*

Chapter Thirteen: RELOADING THE T/C CONTENDER **180**
    It Has The Most Omniverous Appetite For
    Ammo of All The World's Firearms — *Dean Grennell*

Chapter Fourteen: HANDLOADING THE OLD-TIMERS **232**
    Lots Of Elderly Cartridge Handguns Can
    Get Back To Work With Handloads — *Mike Venturino*

Chapter Fifteen: SPECIAL-SITUATION RELOADING **240**
    Winding It Up With Observations On What
    To Do With Those Knotty Problems — *Dean Grennell*

WARNING: This book contains suggested load data for use in the reloading of centerfire metallic cartridges. Developed under controlled conditions, no obvious signs of excessive pressure were noted when fired in the test guns, unless otherwise noted. Loads were developed in accordance with established procedures of reloading safety.

HOWEVER — as neither the publisher, authors nor production personnel have any control over equipment, components and techniques used in reloading by others, they do not and cannot assume any liability, either expressed or implied, for any injuries, damages or other undesirable consequences arising from or alleged to have arisen from use of this load data by others. Any and all such use is clearly and specifically at the risk and discretion of the reloader and/or the shooter who fires such reloads.

ACKNOWLEDGEMENTS

THERE are so many people who contribute so much to any of these DBI books that it's difficult for the guy who is termed the "author" to properly thank all of them. Nevertheless, it would be grossly improper to not give it a try.

When you work with the production people as closely as we do at Gallant/Charger, you come to have vast respect for their problems and products. So for artists Gary Duck, and Paul Graff, Kelly Grant in Photo Services and Pepper Federici and Shelby Harbison in typesetting, we have the greatest respect. Thanks for your efforts! But the one person who is most responsible for the character and tone of this book is unquestionably the effervescent Denise Comiskey. She bears the title of Art Director — and the burden of making it look right.

Editing is the province of Jack Lewis, who can almost do it in his sleep after so many years in the business. But a "Lewis-ized" manuscript always reads a little smoother then the author wrote it and they are sometimes saved from oblivion by virtue of his attentions.

You can't write a book about handloading without the cooperation of the reloading industry and we got the full support of all of the major companies. Speer, Sierra and Hornady contributed lots of bullets; Powder came from Accurate Arms, Hercules, IMR (formerly DuPont), Hodgdon and Winchester; Brass and primers from Remington, Olin-Winchester, Federal and CCI-Omark. Actual tools were from RCBS, Hornady, Lyman, Redding-Saeco, C-H, Lee Precision and Ponsness-Warren.

SO LONG, DAVE

WE AREN'T much for bleary-eyed testimonials, but we can't let a particular event go unrecorded. For many long years, one of the most respected and loved guys in the business was Dave Andrews, Marketing Services Manager for Omark. Dave retired this year and he will be sorely missed.

The nation's reloaders owe Andrews more than they realize. He was the guiding force behind many, many innovations in equipment and products. Over the years that he was up there in Lewiston, Dave Andrews was an active participant in the phenomenal growth of the reloading industry. Throughout it all, he was the most genial, easy-to-deal with guy that you could possibly imagine.

So...enjoy your retirement, Dave. We are all going to miss you.

CONTRIBUTING AUTHORS:

KEN HOWELL

KEN HOWELL has been a pistol competitor and gun collector, forester and wildlife biologist in Montana and Alaska, technical editor-writer for ordnance and aerospace companies, pastor and gunsmith, and Editor of *Handloader* and *Rifle* magazines. A Navy photographer during the Korean War, he has testified in more recent times as an expert witness on firearms and handloading. He has hunted widely in the United States and abroad. Roping brown bear on Kodiak Island, he was mauled by a trophy-sized brownie. Currently, he resides in Arizona where he runs a computerized editing service, writes gun articles and computer software.

MIKE VENTURINO

IN THE SPAN of a relatively few years, Mike Venturino has become one of the nation's most respected and prolific firearms writers. Originally from the east, Mike settled solidly into a Montana lifestyle from which it is unlikely that he will ever be uprooted. His gun interests are generally wide, but his personal preferences are for the old-timers, the guns of the American frontier. At the present time Mike contributes to both *Rifle* and *Handloader* magazines and to DBI books. He is a Contributing Editor, *Guns & Ammo* magazine.

CHAPTER 1

BASICS OF

A Brief Discussion Of Pertinent Steps In The Process

HANDGUN AMMUNITION RELOADING

AS ORIGINALLY manufactured, the neck area of a handgun cartridge grips the adjacent walls of the bullet in a tight friction fit. The case mouth may be crimped into a mating groove of a lead bullet — the *crimping groove* — or into a serrated groove of a jacketed bullet — the *cannelure*. This is done to serve several useful purposes. It prevents the bullet from falling out of the cartridge case, or from remaining lodged in the bore if the cartridge is extracted without being fired. It also prevents the bullet from being driven back into the case mouth under stresses of going through the action of the given firearm. Such deeper seating would result in higher pressures and erratic ballistics.

At the instant of firing, the walls and mouth of the brass cases expand outward, making firm contact with the walls of the chamber, to prevent the rearward leakage of high-pressure powder gas. The term for such a sealing action is *obturation*.

If we're going to reload the .357 magnum, it takes a #6 Pacific shell holder: the little black gizmo just to the left of the empty case near the press ram on the facing page. In the photo at right, the shell holder has been slipped into the mating flange on the ram and a .357 magnum case is positioned in the shell holder.

GUN DIGEST BOOK OF HANDGUN RELOADING

While the exact makeup of a set of reloading dies tends to vary somewhat from maker to maker, this set of Pacific dies, from Hornady, is fairly typical. At left is the full-length resizing die and its sole function is to reduce the OD of the case to slightly less than the desired diameter. The next two items are the expander/decapping stem and expander/decaper die body, disassembled here for clarity of illustration. In use, the stem is turned into the die body, adjusted and locked in place with the locking ring. The two items at right are the bullet seating stem and seater die body, disassembled for the photo. The seating stem is designed to seat most bullet designs without distorting the bullet nose, but occasional bullets may require a specialized seating stem.

As the bullet leaves the muzzle and internal pressures equalize with those of the surrounding atmosphere, the brass case — being somewhat elastic — tends to spring back toward its original dimensions. Even so, the inside diameter (ID) of the case mouth will be somewhat greater than it was before firing. The neck no longer will grip the base of a bullet of the given diameter, as it did in the original load. The base of a bullet of suitable diameter for reloading the case usually can be inserted into the neck of the fired case and withdrawn with little difficulty.

Certain steps are thus necessary in preparing the fired cartridge case for reloading: The spent primer must be forced out of the primer pocket before a fresh primer can be seated and the entire case must be reduced to a diameter permitting it to be chambered without difficulty. At the same time, the case mouth must be brought back to an ID that provides a frictional grip upon the base of the bullet, similar to that of the original load and for the same purposes.

Prior to the actual reloading, the fired cases must be

inspected for such defects as split sides, split necks, snagged necks, foreign matter within the case and other irregularities that may occur. Certain calibers of empty cases, received as a mixed lot of once-fired brass, may include a percentage of cases having the Berdan-type primer. If you look down into the mouth of a Berdan-primed case, you will note two smaller, off-center flash holes, rather than the larger, centered single hole of the Boxer-type primer. Only Boxer-primed cases — in common use in the USA — are suitable for reloading by the techniques described here.

The Berdan cases should be segregated and disposed of. If you attempt to deprime a Berdan case, using standard loading dies, you are apt to break the decapping pin, if undue force is exerted on the press handle.

The exact design of die sets varies from maker to maker. In some die sets, the first die resizes the case, full-length and decaps (deprimes) it at the same time. In other sets, the first die merely resizes the case, full-length. In such sets, the decapping pin will be located on the end of the expander plug in the second die. The exact state of things makes

Here, the ram has been run to the top of its stroke, the full-length resizing die has been turned down to make snug contact with the upper surface of the shell holder and the locking ring on the die body has been snubbed down to hold the die in the desired adjustment. Now, with a fired case inserted in the shell holder at the bottom of the ram stroke, the handle is operated and the case is resized from mouth to rim. The press, incidentally, is fastened to the small piece of two-inch plank by means of bolts, with the heads and washers countersunk to be flush with the lower surface of the plank. This makes a handy approach for mounting to any flat surface by means of C-clamps. A small snap-top plastic box is fastened to the plank to keep track of small parts.

Pacific
DURACHROME
DIES
PART NO.
SERIES
USES SHELL HOLDER
040762
II
#6
DESCRIPTION
38-357-357 MAX CARB
WRITER'S CLAMP

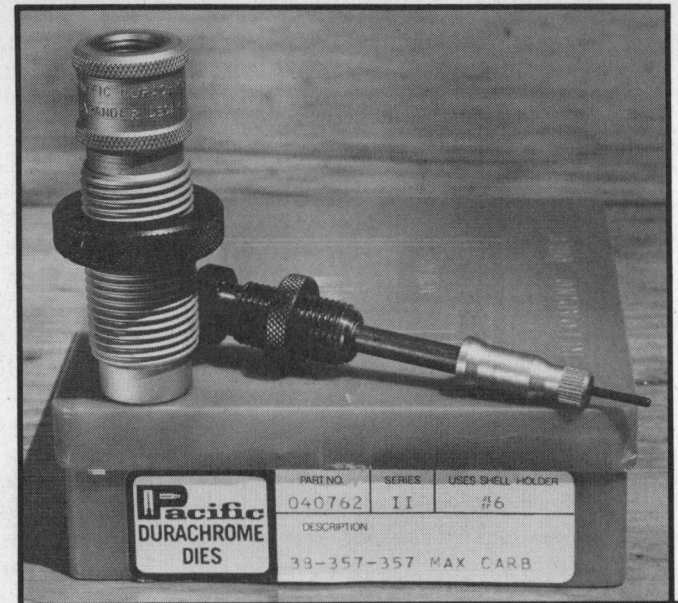

A closer look at the expander/decapper die shows the decapping pin, held in the expander plug. By passing the case mouth over the plug, the mouth is expanded to the desired ID, the case is decapped and the mouth is flared slightly by the tapered area at the top.

Facing page: Press is now set up for the first step, case resizing. Perform the operation on each case in turn. Right, the die set includes a tungsten carbide insert in the resizing die, thus requiring no lube on the cases for resizing. The same set can be used for reloading .38 Special, .357 magnum or .357 Remington Maximum.

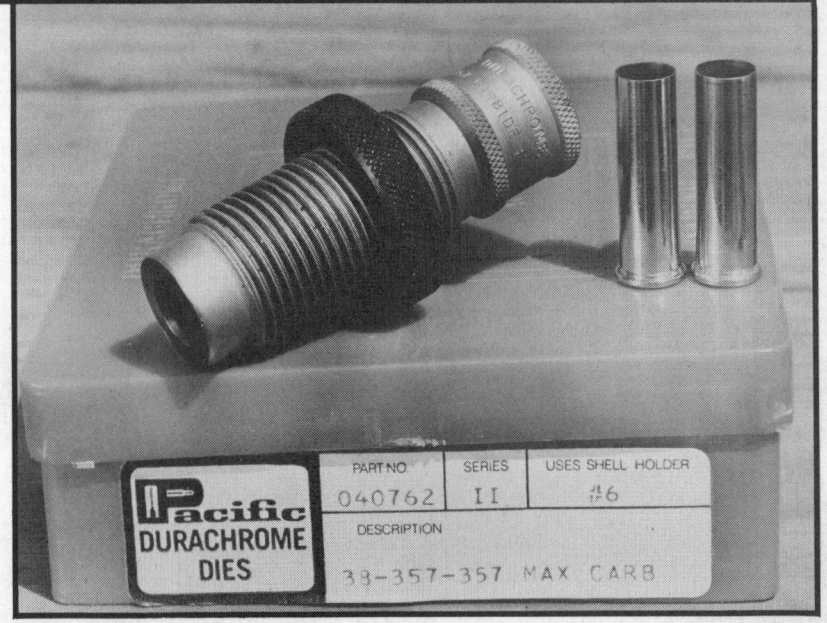

little difference, beyond complicating the description of the procedure.

If you are using a standard set of reloading dies, with the resizing die made of hardened steel, it will be necessary to apply case-resizing lubricant to the outer walls of the cases before resizing. If this vital step is neglected, the bare metal case wall will gall against the inner surfaces of the resizing die, depositing bits of case metal that cling stubbornly to the die surface. Small inclusions of dust and grit will embed into the inner die surfaces and, rather quickly, the die will be ruined beyond much hope of salvage. At the same time, cases may seize within the die, held so stubbornly that the case head may be wrenched off in a vain effort to pull the case back out of the die. Obviously, use of sizing lube is the better way to go.

Resizing lubes are highly specialized compounds and just any old oil will not perform properly. Suitable lubes should by applied sparingly. It can be done by putting just a bit on the tips of a finger or two and rubbing. Uninked rubber stamp pads are obtainable through suppliers of reloading equipment. A small amount of lube can be applied to such a pad, rubbed in well and the cases can be rolled back and forth across it with the palm of a hand to apply the necessary lubrication.

Somewhere down the chain of reloading operations after resizing, the lube must be removed from the outer case walls of the cartridges. In the normal firing cycle, the expanding case grips the chamber wall by friction. Thereby taking some of the strain off the recoil plate or standing breech of the gun. Leaving the lube in place puts unacceptably high stress upon the action of the gun.

Full-length resizing dies with inserts of tungsten carbide or titanium carbide are available for several of the straight-sided handgun cases, or for those with a slight taper, such as the 9mm Luger. With such carbide sizing dies, it is not necessary to apply lubricant to the cases, likewise eliminating the need to remove the lube after sizing. Although carbide dies and die sets that include a carbide resizing die cost more in the first place, they tend to conserve a lot of time, effort and bother, thereby more or less paying for themselves, if any great quantity of reloading is done.

For several bottleneck cartridge cases, neck-size-only (NSO) dies are available as an option. They can be used if the case was fired previously in the same gun in which the reload will be fired. Such cases require but little if any lubrication.

On bottleneck cases, the expanding plug is carried on the end of the decapping stem and restores the case neck to the proper ID as the case is drawn down out of the die. It is helpful to apply a sparing amount of case lube to the inner

Primer seating arm is installed in the ram and, when decapping/expanding, the pin punches out the spent primer, which drops down a hollow in the center of the ram to be deflected into the white plastic box that clips to the rear section of the press frame.

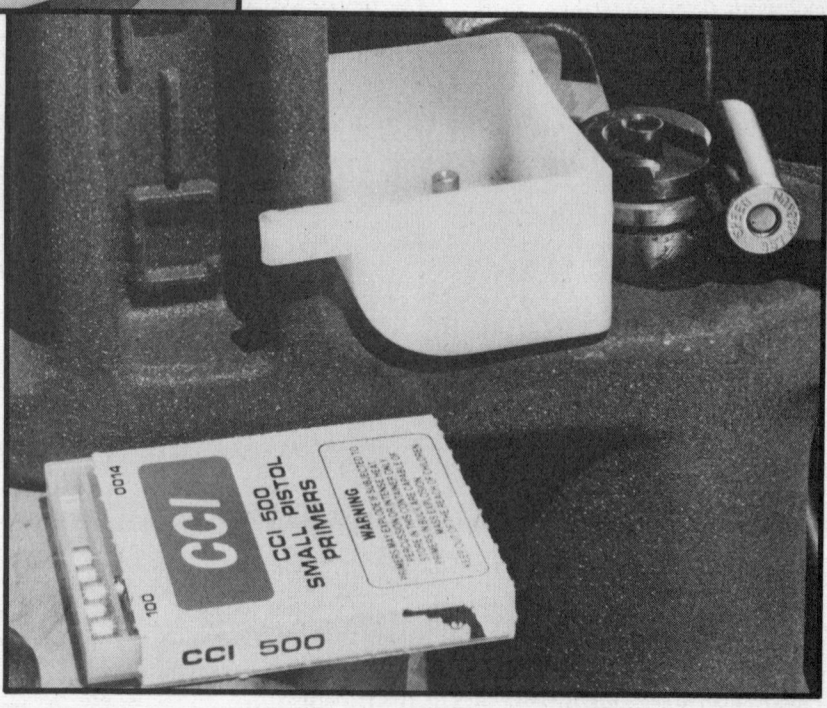

In this photo, the new primer has been seated in the case at upper right. You can see the spent primer in the catch-box. The .357 magnum takes the small pistol size of primer and it's a good idea to leave primers in their box until ready for the next one, for safety.

surface of the case neck by means of a small brush, before putting the case through the resizing die, be it full-length or NSO. That reduces the effort required and minimizes stretching of the case.

Thickness of the brass at the neck of the case varies from maker to maker of the cases and even to some small extent in cases from the same maker. In order to obtain the most uniform case neck possible, the reloading dies are designed to size the case neck down to an ID slightly smaller than it needs to be, as a preliminary, then expand it back up to the just-right ID as a final step. The ID thus produced will be some few thousandths of an inch smaller than the outside diameter (OD) of the bullet, itself.

At the time of expanding the case mouth, a slightly larger area at the top of the expanding plug produces a small funnel-effect or flare at the inner periphery of the case mouth. This serves to ease the seating of the base of the bullet into the case, without risk of snagging the mouth, which would ruin the case for further use.

Loading die sets for use with nominally straight-sided

An arm on the primer seater pivots the seater inward by making contact with the press frame. You can see the primer moving into position beneath the case.

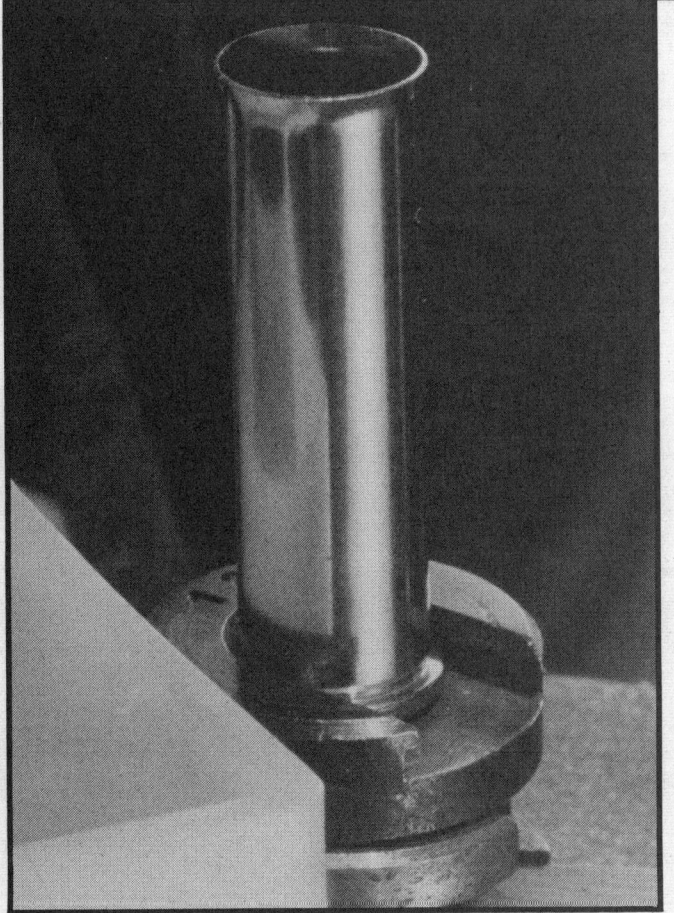

Expander/decapper stem should be adjusted to put a slight flare in the case mouth. The case in the photo at left has too much flare, indicating incorrect adjustment. You want just barely enough flare so the base of the bullet can be started in the mouth without snagging. Too much flare tends to work-harden the case.

handgun cartridge cases usually consist of three dies: The first die is a full-length resizer; the second die is a case mouth expander and the third die is the bullet seater. Each of the dies has an outer thread in the standard thread size of ⅞-14NF: ⅞-inch outside diameter, 14 threads per inch (tpi) National Fine. There is a lock ring, also ⅞-14, with a female thread, to anchor the die in its suitable position.

Let us review the technique for adjusting the first two dies in the press, reserving the seating die for discussion a bit later. The full-length resizing die is adjusted by turning it down into the top of the press until the lower surface of the die makes snug contact with the shell holder on the

Pacific die sets come packed in a sturdy red plastic box, with lift-off cover. Such boxes are an excellent way to store the dies when not in use, but it is not a good idea to store the shell holder with the dies, if you need it for use with other calibers. A dummy round with the dies helps to adjust seating die.

press ram, with the ram at the top of its stroke. You may wish to work the press handle a few times as you turn the die downward, until you can feel a small but detectable *click* as the ram goes over top-dead-center (TDC) against the bottom of the die. At that setting, turn the locking ring down snugly to hold it in place.

The exact setting of the die body for the second — mouth expanding — die is not as critical. It can be at any convenient position in the press, so long as it does not interfere with a full stroke of the ram of the press. What counts here is the positioning of the case mouth expanding plug. That is determined by backing the stem holding the plug up a bit higher than it needs to be, then turning it down a partial-turn at a time, with a resized case in the shell holder. As you do so, pull the case out of the holder after each trial and see if you can start the base of the bullet into the case mouth. You are looking for the setting that produces just a little case mouth flare, but not too much. If you end up put-

These heads were sectioned from 9mmP cases to show the single central flash hole of the Boxer primer, left, and the smaller twin flash holes of the Berdan primer, right. Reloading Berdan-primed cases requires special gear.

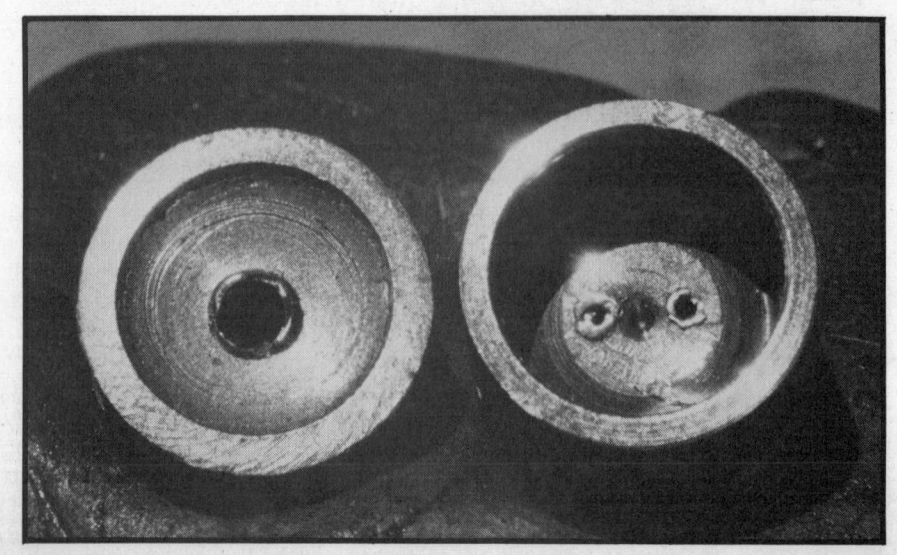

ting too much flare on the case mouth, readjust the expander plug and give the slightly over-flared case another pass back through the resizing die — as time affords before going on to re-expand and load it.

It remains to seat a new primer, dispense a charge of powder, seat the bullet and crimp it in place, as may be desired. Many presses have a provision for seating the primer, although exact procedures vary with the make and/or model. The primer must be seated firmly, with the exposed face of the primer cup from .003- to .005-inch below the rear surface of the case head.

Mindful of the cartridge caliber and weight of the selected bullet, the appropriate weight of a given powder must be determined by reference to a reliable reloading manual or handbook. Examples include: Speer Manual, Sierra Manual, Hornady Handboook, Lyman Reloading Handbook, Hodgdon Manual, Reloaders' Guide for Hercules Smokeless Powders, Handloader's Guide for Smokeless Powders (by IMR, formerly Du Pont), Winchester Ball Powder Loading Data, Nosler Manual and the data pamphlet from Accurate Arms.

If you have several such works at hand and compare them thoughtfully, you quite probably will encounter some

Primers should be seated slightly lower than flush with the case head. Primer depth gauge from RCBS had a GO and NO GO button at opposite ends for checking depth.

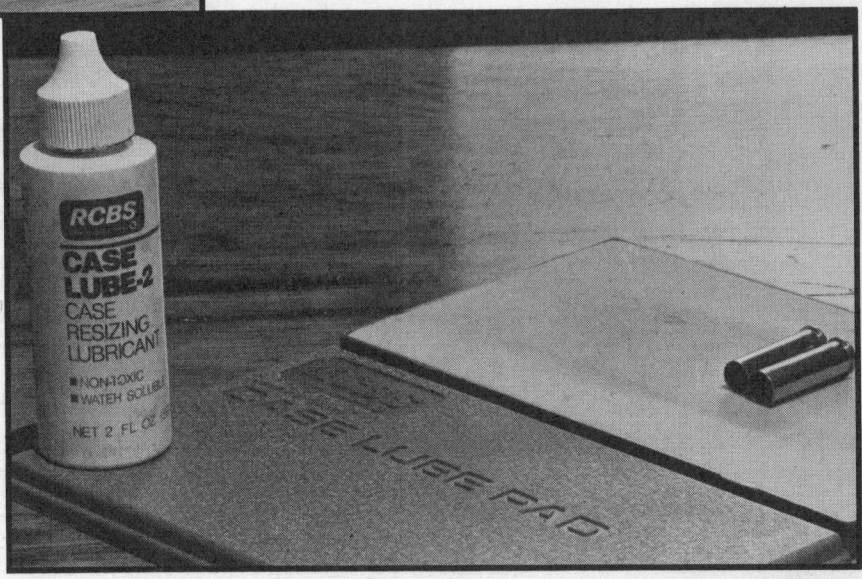

If the resizing die is of standard hardened steel, rather than carbide, it is necessary to apply case resizing lube to the cases before resizing. Many makers offer lube pads for application by rolling cases across lubed area.

GUN DIGEST BOOK OF HANDGUN RELOADING

The weights of bullets and powder charges are given in grains. A "grain" is 1/7000th of a pound; 437.5 grains to one ounce. A reloaders' scale is sensitive to plus/minus one-tenth of a grain and, as you'd rightly assume, gets quite a lot of use!

You can acquire a supply of cases by saving the empties from fired factory loads but, if that isn't quick enough, you can buy virgin brass by makers such as Winchester, Remington, Federal. Usually, it's supplied unprimed.

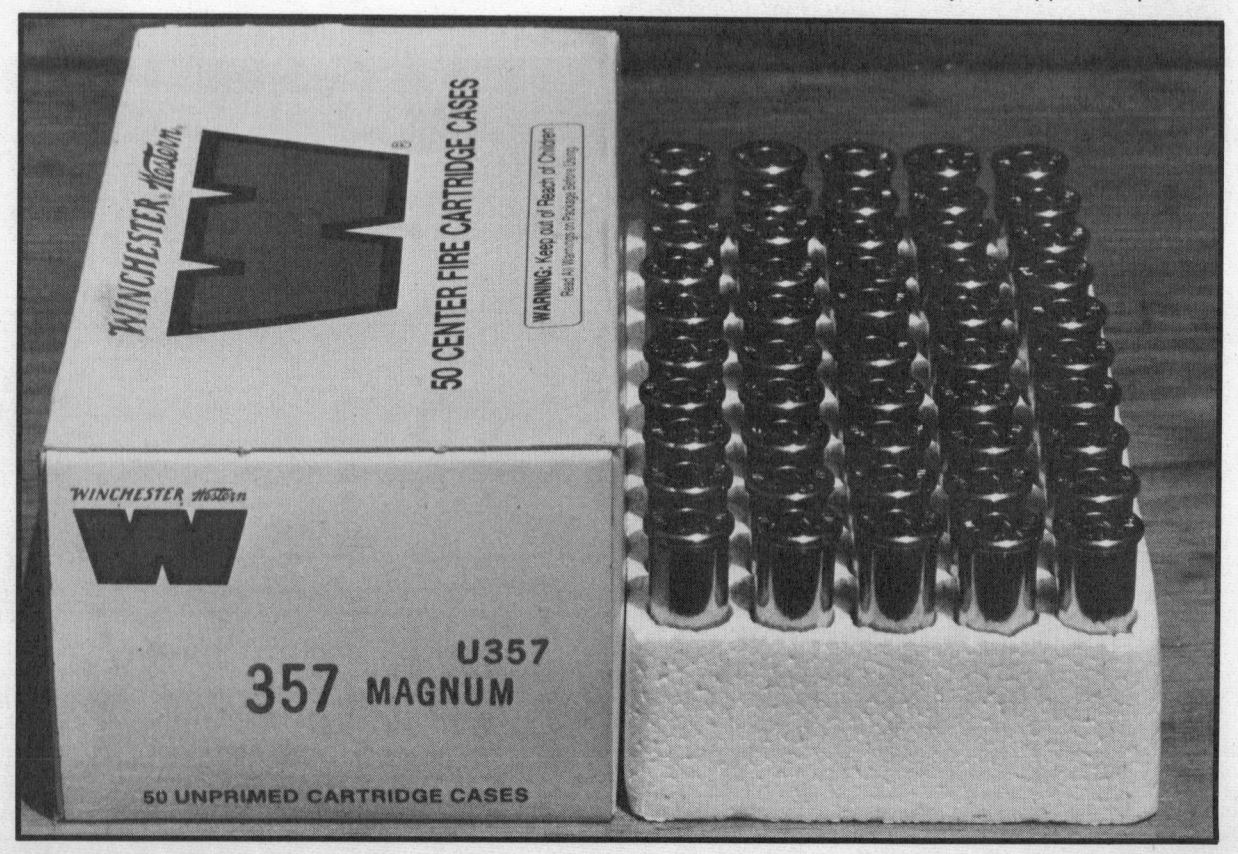

Die sets from Lee Precision come in these neat red plastic containers, complete with dipper-type powder measure, shell holder and sheet of suggested load data.

Powder comes in cans, usually smaller than this five-pound keg of Hercules Blue Dot and, being extremely flammable, requires safe, thoughtful storage. The RCBS "Little Dandy" powder measure uses a fixed-cavity rotor for dispensing charges and 26 rotors can be had.

amount of variation in the charge weights they list for a given combination of caliber/bullet/propellant (powder). You have to understand that tables were compiled by different people at different places and different times, using different guns. Realizing that, it is hardly surprising that the specifications differ somewhat. What *would* be surprising would be if they agreed, to the tenth of a grain, in every instance.

The prudent, sensible course is to start out with a charge weight of powder below the listed maximum. You can always increase the charge slightly next time, should conditions appear to warrant it. However, if you blow up the gun on the first try, you are back to Square One, after you get out of the hospital, hopefully assuming you do so. What's meant to be said is: Be on the conservative side, at the start and for a long while after starting.

The selected powder charges can be dispensed in any of several approaches. You can weigh each charge on an accurate reloaders' scale, going on to pour it into each case mouth in turn, with the cases resting in a loading block. Or you can dispense the charge by use of a fixed or adjustable rotary measure, by use of a dipper-type measure or other methods. In any event, it is a good idea and a great mental comfort to check the charge weight against an accurate reloaders' scale, before going on to seat the bullet and fire the reloaded cartridge.

While the charged, load-ready cases are still resting in the loading block, take a keen and thoughtful squint down into the neck of each case, comparing it with all of the other charged cases. You are checking against the presence of any case(s) with a level of powder detectably higher or lower in the case than all the rest. If they pass that test, go

38 CALIBER (.357" Dia.)
160 GRAIN FULL METAL JACKET
(Flat Point) #3579

CARTRIDGE OVERALL LENGTH 1.580"

POWDER	1000 fps	1050 fps	1100 fps	1150 fps	1200 fps	1250 fps	1300 fps
BLUE DOT			9.8 gr.	10.2 gr.	10.6 gr.	11.0 gr.	11.4 gr.
HS-7	10.2 gr.	10.6 gr.	11.0 gr.	11.4 gr.	11.8 gr.	12.2 gr.	12.6 gr.
WIN 630			12.1 gr.	12.7 gr.	13.3 gr.	14.0 gr.	
2400	11.3 gr.	11.9 gr.	12.6 gr.	13.2 gr.	13.9 gr.	14.5 gr.	
R-123	12.3 gr.	12.7 gr.	13.2 gr.	13.6 gr.	14.0 gr.	14.4 gr.	14.8 gr.
H-110	10.9 gr.	11.9 gr.	12.9 gr.	13.9 gr.	14.9 gr.	15.9 gr.	
WIN 296	13.4 gr.	13.9 gr.	14.5 gr.	15.0 gr.	15.5 gr.	16.1 gr.	16.6 gr.
IMR 4227	13.5 gr.	14.2 gr.	15.0 gr.	15.8 gr.	16.5 gr.		

Indicates maximum load • use with caution

(Handwritten energies: 346, 355, 392, 430, 470, 512, 555, 601)

Here's a typical block of load data, as given in the third edition of the Hornady Handbook. Vertical columns are headed for the typical velocities that the given charge weight of powder can be expected to deliver and the owner of this copy has inked in the equivalent energies in foot-pounds at the bottom of each column. Shaded listings at the right end of each line are maximum charges and should be used cautiously or, much better, avoided!

---

on to seat a bullet in each case, as planned.

The seating dies should carry a seating punch that will seat the given bullet without undue distortion. Seating punches of various specialized configurations usually can be obtained from the maker of the die set. Many seating dies have punches that do a pretty good job of handling most bullets.

If you have a problem bullet for which a custom punch is not apt to be available, it is often possible to help yourself out of the predicament by adapting an existing punch. Choose a punch that has a fairly generous lower cavity — the part that makes contact with the bullet — and pack the cavity with wadded-up facial tissue, paper towel, Saran Wrap, aluminum foil or a similar material. The tip of the bullet will custom-form the packing to fitting contours. When finished, the wadding can be removed with an ice pick or scratch awl and re-packed at a time of future need. My preferred medium for such use is paper towel and I mix a few dabs of bullet lube with it to help it remain in place and hold its shape.

My usual procedure is to apply a taper crimp, rather than the usual roll crimp. The ostensible purpose of the roll crimp is to prevent forward bullet migration in the chambers of revolvers. It's been my general experience that, if the case neck grips the full-diameter portion of the bullet snugly — as it should and must — no further holding arrangement is required.

My objection to the heavy roll crimp, as recommended in many books, is that brass is harder than lead and, in emerging from the case mouth, the roll crimp may tend to scrape and distort part of the full-diameter portion of the bullet. That does nothing at all to help the potential accuracy, particularly if the bullet is of a rather soft alloy.

If a savage roll crimp has undesirable effects — and I think it does — a savage taper crimp can be almost equally harmful, particularly when loading jacketed bullets. If you

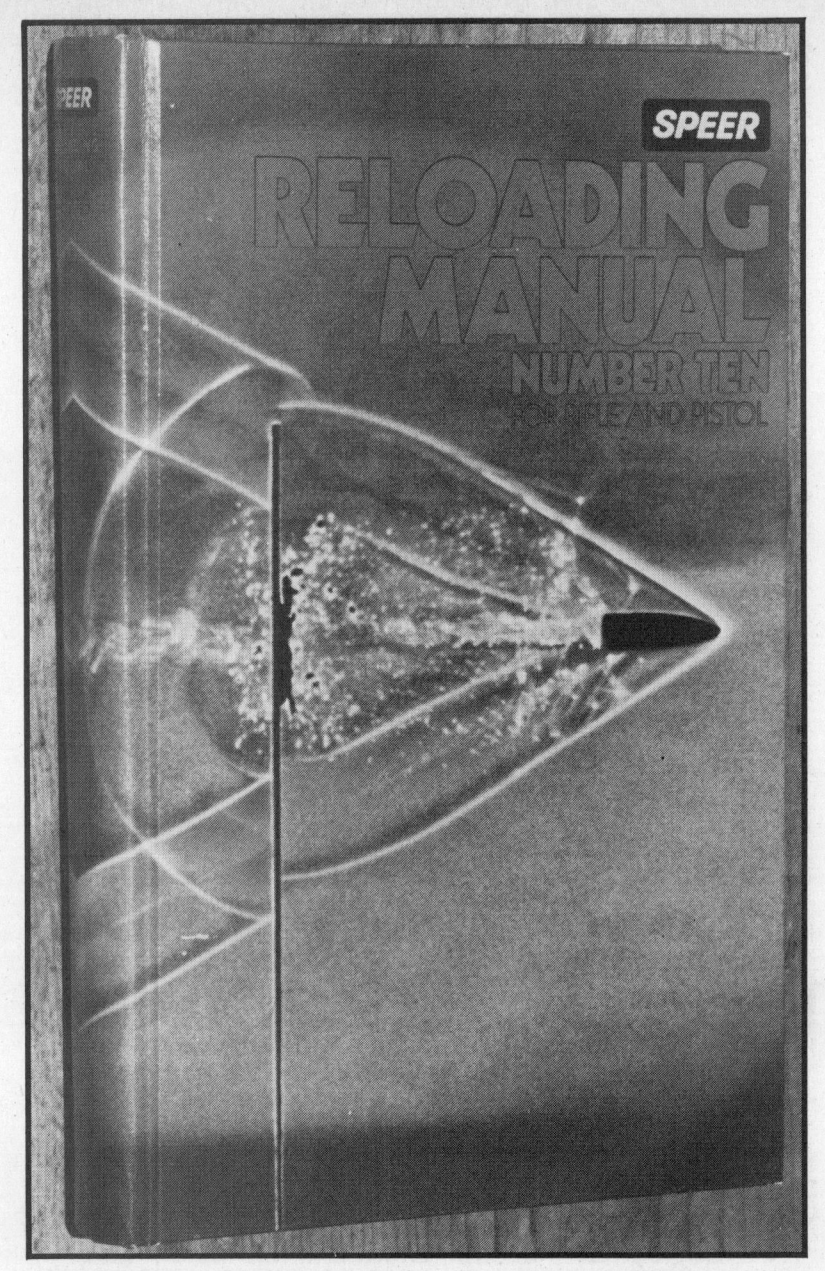

Although it's a little hard to make out the title, this is a copy of the Number Ten Speer Reloading Manual, originally published in 1979 and reprinted many times since. The Number Eleven should be appearing, almost any year now. An incredible amount of work goes into one of these, but they offer invaluable aid to reloaders.

taper the crimp too hard, you will squeeze in the full-diameter portion of the bullet to some extent. When that is done, the jacket metal, being somewhat elastic, will spring outward slightly as the cartridge comes down out of the die. The lead core, being quite inelastic, will not spring back and you will lose the intimate fit of core to jacket that is quite vital to accuracy.

Adjust the taper crimp die so that there is a slight "feel" on the operating handle of the press when the cartridge goes up into it. That's all you need. The purpose is to remove the flare you put into the case neck for the sake of easy, gouge-free bullet seating. Anything beyond that contributes nothing useful to cartridge performance.

You can seat the bullet as one step, then readjust the die,

GUN DIGEST BOOK OF HANDGUN RELOADING 19

Here's the Third Edition of the Hornady Handbook, from which the data block on page 18 was reproduced. It covers load data for a great many rifle cartridges, as well as handgun calibers and it's a good idea to have a copy.

turning the die body downward and backing the seating stem upward, to run all the loads through again to crimp the case neck into the crimping groove or cannelure, as may be available. Do not try to crimp the case mouth in the absence of one or the other. To do so is apt to produce cartridges that cannot be chambered. The brass has to go somewhere. If it has no place else to go, it may buck up into a small ridge behind the case mouth, too large in diameter to allow the cartridge to fit into the chamber.

It is a good idea to have the gun on hand in which you propose to fire the reloads and use that — with suitable safety precautions! — as you go along, by way of making certain that the reloads will load easily into the chambers of the revolver, for example, without the nose of the bullet projecting ahead of the front surface of the cylinder.

Reloads for an autoloading pistol must be short enough in overall length to load into the magazine and of such a length overall to assure reliable feeding out of the magazine and into the chamber. When performing any such checks, take every possible safety precaution and, above all, keep the muzzle pointed in a direction so that, should an accidental discharge occur, the resulting damage and consequences will be of minimal nature, with which you can continue to live without undue emotional, legal or financial discomfort.

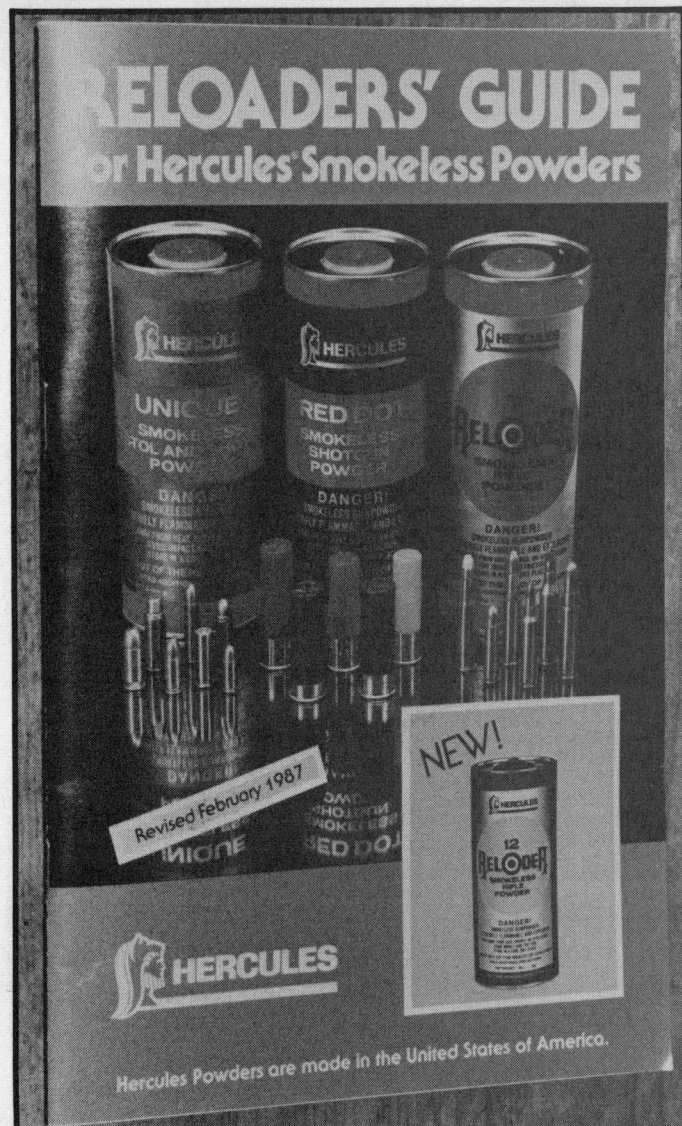

Hercules publishes updated editions of their Reloader's Guide annually and it's often possible to obtain free copies from your local dealer. It covers data for rifles, handguns and shotguns, but only for Hercules powders.

Just as the .357 magnum was an elongated version of the earlier .38 Special, the .357 Remington Maximum, at left here, next to a .357, is a further stretch of the .357 case. The Pacific die set shown earlier will handle all three, by suitable die adjustment.

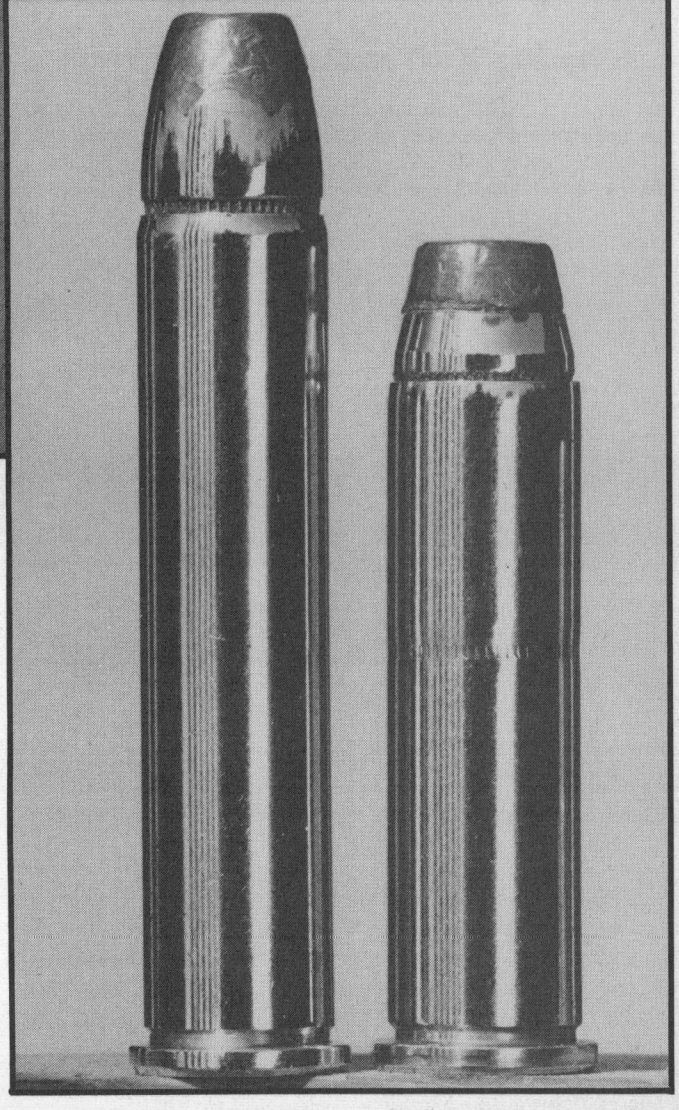

What you install "up front" governs the performance and capabiity of the reload to a great extent. Factory-made bullets usually come packed 100 to a box and the 140-grain JHP is an excellent general-purpose choice for all-around use in the .357 magnum cartridge.

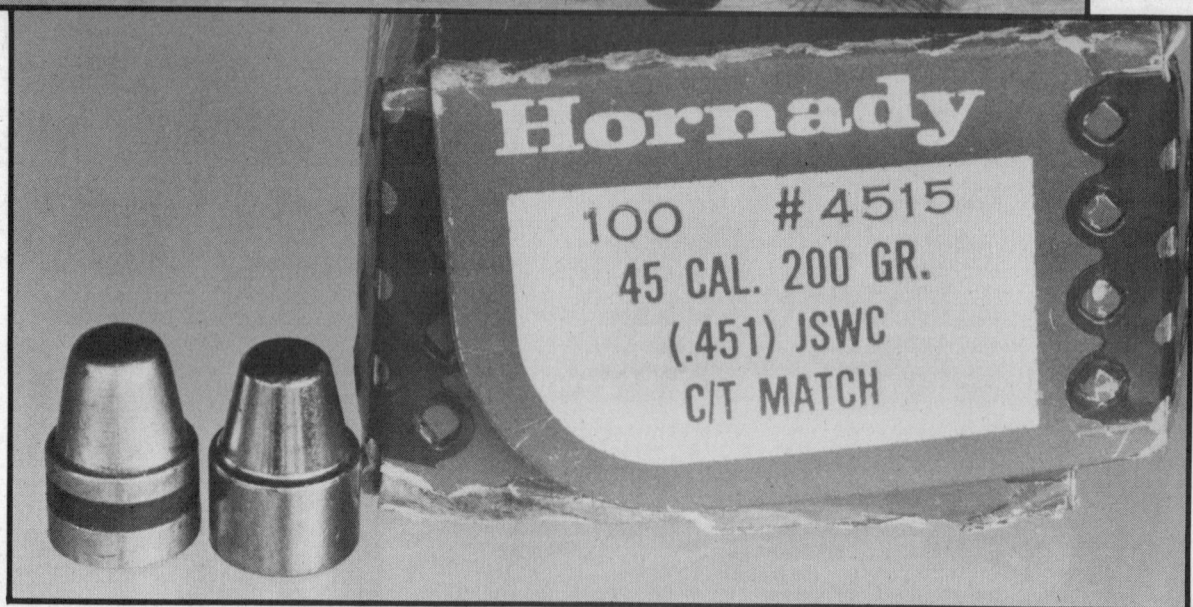

Many reloaders cast their own bullets and, for the .45 ACP, bullets from the Hensely & Gibbs #68 mould (at left, here) are exceptionally accurate. That led Hornady to bring out their #4515, essentially a jacketed version of the same basic design and it groups well, too!

Reloads should be stored in an appropriate container. The MTM Case-Gard ammo boxes work exceptionally well, but there are satisfactory alternatives. Be sure to record the pertinent details of the reload somewhere, against the time you will wish to retrieve and consult them. You can keep the details in a spiral-bound notebook, or affix them to the container of reloads, enter them in your personal computer, whatever course you prefer.

The thing is, if the reload performs well, you may wish to duplicate it. If it does not perform well, you will wish to avoid the waste of time and components in making up more of the same. In either event, having the pertinent details readily at hand for consulting can be of immense value. There is hardly anything so gallingly frustrating as to come upon a quantity of reloaded ammo with no slightest clue as to its inner makeup.

For target work or casual plinking, lead bullets of the wadcutter type work quite well, but they must not be loaded to high velocities, as it causes bore leading and, in the example of the hollow base wadcutters here, the risk of a separated skirt stuck in the bore.

Another specialized design is Hornady's JTC-SIL, available in various weights and calibers. It's designed expressly for the best possible performance against the steel targets used in silhouette competition.

Today, there are many handguns that work with bottleneck cartridges. Most die sets for reloading bottleneck cartridges have but two dies. There is a full-length resizing die, with an expanding plug on the stem for purposes of sizing the case neck to the proper ID as it's dragged back down out of the die. The second die is for seating the bullet. On such cases, you must keep attentive eyes upon the overall case length and trim the cases if they exceed the length for that caliber, as given in most of the sources of load data cited earlier.

After some number of loadings, a bottleneck case may require inside-reaming or outside case neck turning to remove excessive thickness of brass at the case neck. If you cannot insert the base of a suitable bullet by hand into the neck of the fired case, such a condition may be in existence. To make up further reloads in such a case is hazard-

There are all manner of other gizmos and widgits used in reloading, helpful, but not necessarily all-out vital. Here, for example, are an inside/outside case neck deburring tool, installed in a really vintage crank stand, along with the stuff used to refuel both the dry-type and wet-type case cleaners, which are discussed in considerable depth, along with a great many other topics in...

...books such as the one above. A fourth edition is scheduled for publication in late 1988, as currently planned. ABC/R carries no load data; all it does is tell you how to load and use data!

ous, as the too-thick neck may lock the bullet in place and resists its release at the time of firing. If that happens, peak chamber pressures will go up and up — quite probably to hazardous levels.

A handy tool for checking case length is a caliper — either vernier or dial-type. Several makers of reloading equipment offer case trimmers, as well as inside-outside case neck deburring tools for removal of the burrs left after trimming.

Accessories for the case trimmer available from Forster Products include inside neck reamers of suitable diameter and outside case neck cutters. If it is impossible to insert the base of a suitable bullet into the neck of the fired case, the reamer can be installed and used to remove excessive metal from inside the case neck. Note that this is done *before* resizing. If it is done after resizing, it will remove so much metal that the case will be unusable for further reloading.

As a broad generality, it is rarely if ever necessary to trim straight-sided handgun cases, such as the .38 Special, .44 magnum and the like. Several of these — notably the .45 ACP and 9mm Luger — rarely come up to the length specified in the manuals and, if anything, seem to become slightly shorter with extended reloading.

The topic of reloading does not lend itself to comprehensive coverage in a few pages. If the reader feels need of more intensive discussion, a companion book from the same publisher — *ABC's of Reloading, Third Edition* — is suggested as a most helpful supplement to the foregoing discussion. — *Dean A. Grennell*

Chapter 2

A Veteran Handloader Offers Advice, General and Specific, On The Knotty Topic Of Selecting Tools Where...

...GOOD ENOUGH ISN'T.

COMPARED TO "the good ol' days," these are marvelous times to shop for handloading tools. Selection and quality are worlds better than they were 25 years ago. Only a few well-known tools are junk, made and advertised to snare the unwary. But not all top-quality tools are equally good for loading handgun ammo, so there's still plenty of good reason to shop carefully — and not just for the best prices.

South of the border, *Gato escaldado del agua fria huye* — the scalded cat flees cold water. That makes sense to me. By the late Fifties, I'd formed the habit of buying from the one outfit whose tools had been consistently good. Other brands — usually bought because the prices were better — had scalded me more often than they'd been satisfactory. After a bad time with any maker's junk, I shunned that brand with the caution of a scalded cat.

For example, I had nothing but trouble with one name-brand H-frame press and .357 magnum dies. The vertical guide bars of the press were not parallel, so the assembly bolts had to be rattly loose before the shellholder could come up to the bottom of the die. The shellholder didn't line up properly with the die. The body of the seating die was smaller (inside diameter) than the sizing die. It sized cases down smaller than the sizing die had made them. Trying to seat a bullet crumpled cases, since the cases were about the same diameter as the bullets.

Seating a primer in the first .357 case pressed the priming punch deeper into the priming post. That made it too short to seat another primer. The manufacturer replaced that priming assembly — but after the new one had primed one case, it too was too short to prime the others.

At the plant of another "name brand" in the early Sixties, I tried the tools exhibited in the lobby. They were well designed and carried a respected old name, but were crudely made, poorly aligned junk. A friend bought one of their presses. He had to squeeze with all the strength of both hands to push the priming arm into the slot in the ram and needed a small crowbar to pry it back out. Other products with that brand were equally crude — yet advertising had made that brand a "name."

Since then, friends have reported satisfaction with later tools bearing the brands of that earlier junk. So I've checked

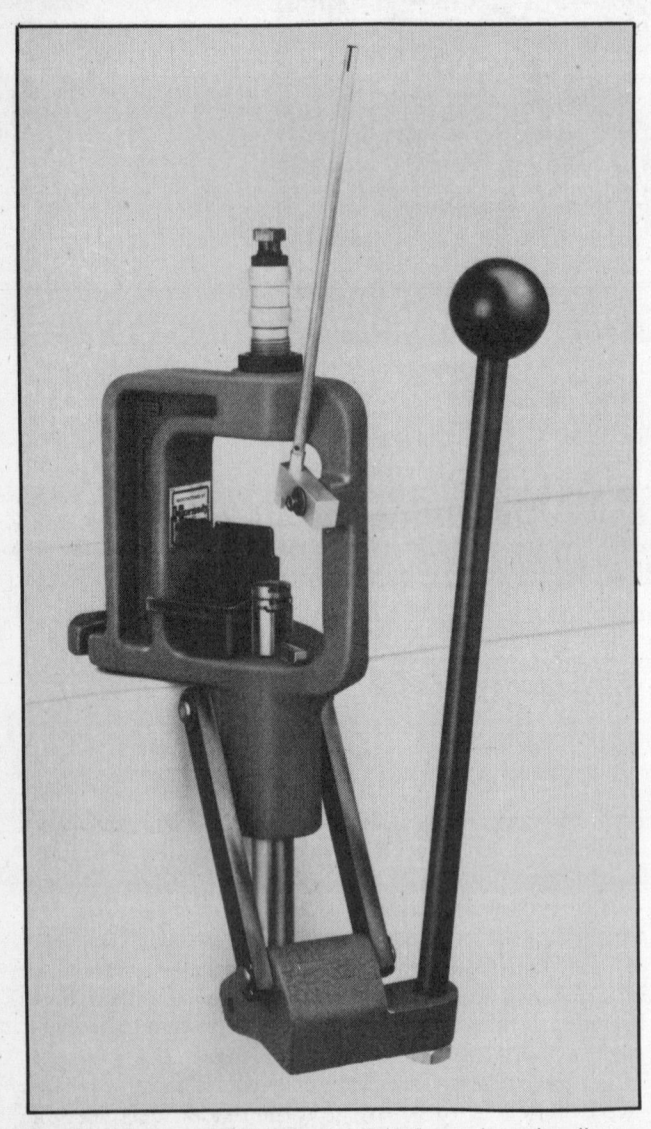

While they are available from all of the major reloading toolmakers, they differ in details of shape, leverage and color. The basic tool is usually a hefty "O" frame type.

26   GUN DIGEST BOOK OF HANDGUN RELOADING

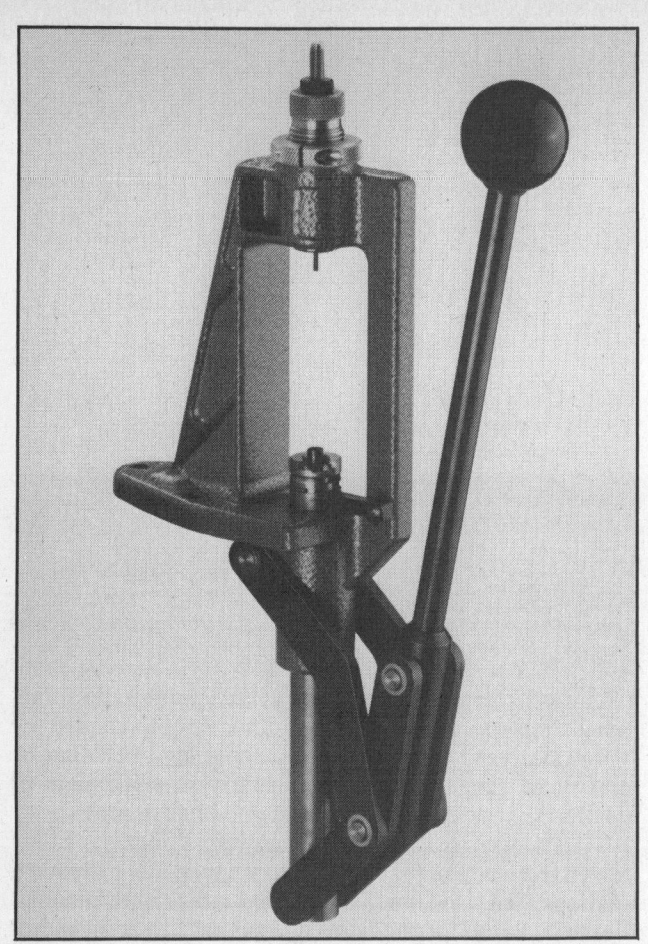

The newest thing out is a smaller version of the basic "O" press. This one is green — from RCBS — and they'll market it as the Partner. It's light enough to take along.

them out by testing samples and visiting their factories. Those brands have changed ownership several times, some factories have been moved or retooled and current owners of those old companies now make good stuff. I've been pleased with recent products made under brands that I used to shun. My loading bench and shelves still reflect my long-established trust in a certain brand of green tools — but the last ten years have added red, orange, blue, gray and black tools of equally superb quality.

Shopping advice for the handloader used to be simple: "If RCBS makes what you want, buy RCBS." Now, brand makes far less difference, so smart shopping depends more on your handloading requirements, tool design and manufacturers' quality control. When I started, it was often necessary to make some tools myself or modify (sometimes even overhaul or rebuild) some of the best "store-boughtens." Today, there's far less need to cobble up something of your own. It's wiser now to shop carefully, then get plenty of experience with good ready-made tools before you consider making or modifying your own.

Quality and dependability cost less then junk in the not-so-long run, so the purchase cost of a tool isn't the first thing to consider. The best economy lies in buying the best you can manage. The best you can afford is second most economical. Junk never saves you money. It has to be replaced sooner or later and, the worse it is, the sooner you'll have to toss it out.

If you merely upgrade from junk to "junque," you simply add to the overall cost of the quality tools that you eventually have to buy to get even minimum satisfaction. Don't postpone buying the best. If you have to squeeze your shekelim and use a shoehorn to fit loading tools into your budget, start with the bare minimum of tools of the highest quality. Get the essentials first and add the frills later.

Some tools are basic necessities; others are merely nice — conveniences or time-savers. If you have to economize, don't buy anything you can get by without, stay away from junk and don't consider "good enough" good enough. First determine the best tools for your needs, then shop to find where you can get those tools for the best prices. You can conveniently load hundreds of good rounds at a crack with even the minimum of good tools.

By all means start with a good bench-mounted press. Whatever else you have to scrimp on or do without for a while, get the best press and dies you can find. This means an RCBS, a Redding, a Lyman, or a Hornady, most likely, unless you plan to go into moderate mass production with a progressive loader. But if you're new at handloading, I strongly advise against starting with a progressive loader. (However, I'm not the bird to chirp about progressives. Mike Venturino covers these special tools fully in Chapter Three.)

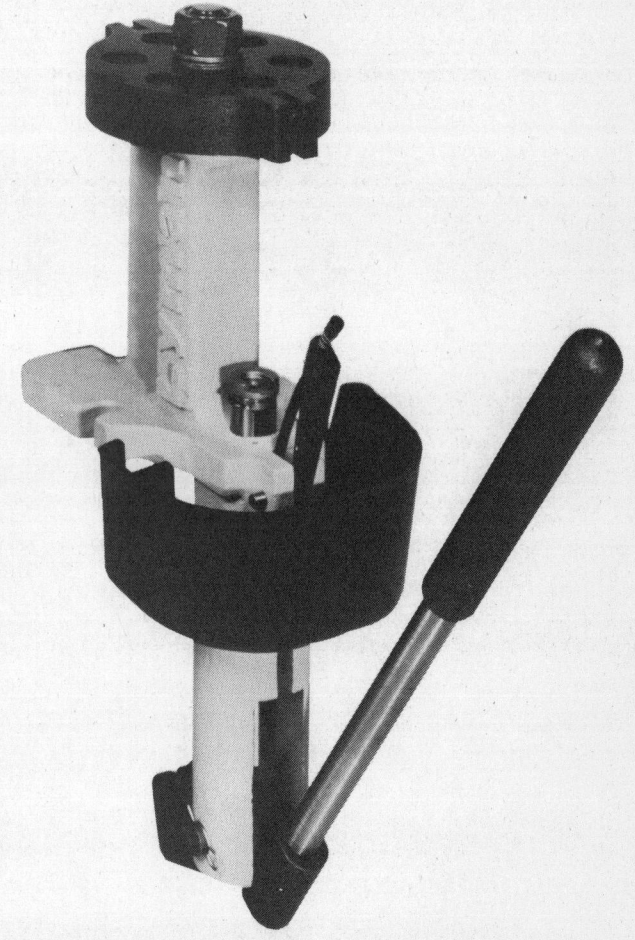

One of Lyman's newest is the T-Mag. It's a turret press and a good one. This class of loading press is one that the handloader needs to be careful about purchasing.

GUN DIGEST BOOK OF HANDGUN RELOADING 27

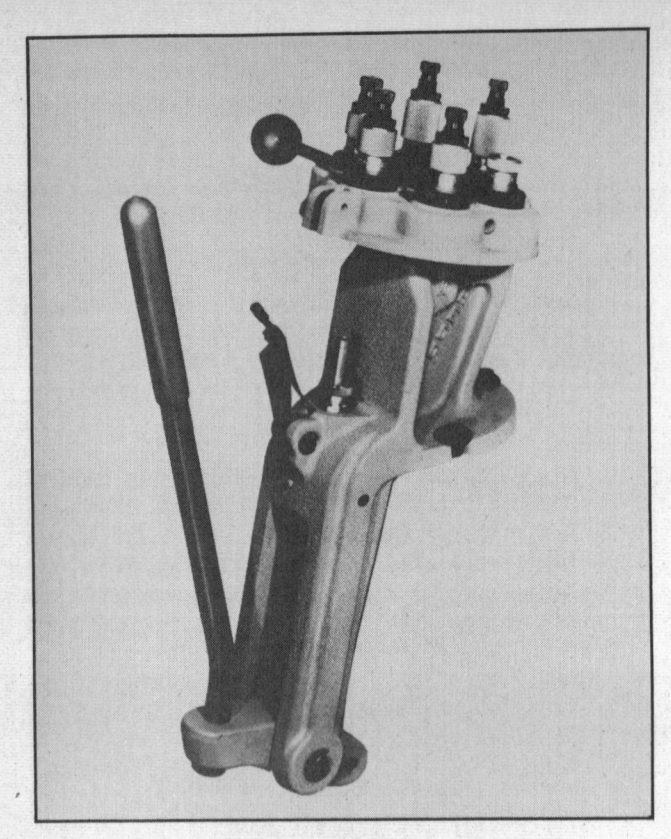

Author Howell's favorite turret press is this one from Redding. It's exceptionally well made. Below: This tiny hand press from Huntington Die Specialties is the heart of many compact "take to the field" reloading outfits.

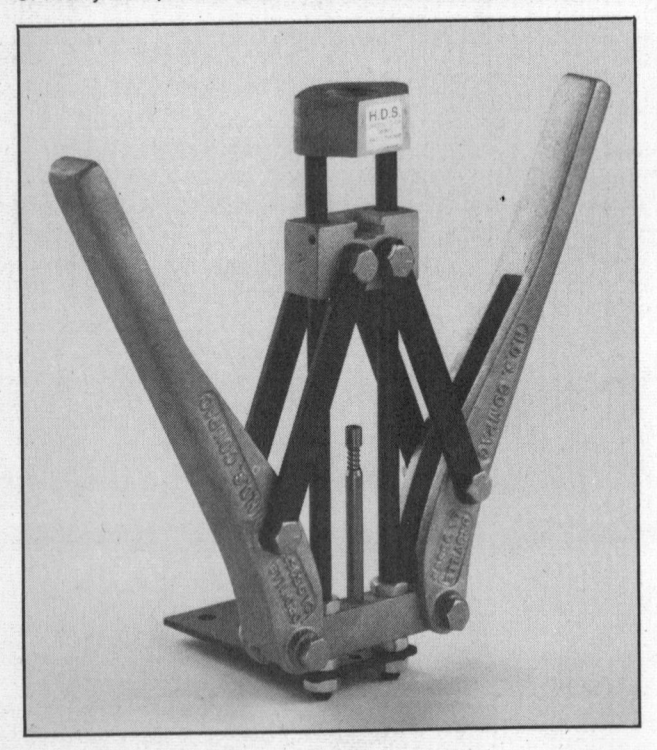

The old "grasshopper leg" linkage of the press on left has given way to more modern and efficient compound leverage systems pioneered by Fred Huntington at RCBS.

Compac in the shop frequently when I want to do something simple to a few cases and don't want to change the setup on any of my bench presses. Some hand-helds can be downright dangerous. None is nearly so convenient or comfortable as even the simplest good bench press.

Handloaders generally load many more rounds at a time for a handgun than for a rifle — so a handgunner's loading sessions tend to be much longer than a big-game hunter's. But then ammo for handguns doesn't have to be as microscopically precise as loads for bench-rest rifles, so you don't have to fuss quite so much over the preparation and loading of each case.

The vitals of a good press are its frame type, material, linkage, alignment and finish (on its sliding surfaces, screw threads and such — not where it's enameled). A good C-frame press is good enough, but an equally good O-frame press holds it alignment better under the extreme force of the best linkage systems. (H-frame presses are structurally modified O-frames.)

Now that the patent has expired on Fred Huntington's superb compound-leverage system, several makers use it on even their most economical presses. There's no longer any reason to settle for the old grasshopper-leg linkage. The old linkage is good enough, but the Huntington linkage is stronger and much more comfortable to use.

Cylindrical cases size harder than tapered cases — and most handgun cases are cylindrical. The more sharply a case tapers, the more of its length enters the sizing die before the die begins to swage it down. The leverage of the press linkage therefore produces a greater mechanical advantage — meaning less effort on the end of the handle. But a cylindrical case must be sized from its mouth down; this sizing begins while the leverage of the press linkage is nearer its lowest mechanical advantage.

This means more effort on your end of the handle. The difference may not be much, per case, but it makes a big difference in time, comfort and convenience by the time you've loaded a good-size batch of ammo. The compound-leverage linkage is worth more than the difference in price.

Few hand-held tools are worth the bother; their best use is for backing up your basic outfit — to use on a trip, for example. The only "hand press" still in my outfit is Fred Huntington's strong and handy little Compac. It's a hummer, but nothing else I've ever seen comes close. I use the

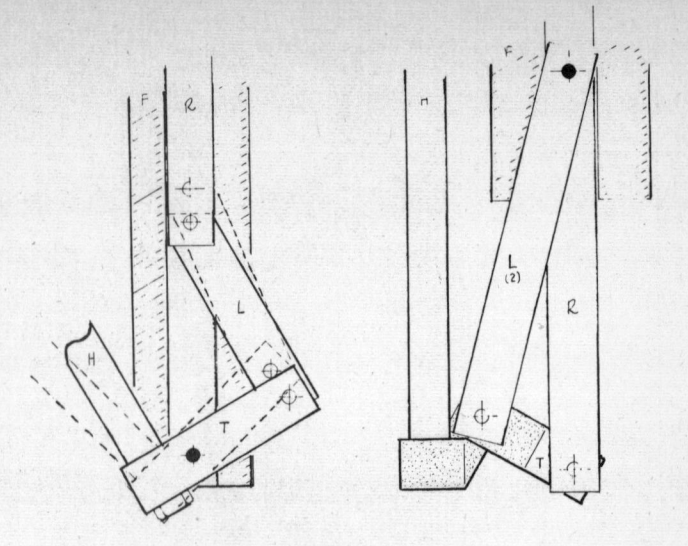

GUN DIGEST BOOK OF HANDGUN RELOADING

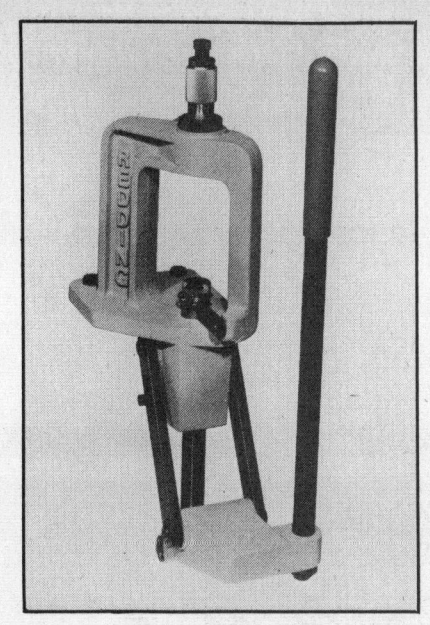

Redding makes good presses whose quality exceeds their reputation. This is their simplest with compound linkage and it's a combination of ease of use, value and power.

One warning: Don't fall for a cheap press with the Huntington linkage but made of pot metal. I've seen ads for neat-looking economical presses made with this linkage, with the claim that these presses develop more applied force than any other. I got one and checked it out. This press developed greater leverage, all right — but only because the handle was longer than those on other presses.

Its greater leverage did indeed apply more force to the frame — but the frame wasn't strong enough to withstand so much force. In another examiner's tests, the frame of this pot-metal press broke before his testing device applied enough force to give him a reading. This press is far weaker than any other now on the market, because its frame is flimsily made of weak material.

Zinc alloys are cheap but weak; accept no economical material other than iron, steel, or *hard* aluminum alloy. Shun pot metal like the plague that it is. Strength is a greater concern than just whether your press breaks or hangs together. The mechanical advantage of even the old linkages require very strong frames and other tool parts. Strength is necessary not merely to avoid breakage or deformation of the tool, but for the solid rigidity required to keep cartridge dimensions uniform.

Nearly perfect alignment is mandatory; no tool is usable without it, however good that tool may otherwise be. The ram or stage that raises the shellholder to the die must align each case with the die that it's to enter. The long axis of the die and the line of the shellholder's travel must coincide — they must not be offset or at an angle to each other. All the better manufacturers check alignment with machinists' dial indicators; you can weed out the worst offenders with a simple bench test.

Set a fired case — the longer the better; a .357 or .44 magnum — in the shellholder. Make sure the shellholder is properly seated in the ram or stage and raise it slowly toward the base of the die. If the mouth of the case enters the rounded mouth of the die evenly, you're in like Flynn

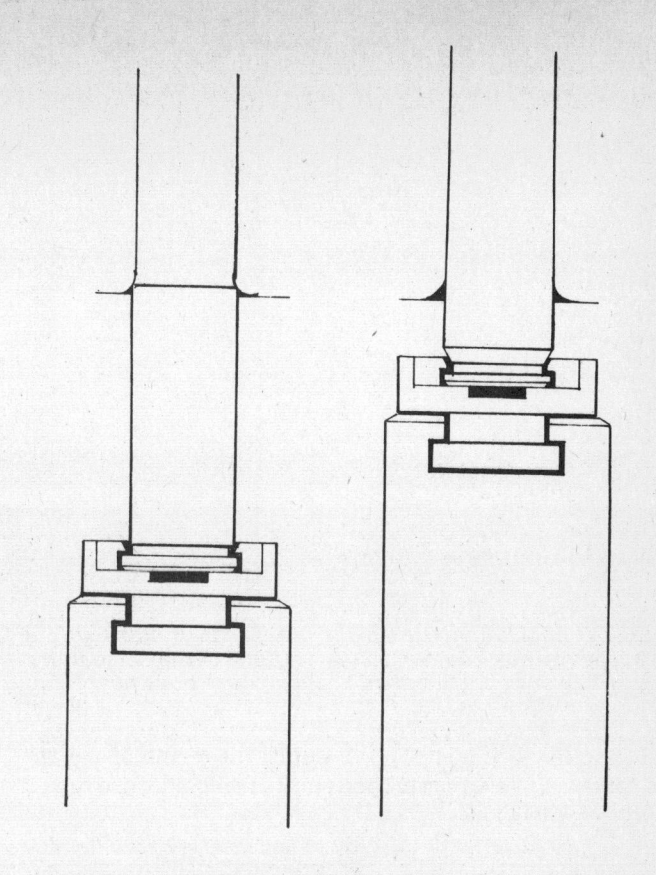

Above: The .44 magnum case (left) is harder to size than the '06 case because the actual work occurs when the mechanical advantage is lowest. Below: Before buying, it's wise to check out the alignment of the press with dies installed. Left view is correct; other two aren't.

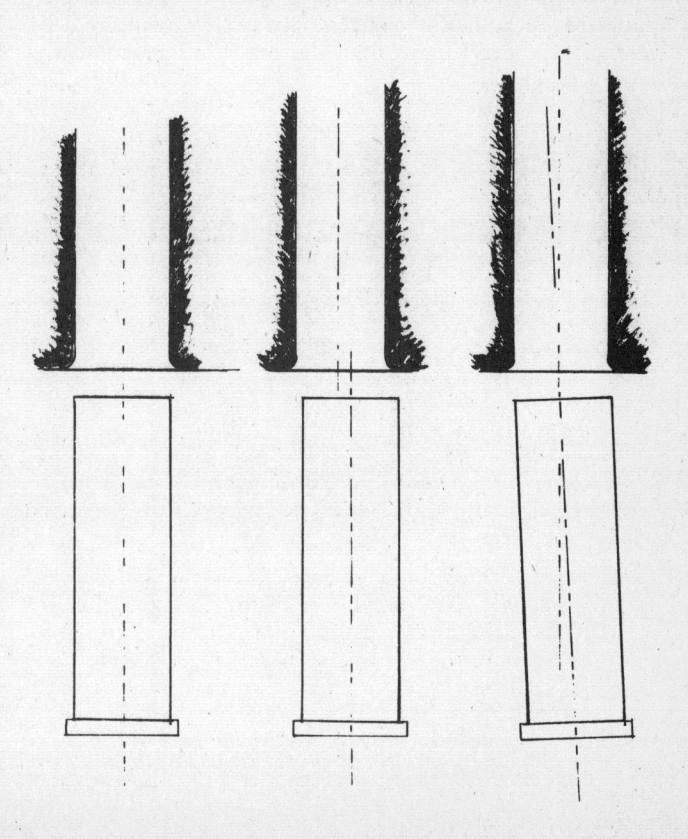

GUN DIGEST BOOK OF HANDGUN RELOADING

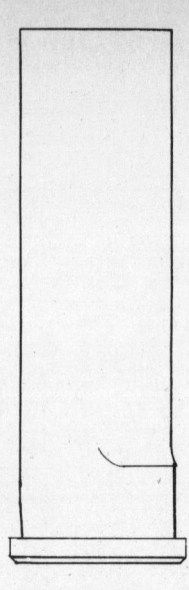

This is an extreme example of the results of press-die misalignment. You can see indications of the same trend by checking for burnished spots on case body near head.

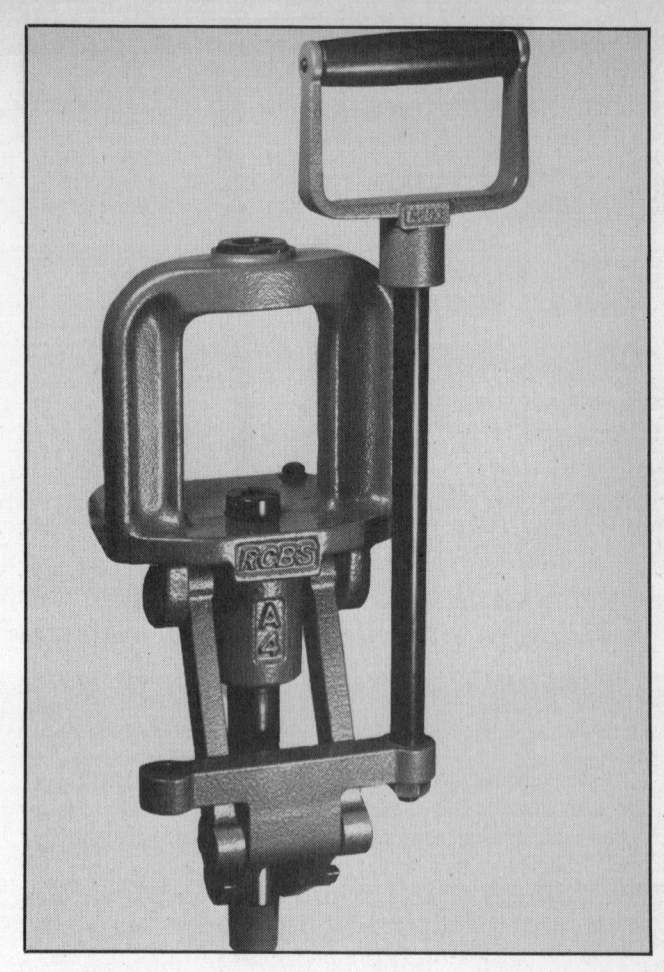

— so far. If it bumps hard against the mouth of the die, make sure it's seated properly in the shellholder and raise it to the die again.

If you have to tilt the case to line it up with the die, the press may be poorly aligned. A little misalignment is tolerable, but if you have to push the case far over to line it up, smile at the sales clerk and try another press, or go shop somewhere else.

If this alignment is good or tolerable, size the case full-length and withdraw it. Near the rim, you should be able to see how far back the die sizes the body of the case. Turn the case to see whether it has been sized evenly all around. If one side is sized more than the other, alignment isn't good enough.

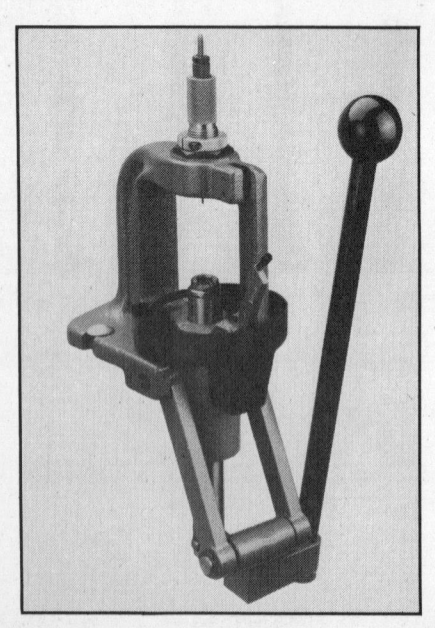

It's best known as the Big Max, but sometimes called the A4. A big sturdy brute that makes handloading a breeze, the A4 is about as hefty as single stage presses come.

One of the all-time best buys in reloading tools is the great RCBS Rockchucker. It's slightly smaller and a bit more economical than the A press which preceeded it.

You want specific recommendations? All right. The least that I'd consider as a bench press for handgun ammo is the RCBS Reloader Special, but I'd rather have a Rock Chucker or a Big Max. But then I'm obviously still biased toward the brand of green tools that have consistently given me superb dependability for decades. I'd also be content with — and happily recommend to you — a Lyman Orange Crusher, a Hornady 00-7 or a Redding Boss, all high-quality presses.

The Hornady 00-7 and Redding Boss have the best built-in priming devices I've seen on this class of press. With their excellent C-frame Ultramag press, Redding has even improved the basic Huntington linkage and reduced the preference for an O-frame. In the new Redding design, the upper ends of the leverage arms are attached to the upper part of the frame at the base of the die. This is an excellent design, and it's well executed.

For the most dependable performance, primers must be seated evenly and consistently. All must be seated with the same slight stressing of the priming pellet between anvil and cup. This means that all must be seated to the same depth in the primer pocket without any cocking or tilt in the cup. Ideally, the primer seater should be supported directly in line with the center of the primer and case. The old-fashioned swing-forward priming arm had its support too

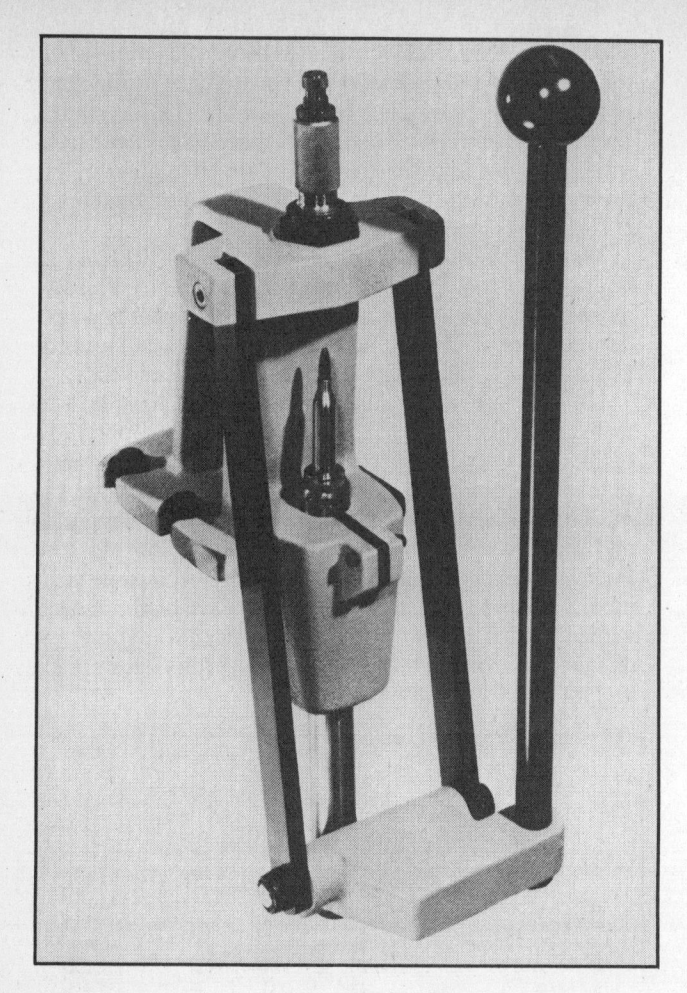

This big brute from Redding takes raw strength a little farther by anchoring the linkage straps at the base of the die, not at the base of the frame. Note open front. Below: The Lyman Orange Crusher Set is solid value.

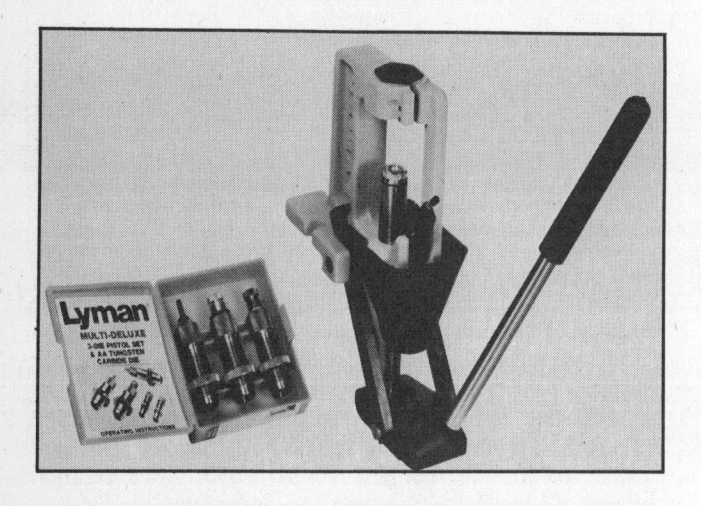

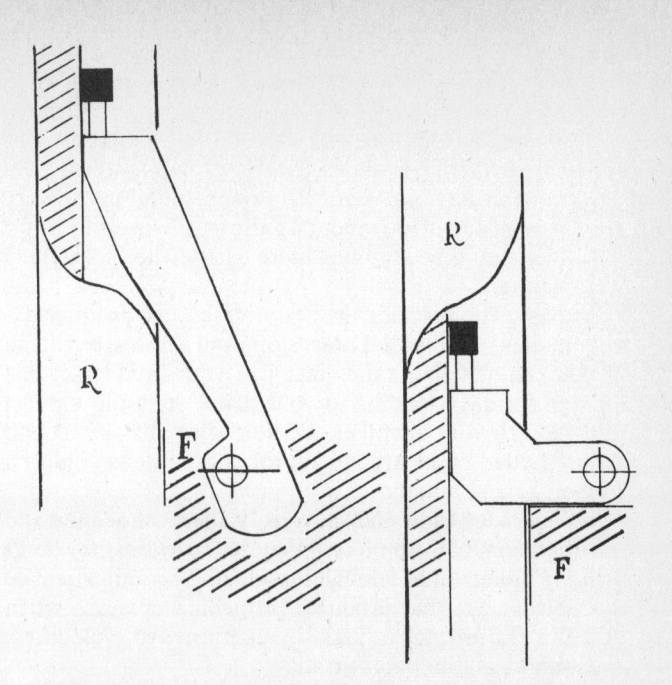

Older press systems use the primer arm seen on the left but the newer ones are as on the right. The older way had primer seating alignment tied to press ram alignment.

far from this axis for the best seating. It's usable, but it isn't the best design.

Proper seating with this old-style priming arm requires the old technique of seating the primer, easing the case up a bit, turning the case 180 degrees and seating the primer again. This is a nuisance. Some purists, however, still use it even with in-line seaters — some even go so far as to turn the case 90 degrees at a time for four seatings per primer.

You notice that a few classic brands of presses aren't on this list at all. One brand that signified quality when its originator made it is now made — by somebody else — with very poor fit, finish and alignment. Ones that I've checked out are of substandard quality in several ways.

Another brand of presses are so atrociously designed and made that I ache to violate my own firm principles and identify them here. But anything that bad doesn't deserve sentence space when I write to recommend good stuff. Besides, I consider only the best good enough, for me or for you. (Napoleon Bonaparte's axiom certainly applied to handloaders and our tools: without regard to your birth or fortune, *la carriere ouverte aux talents* — the tools to him who can handle them. If you agree, you won't get suckered into buying junk anyway.)

One thing that I dislike about the design of most presses is their usual setup for priming cases. I even hate decapping fired cases on the same press that I load on, but that's a subject for another time. I prefer to size my brass and bell their mouths slightly, then clean the ash and grit out of their primer pockets before I prime them. But then, I'm picky.

This procedure is a major reason for my use of presses with only one die hole at the top. Progressive and turret presses don't fit my usual loading scheme, although they're mighty nice sometimes. I prefer also to seat primers with a separate priming tool — but this is more luxury than necessity, an extra tool is not vital to a basic outfit. I dislike and distrust primer feeds and usually have trouble with them the few times I have to use them.

Old advice says not to handle primers with your fingers, because skin oil can mess 'em up. I've been handling them — carefully — with my fingers for over thirty years and haven't seen any adverse effect yet. But then, my skin is so dry, chips and pretzels write to me for beauty tips. For a long time, I didn't have a primer flipper, but used whatever container lid of the right size and shape was handy.

GUN DIGEST BOOK OF HANDGUN RELOADING

With your eyes and fingers, you can better tell whether a primer is upside-down or rightside-up. You can also more easily make sure it's seated square and level in the priming device before you pull a case down over it. If you keep your loading outfit trim and simple at first, you'll quickly develop that crucial gut sense of what's going on at every step. If you depend too much on tools that do everything for you, then you'll be stumped more easily when something goes wrong.

Makers who produce the best presses also produce excellent dies — usually. Dimensions and finish are crucial. If you can, get sizing dies that size your fired cases just enough for easy loading in your pistol or in the tightest chamber in your revolver. Be sure also that sized and mouth-belled cases are neither too loose nor too tight on properly sized bullets.

If they're too tight, they adversely affect the seating and the alignment of the bullets; if they're too loose, they don't grip the bullets tight enough for adequate confinement of the powder gas that produces propellant pressure when you fire. Loose necks make your pressures, velocities, trajectories and impacts erratic.

Roughly finished dies make loading rough — and that's bad enough. They can also damage cases enough to shorten case life. Also, poorly finished dies are frequently poorly made and aligned — so a poor finish can be both a problem and a symptom of other problems that aren't so obvious.

With so many superb brands of dies on the market, there's no reason to settle for less than the best — which includes a bunch of brands, not just one. Once you see a set of the best dies from Hornady, Lyman, RCBS or Redding, you won't put up with anything less.

One reason that I've favored RCBS dies all these years is the internal design of their seating dies — a cylindrical chamber above the mouth of the case aligns the bullet for seating. In earlier days, other makers' dies depended on the mouth of the case and the nose cavity in the seating punch to align the bullet between them. This system has given me several poor loads and ruined cases from time to time. Now some other makers use the alignment cylinder above the mouth of the case. Check for it in any dies you consider buying.

The only other absolute necessities for loading good ammo are a good powder scale and funnel. Nothing else is as necessary for safety and quality, although a good powder measure, case trimmer, micrometer and calipers come awful close to being as necessary as press and dies.

Unfortunately, the simple, economical powder scales from any maker can be excellent, adequate or poor. An excellent one is a triumph, because it's terribly hard to make a precision tool as dependable and durable as a good scale must be, yet make it at a good price. The most dependable and durable scale that I've ever used — the old 1940s Pacific — is no longer made, probably because it was so inconvenient to use. In its place, we have a plethora of convenient scales that can be tricky to use and sometimes even dangerous. Fortunately, the few dangerous ones are easy to spot and to stay away from.

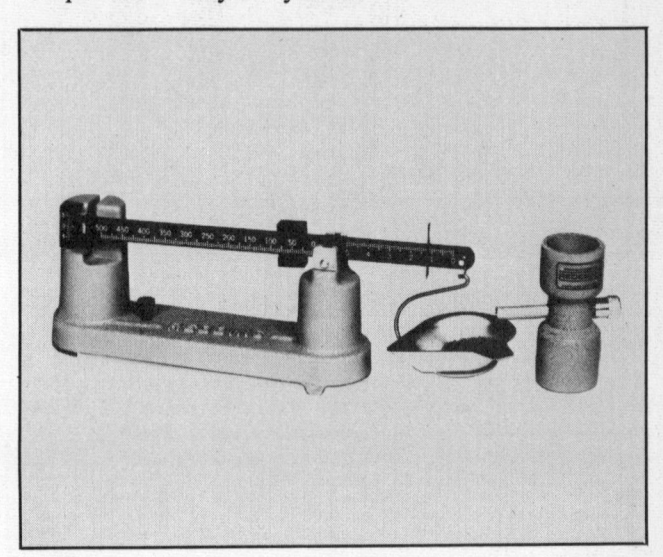

The basic scale, such as this one from Redding, may be different from most others on the market only in color, shape of the base casting and method of setting poises.

There's little difference in either the basic design or the quality of manufacture in today's simple powder scales. You can generally count on them to be excellent, but shouldn't overlook the possibility of getting a flawed one. The doodad that tells you when you have the right amount of powder for each case isn't something that can be off and still be acceptable. A slight error in measuring a maximum charge of a fast-burning handgun powder can be more than slightly dangerous — especially if the amount of error grows as your loading session continues.

Shun a scale that appears erratic in any way. Set one up level, with its pointer zeroed, and weigh several small check weights. Coins are good test weights for checking powder scales. Set the counterpoises in the beam back to zero after each test weighing and see whether the pointer returns to zero dependably time after time. If you have a

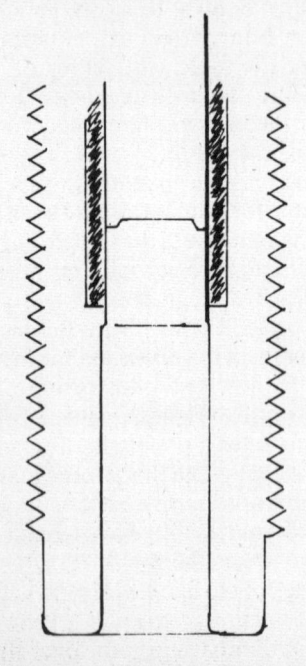

Some makes of seating dies leave an area where there's no support for the about-to-be-seated bullet except the case mouth and seating stem. RCBS dies don't do this.

choice, favor the scale with the deeper teeth along the top of the beam. Beam counterpoises can too easily hop from notch to notch along the beam — changing your charge setting by great amounts as you use the scale.

At light, medium and heavy weight settings (at both ends and the middle of the beam scale), check the consistency of weight readings that you set in two ways. For example, set the beam counterpoise for five grains and the tenth-grain counterpoise for zero-tenths. Weigh five grains of any convenient test substance (salt, sand, birdseed, powder — whatever); then set the main counterpoise for four grains and the tenth-grain counterpoise for ten-tenths. Don't be surprised if your test substance is heavier or lighter now.

You may not be able to find a scale that weighs the same amount of sample at both weight settings; 4 + 1.0 grains

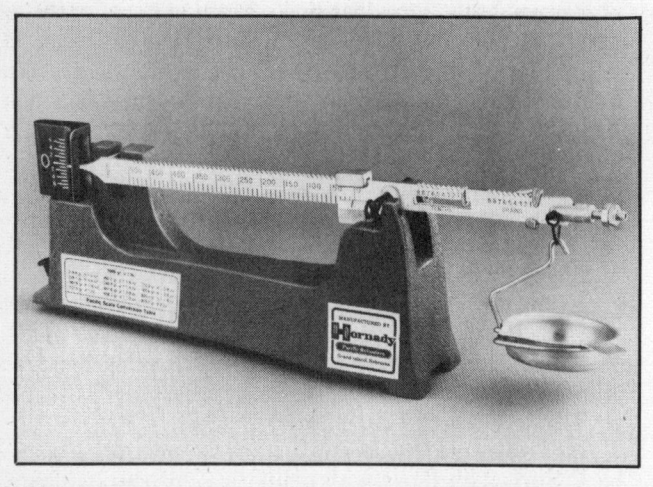

You can't handload for very long or in confident safety without a scale. Choose a good one on the basis of its accuracy and then details of construction, color, etc.

may not equal 5 + 0.0 grains. But don't give up yet — stick a label on your scale with a note to that effect and ever afterward be sure to weigh your charges the same way. It's a good rule never to use the tenth-grain counterpoise for a full ten-tenths (1.0 grain).

Also, don't be surprised if two apparently identical scales weigh the same coin as different weights. And, if you're loading handgun ammo only, errors at the heavy end of the scale beam are less important to you then they'd be to a fellow who carefully handloads maximum loads for big rifle cartridges.

If you want to be very careful and reasonably certain how far off a scale is, weigh three coins of different denominations — a penny, nickel and dime, say — separately, in pairs and all together. Record these weights, then have a pharmacist or other lab scientist weigh them on a dependable laboratory balance. You want to know what equals what!

Convert his gram weights to grain weights and you have calibrated the error range of your powder scale. A 3.3-grain load of Bullseye listed in a loading manual may read as little as 2.9 or as much as 3.7 grains on your scale and the error might be different at other points along the beam. (Multiply grams by 15.43 to get grains. Multiply grains by 0.0648 to get grams.)

Be wary of a scale if the beam swings jerkily or stops sud-

denly at any point in its swing. Accurate, consistent weighing depends on smooth, consistent swinging and stopping of the beam. A jerky swing or sudden stop may be caused by lint, dust, grit, chips or burrs on the beam bearings.

An engineer friend discovered an error of about four or five grains — around a nominal sixty grains — when he weighed the same test weight on two scales. By cleaning the bearings, he reduced the error to a few tenths of a grain. If careful, tender swipes wtih alcohol on a cotton swab take care of the problem of an erratically swinging beam, fine — if not, not so fine. (Be sure you leave no lint from the swab in the works!)

Don't even consider trying to "tune" faulty scale bearings. If a new scale has bad bearings, squall lustily at the maker of that scale if your first gentle request for a replacement doesn't get you a good one. If you may have damaged your scale yourself, get the maker to repair it for you,

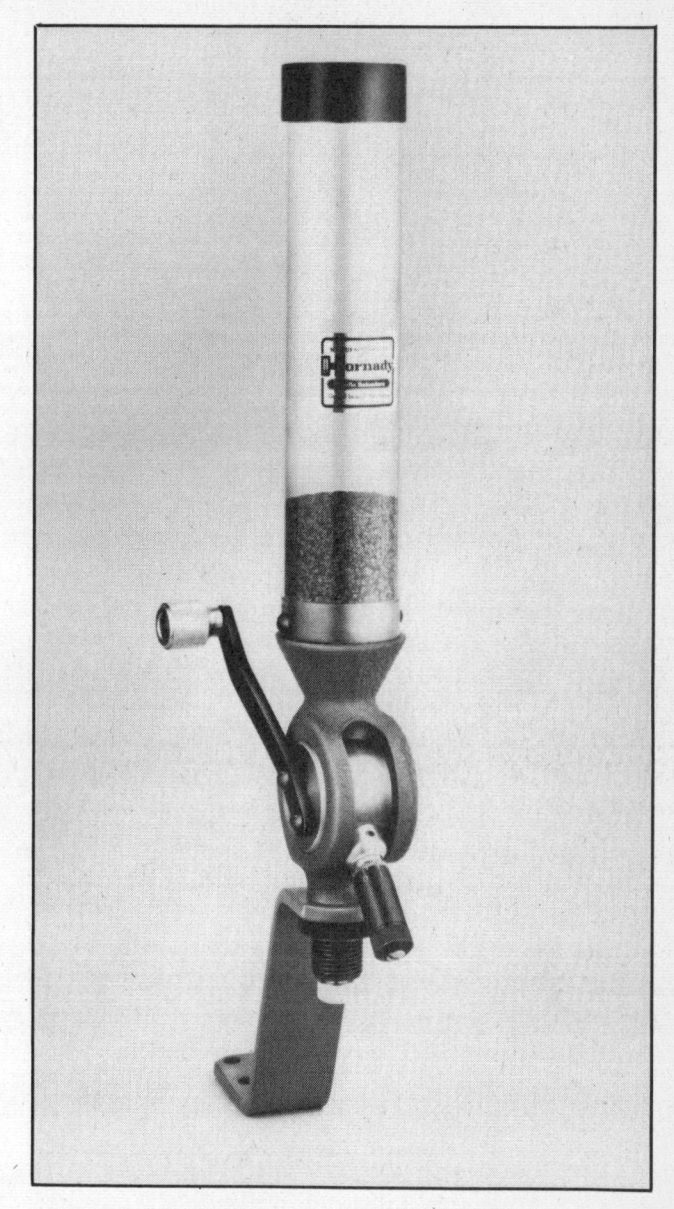

Most good powder measures, like this Hornady, work on the same time-proven formula. Other makes are usually just as good. Your choice may be a matter of loyalty.

GUN DIGEST BOOK OF HANDGUN RELOADING

This is a fine powder measure from Redding; precision ground from suitable metals. There's plastic in the drop tube and hopper. Those are the only places for plastic.

whatever the cost. He may want to replace it instead. But don't rely on the accuracy of any powder scale with a bad bearing and don't try to fix it yourself if it needs more than a simple cleaning.

With other makers using plastics to make powder funnels cheaply, the choice of funnels is simple; avoid 'em all except Hornady's aluminum ones. I hear and read encouraging words about static-free plastic funnels and haven't come across one yet that was truly static-free. Aluminum is static-free; plastic ain't. It's that simple. Get aluminum no matter what else your dealer tries to sell you, even if you have to order direct from Hornady. (Get two — who knows when Hornady will go to plastics, too?)

One tool neatly straddles the fence between necessity and luxury — the powder measure. It isn't absolutely necessary, but it sure does make handloading a passel of cases a lot easier. A good powder measure is hard to make cheaply. One nice, old measure that's no longer on the market lost its maker money with each sale, even when he sold it direct at full retail. But he wouldn't compromise its quality, so he finally had to drop it — despite the howls of handloaders who knew how good it was.

I have several good powder measures, my favorites being an old pre-Omark RCBS Uniflow with large and small measuring cylinders (new ones should be just as good) and the classic Belding & Mull. Others that I've seen lately — from several manufacturers — seem well made. Those with fixed rotors are superb for loading handgun ammo, once you determine what your best loads are and get rotors to match your powder charges.

Avoid measures made with rectangular measuring chambers. Stick with those with cylindrical measuring chambers — they're likely to be more accurate and easier to use. Flee in terror from powder measures with plastic slides or rotors (plastic hoppers or reservoirs are okay).

A friend and I tested prototype and production samples of an all-plastic powder measure and we found them to be wildly erratic. Static electricity held a few kernels of powder from each charge stuck to the sides of the internal funnel below the measuring chamber — until the buildup was heavy enough to be swept along with a later charge. Most charges were too light, but several were hair-raisingly heavy. We were never able to get any consistency, since we never found a way to kill the static electricity that built up so quickly in those plastic parts.

There you have the gotta-haves. Every other handloading tool, however nice it may be to have one on your bench, is a frill to some degree. Add as many as you like, as often as you can afford 'em — but use the same kind and degree of care that you've used to select the best basic tools. The best tools are best in two vital ways: They do the job right with a minimum of mislicks and effort, and they stay with you until you wear them out. That takes awhile!

I know one handloader who has worn out a set of top-quality tools with careful handloading. With only one press, he regularly turns out more rounds per year than anyone else I know. He loads great quantities of test and field loads for an extremely wide range of handgun and rifle cartridges. This great variety means a lot of adjusting and readjusting of his tools — therefore much more wear than he'd get from handloading a pile of ammunition for one gun.

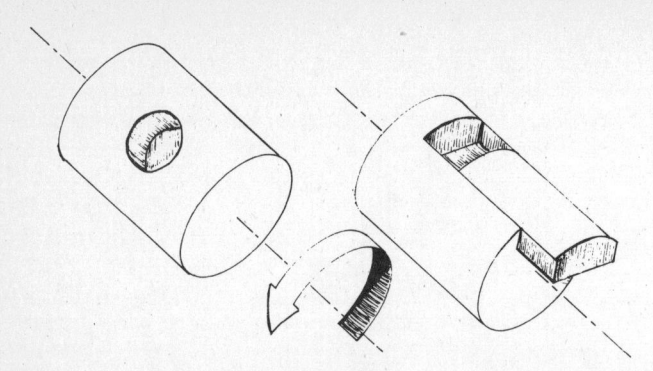

Powder measures are best fitted with drums that adjust by changing the depth of the cylindrical cavity. Slides, as in the right measure, can be often non-repeatable.

High-quality tools can make life at the reloading bench perceptibly happier. Choose quality stuff from leading makers. Hornady is one of several manufacturers who'll stand by their products. What more can you ask of them?

It took this pal of mine a passel of years to wear out those tools. In all those years, he would've gone through a binful of economy tools — enough to pay for several outfits like the one he wore out, with less satisfaction all along the way. He bought wisely at the start and I know he's never regretted paying the initial extra cost of quality tools.

I'm also sure that you can learn, simply by reading and heeding good advice, a vital lesson that my pal and I learned the hard way over a bunch of years and paid a wad of dollars for. And this is it: Get the best (I said that already, didn't I?), but don't depend blindly on the accuracy, alignment, calibration, consistency, durability or strength of the tools you buy. Fancy ads and colorful paint jobs do not a cartridge make.

Go to the name brands first, because that's where you're most likely to find the best quality at manageable prices. The best big makers have a bone-grinding tough time manufacturing top-quality tools that they can pass along to you and make a profit on without taking your arm and leg. Smaller outfits have to cut corners somewhere to produce anything vaguely resembling loading tools at those same prices. The odds don't favor your chances of getting both quality and economy in off-brand tools.

Learn what to look for and keep looking until you find it. "Good enough" isn't — for ammo or the tools you load it with. — *Ken Howell*

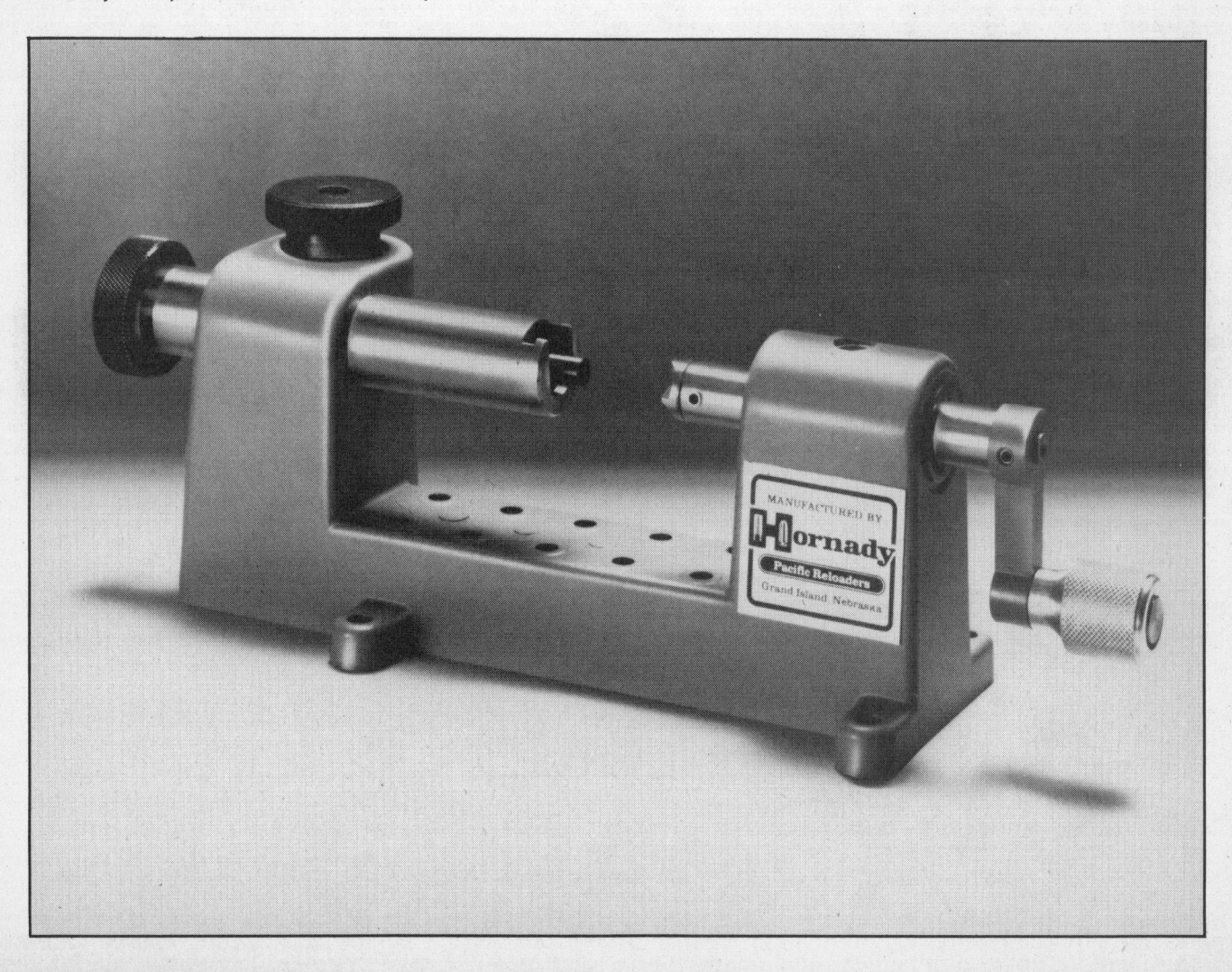

Chapter 3

SPEEDY RELOADING: To Progress(ive) Or Not

Handgunners Shoot More: Reloading Handgunners Reload More Surveying The Tools That Ease The Burden

SINCE THE first bullet was loaded in the first metallic cartridge case, reloaders have been looking for a way to speed up the process. Although some reloaders consider the necessary steps required to produce shooting ammunition as a labor of love, many others consider actual reloading a nuisance. Many just don't have the time in today's busy society to spend more hours at the reloading press; they would rather spend their free time shooting. Then, too, there is the action shooting crowd whose pursuits demand vast quanities of ammunition for both practice and competition.

Due to all of these factors, production-type reloading presses are getting to be the vogue among handgun handloaders. Only a few years ago, companies specializing in such presses had the market to themselves. Now many of the larger companies manufacturing reloading equipment are pushing their progressive-type presses.

The concept of speedy reloading is not new. In fact, I was introduced to reloading on the old Lyman All-American press, which is described best as a forerunner of the progressives. It contained only one shell holder, but the revolving turret head had four die stations. The unit was meant to be used with three dies and a powder measure. In practice, the turret was rotated and the cartridge case run into die or powder measure with each stroke of the reloading handle. An automatic primer feed on a swinging pivot came around and deposited a primer in the punch under the shell holder.

In theory, the old Lyman All-American sounds fast and efficient, but it was not. For one thing, its turret head was stiff and required a strong hand to rotate it. It started that way and remained so even after many thousands of rounds were loaded. Also, long loading sessions tended to leave blisters on one's hand where the turret was grasped for turning. After many, many thousands of reloads with the Lyman All-American, I decided the fastest, easiest method was to run several hundred cases through each die before pivoting the turret. That left me with the equivalent of a single-stage press.

There are still turret types around. While not true progressive reloading presses, they are intended to speed up the reloading process. However, with the advent of near-automated progressive presses, the days of the simple turret press must be numbered.

Facing page, Lee Precision offers presses at several levels of sophistication. Clockwise from lower left, there's the single-station 2001 Challenger; above it is the Lee Turret Press with Auto-Index; the Lee Turret Press for manual indexing and the Lee Progressive 1000, set up for fully automatic operation. Right, the 1000 now includes a case sensor to prevent feeding of primer unless a case is present in the station.

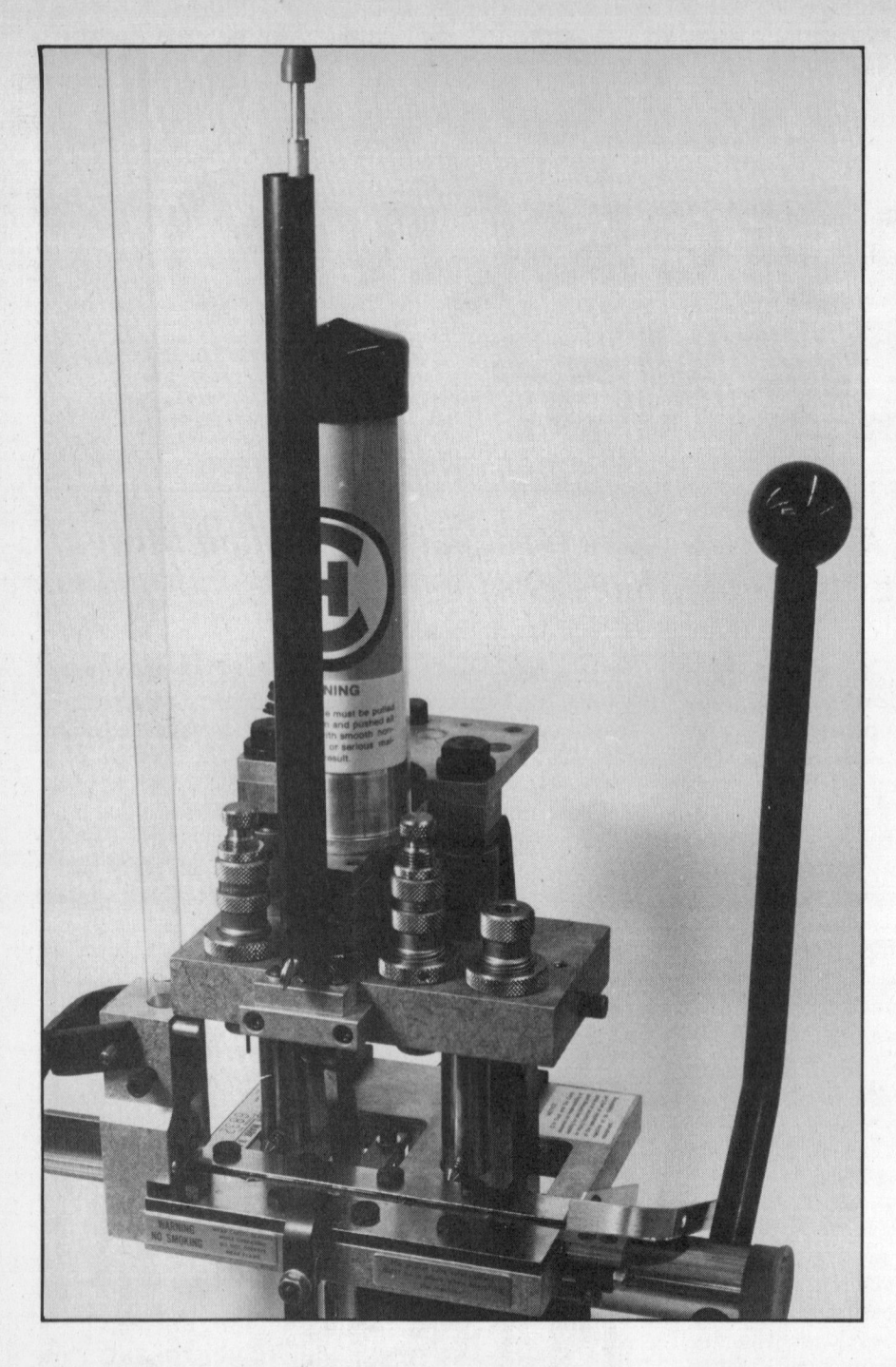

This is the Mark V-A Auto CHampion, from C-H Tool & Die Corporation. It can be had for .38/.357; .45 ACP; .44 magnum or 9mmP. Empty cases feed automatically in a straight line through the four stations. Station 1 resizes and decaps; station 2 primes, expands and bells the neck as well as charging the case with powder. At the user's option, the bullet can be seated in station 3 and given a roll or taper crimp in station 4. Alternatively, the die in station 3 can be set to seat the bullet and crimp the neck simultaneously.

Lyman still makes a turret press called the T-Mag. It is much advanced over my old All-American, but still uses a single shell holder. The turret has six stations instead of the four I grew up with. Another is the Lee Turret Press which has several interesting features such as optional automatic indexing of the turret. That would eliminate the sore hands I was plagued with. In fact, the Lee Turret Press can be upgraded to fully progressive, if the owner desires to go to the expense. It is the basis for their progressive model which will be discussed. However, as a simple turret press, it uses a single shell holder.

Another attractive feature is interchangeable turrets. You can keep seperate turrets set up for favorite calibers, complete with powder measure, and switch calibers in only a matter of seconds. That's a most attractive feature for those of us who load many different handgun calibers.

Another turret-type press is the Redding Model 25 which has a full half-dozen die stations and an operating handle to rotate the turret. The Redding Model 25 also is sturdy enough to handle such tough reloading operations as case forming or bullet swaging.

An interesting variation, still one step down from being a progressive press, are the Model 444 and Model 444X presses from C-H Tool & Die Company. Instead of a rotating press turret, these heavy-duty pieces of equipment have four die stations and four shell holders. A case can be processed through each step one at a time, or the operator can develop a rhythm and insert and progress cases with every pump of the press handle. C-H also sells push-button powder measures and automatic primer feeds for these

The Dillon Model 550 has four turret stations and all operations are automatic except for inserting the empty case, inserting the bullet in the neck of the charged case and manipulating the press's operating handle.

tools. The only complaint I've ever heard from 444 and 444X users is that one must buy four shell holders for each caliber loaded.

Getting into the true progressive reloading presses, we find at least a half dozen by popular makers. All have one factor in common. They use rotating shell holders or plates, while the dies remain stationary. However, in operation they differ greatly. Some of the more basic models rely on the operator to see that a primer is seated and that powder is dispensed. Some of the newer versions carry matters farther, because they do the priming and powdering automatically. And, in a couple of instances, they even have automatic case feed tubes.

One of the big names in modern progressive reloading presses is Dillion Precision Products, Incorporated. Dillion's sole business is high-production reloading equipment. Their earlier #450 press gained wide acceptance among handgun reloaders even though it had manual indexing and priming with four die stations.

The new Dillion Model 550 has automatic powder and primer systems and interchangeable die holding tool heads. This outfit uses standard ⅞x14 die sets and is advertised to handle over 115 different calibers; I know one fellow who loads the .50 Sharps with a Dillion.

The Model 550 has four die stations, one being for a powder measure. It has an automatic loaded round ejection system, and auto priming and powder measuring at the pull of the handle. All the operator must do is insert empty cases and bullets manually and, of course, pull the handle. It is as easily used for rifle reloading as for handgun. A conversion kit for additional calibers is $23.

Even newer than the Model 550 is the new Dillion Square Deal which retails for only $135, complete with adjustable carbide sizing die. The Square Deal has the same automatic features of the Model 550, including automatic indexing, powder measuring, priming and ejection of load rounds into the collection box. Available calibers are .38 Special, .357 magnum, 9mm, .44 magnum and .45 ACP. As the Square Deal arrives from the factory, it already is adjusted. All the operator need do is select the proper bullet seating depth and the correct powder charge. As with all other Dillion products, there is a thirty-day return policy, if the buyer is not satisfied, and a lifetime warranty.

The RCBS 4x4 was introduced in 1985. The green progressive has four die stations, again, with one intended for powder measure, and a shell plate for four cartridges. It does not rotate automatically, but that can be a blessing, because it allows the press to be used just as a single-stage reloader, if smaller batches of ammo are required. It does have a primer feed, but it is not automatic. The 4x4 uses standard dies and special shell plates available for most calibers.

Hornady's newest entry in the progressive market is the Pro-Jector. As with the other progressives, this one uses standard dies, but has its own five-station shell plate. With the Hornady tool's automatic indexing and priming, the

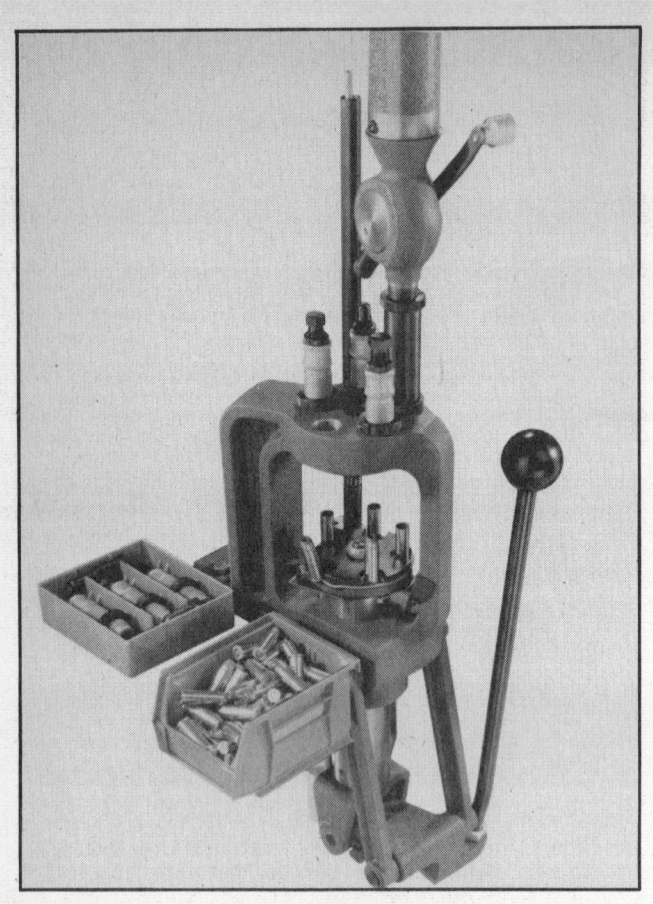

Hornady's Pro-Jector press is highly automated, with completed cartridges flipped from the shell plate into the catch box by the "Brass-Kicker" arm at the bottom of the ram stroke. More photos start on facing page.

and all the user has to do is place a bullet in the belled mouth of the case. All other operations are done for him, including case feeding and ejection.

As C-H states in its catalog, the Auto-Champ is for the professional, and is not intended for the hobbiest. It comes complete for one caliber and weighs about thirty-four pounds. Calibers available are .38/.357 magnum, .45 ACP, .44 magnum, and 9mm Luger.

Actually, there are some pretty fancy speedy reloading presses. C-H also makes a Manmatic that weighs 525 pounds for reloading .223 Remington rifle ammunition. It costs $13,700. RCBS makes a nearly fully automated progressive billed as the Green Machine for .38 Special/.357 magnum, 9mm Luger, .44 Special/.44 magnum and .45 ACP. It can load up to six hundred rounds per hour. They don't even picture it in their catalog anymore, but there is a toll-free number for those seriously interested. Those who

The Lee Progressive 1000 is set up with the automatic case feeder and a four-column magazine for empty cases. As with most presses that seat the primer at the end of the ram down-stroke, it needs to be mounted on a solid bench to resist movement.

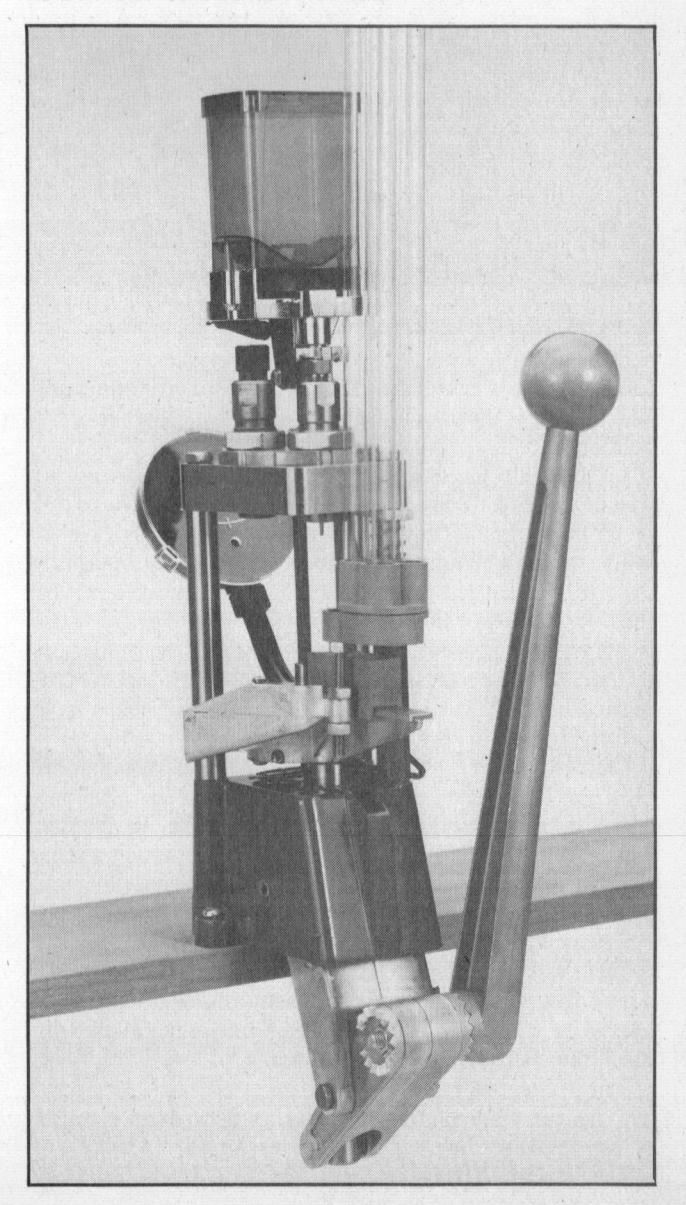

user inserts the cases, drops powder and, of course, pulls the lever handle all the while. Extra shell plates for both RCBS and Hornady tools run about $25 each. As with the RCBS 4x4, Hornady shell plates are available for most all handgun and rifle calibers.

Lee Precision has made quite a name for itself over the years. Interestingly, the Lee name became known to reloaders with a kit with which the user pounded cases into dies with the aid of a wooden mallet. Such loading was slow, but the resulting ammunition functioned just fine. Those kits still are being produced, but Lee has come full circle now with an offering in the speedy progressive tool area.

Lee's Progressive 1000 has automatic indexing, case feeder, automatic priming, automatic powder charging and loaded-round ejection. One interesting feature of the Lee tool is that the shell plate carriers are sold as a total unit for $52.98. Each shell plate carrier comes with the shell plate itself, Auto-Prime, case ejector, Auto-Index, plus some spare parts. Additionally, the dies and powder measure are contained in a die head which can be changed in mere seconds. The Lee Progressive 1000 is available for almost all handgun calibers, plus such rifle rounds as there .222 and .223 Remingtons and the .30 M1 carbine.

Probably the most sophisticated of the progressives is the Auto-Champ sold by the one of the oldest and most respected names in reloading equipment, C-H Tool and Die Corporation. The Auto-Champ is nearly fully automated

The vertical black tube, seen here, is a steel housing that encloses the primer feed tube to help contain the force of a possible detonation of the primers within the tube. Press is the Hornady Pro-Jector and Vari-Air can is handy for cleaning spills.

With dies and powder measure in four of the five stations, we're ready to start loading .44 magnum cases. It's a good idea to wear shooting glasses when reloading and you'll note a pair's on hand. Below, the ram on the Pro-Jector is a husky one!

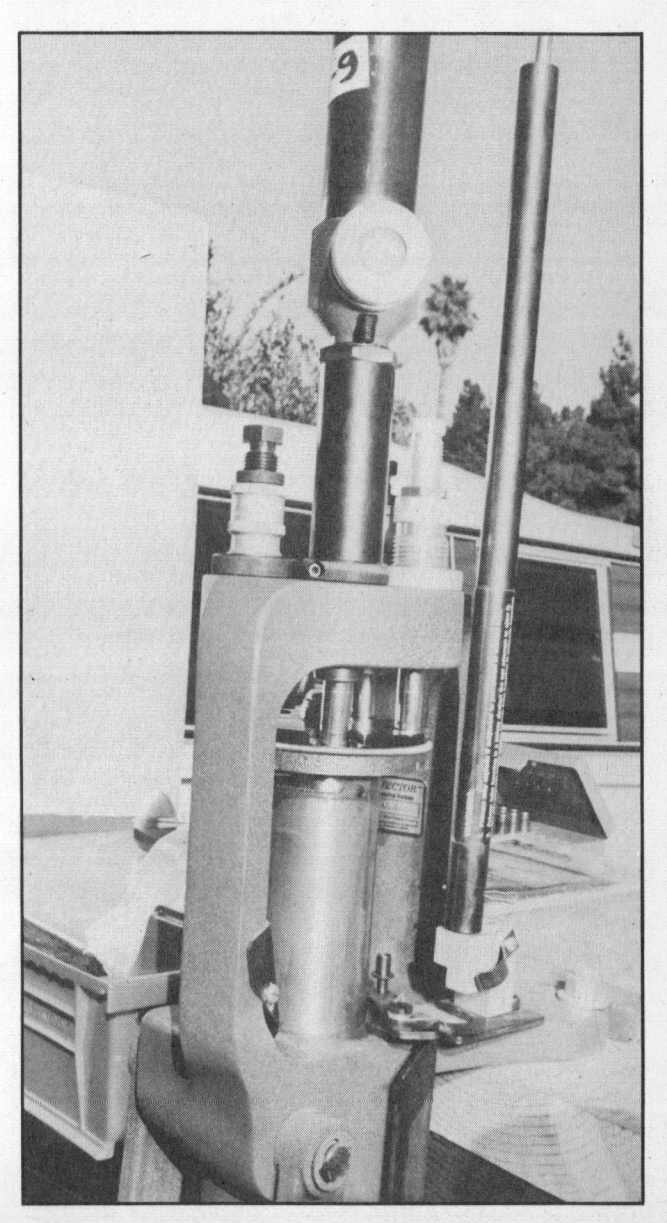

read the C-H catalog for the first time might be surprised to find a black bordered full-page warning asking, "Is The Auto-Champ For You?"

This brings us to a serious question concerning safety with the progressive presses. The C-H catalog statement is a warning against primer detonation in the automatic tube feed. They say it is unlikely, but possible. They supply a strong safety shield to go between the user and the tool, just

Arrow points to the small plastic cup at the end of the brass tube; it's designed to catch the spent primers and needs to be emptied occasionally.

GUN DIGEST BOOK OF HANDGUN RELOADING

Above, from left, resize/decap die, expander die, powder measure and seating die. Fifth station permits seating the bullet and applying the crimp in two stages, should the operator prefer. Left, marked tape on powder measure reservoir identifies powder.

in case. They also request that all buyers read the instructions thoroughly and keep untrained operators away from the machine.

Worry about the primers in a tube is genuine. A single primer going off by itself may not amount to much, but all of them going off together in a tube stack can be totally dangerous. I know of experienced handloaders who refuse to use automatic priming systems for just that reason.

Another safety factor to consider is user error. This is a real consideration, as I found out personally. When reloading with any tool, you should give it your total concentration. This is doubly true of progressive-type presses, especially those whose powder is dumped manually. It is extremely

A fired .44 magnum case has been inserted in the shell plate at station 1. The plate is indexed automatically as the handle is operated.

42

GUN DIGEST BOOK OF HANDGUN RELOADING

Down comes the handle and up goes the ram to start the first case on its way to the resizing/decapping die. A carbide die is used here and case needs no lube.

At the top of the ram stroke, the spent primer has been punched out of the primer pocket, to head down the tube to the plastic container shown on page 41.

The handle goes up, the ram comes down, indexing the first case to station 2 and leaving station 1 open to receive the next empty case to be processed.

GUN DIGEST BOOK OF HANDGUN RELOADING

Second case has been inserted and it's ready for you to pump the handle again. Primer was seated in the first case at the bottom of the ram down-stroke.

Now we have three cases working. At the top of the ram up-stroke, this time, actuate the rotor on the powder measure to drop a charge into station 3 case.

At the next cycle, with four cases in play, we start the base of a bullet into the neck of the charged case at station 4.

GUN DIGEST BOOK OF HANDGUN RELOADING

If you look closely, you can see that the bullet we set in the case neck at station 4 has been seated and crimped.

Coming down the home stretch, the completed cartridge is being approached by the arm of the Brass-Kicker.

easy to forget to charge a case or, conversely, to double charge one.

The day I first used my RCBS 4x4, a friend dropped by just as I started the first cases into it. Stupidly, I visited with him while working the press handle. Later, at the range, when I started to fire those first fifty rounds of .38 Special, I had three failures due to no powder in the case. That taught me a lesson — and I mistrust progressive presses to this day. Of course, that does not keep me from using one; it makes me a more careful handloader when operating it.

Consider this: Progressive presses are intended for those who shoot a lot, such as action-type competitors. When shooting fast, it is virtually impossible to stop, if a round sticks a bullet in the barrel because of a lack of powder. If a round is fired behind it, the best one can hope for is a ruined barrel. This is a serious consideraton when indulging in speedy reloading. Fortunately, most of today's automated machines that dispense powder automatically are reliable, if the user remembers to keep the powder hopper full.

Progressive presses have their place. Some competitor

Ejection of the completed cartridge takes place within the flick of an eye, over the course of just a little ram travel. In the four-photo sequence, here and on the facing page, you can watch it making the final dive to the box.

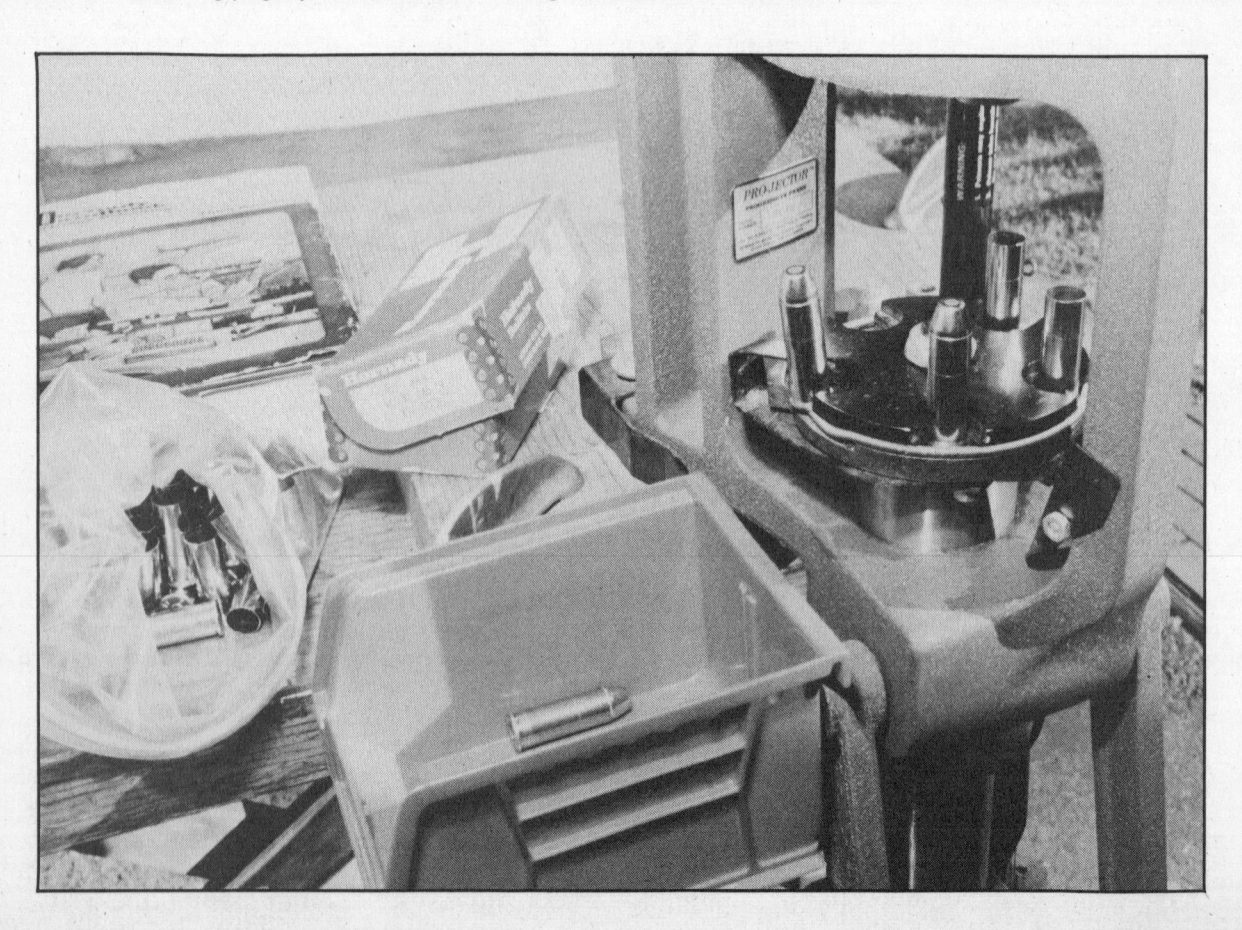

46 GUN DIGEST BOOK OF HANDGUN RELOADING

Fairly well out of the slot and heading to join the rest of the herd — no turning back now...

...and you can take it from there, right? If you think it's an easy project to manipulate a Hornady Pro-Jector and a Nikon F3/T equipped with a flash, simultaneously, try it sometime — a real challenge!

types would be hard put to do the practice necessary without one. However, the average reloader should do some deep thinking before investing his money in such machinery. One factor is that progressives can be expensive. They run from $135 for the Dillion Square Deal to about $700 for the C-H Auto Champ. Additionally, shell plates, holders, carriers, and so forth are likewise expensive for changing calibers.

Progressive presses are nice. In their many forms they are the state of the art as far as reloading tools go and anyone who appreciates fine machines cannot help but admire them. Before diving into one, however, the handgun handloader should ask himself a few questions. Is it really necessary or is he just buying it for prestige? Also he should ask himself whether he has the dedication necessary to operate it safely. And he should consider whether he intends to load hundreds of rounds of the exact same handgun handload or will he tend to bounce around from bullet to bullet, powder charge to powder charge. If the latter is true, a less than progressive press still may be the best choice. Does the shooter actually expend hundreds of rounds of handgun ammo a week? Or is that the figure expended in many months or a year?

If, after due consideration, the shooter still feels a progressive is for him, he has an astounding assortment to choose from nowadays. Like computers, they are progressing so rapidly that it is difficult to keep up with what is currently state-of-the-art. And who knows what is on the drawing boards for reloaders in the 1990s? — *Mike Venturino.*

*The Process Is More Efficient —
Easier, Quicker And Safer When You...*

...ORGANIZE YOUR RELOADING

It's a good gun — a Colt Commander in .38 Super with a bunch of custom features — but if you want to get a really good performance from it, you'll want to produce quality ammo. And that means organizing your reloading.

CHAPTER 4

RELOADING WRITERS don't spend a lot of time dwelling on the subject, but you can get hurt at the loading bench. Moreover, you can get hurt at the range when you shoot ammo hastily loaded without concern for the normal commandants of safety. That it doesn't often happen is testament to the fact that most handgunning handloaders are pretty careful. But once in a while, someone misses a step, doesn't check a powder level, substitutes one powder for another or mishandles primers. The results can be tragic. More often, the result of carelessness is simply a waste of time.

The only logical preventive measure is care — careful attention to correct procedures, careful record-keeping, careful handling of components — just plain watching what you are doing. We all screw it up once in a while, but the possibility of a catastrophe can be reduced with the material we'll discuss in this chapter. The matter at hand is organization of the loading process, because organized reloading is probably going to be safe reloading.

Still, there are other reasons for loading bench organization. For one thing, the process is more efficient and more likely to produce quality ammunition when it's organized. It's easier, too; organized reloaders don't waste energy in correcting mistakes. And economical use of available time means the the reloading is quicker.

This is what we are after: procedures; ways of doing things that result in better use of time and make the handloading process safer and more likely to produce quality ammo. I am not going to go back through the reloading process — Dean Grennell handles that one beautifully in Chapter One. What I'll talk about in this chapter is organizing your loading bench and equipment to get the most out

This is one example of a well-planned, well-organized loading bench. It has served the author well for better than ten years and shows no sign of serious wear. With the exception of case tumblers, which are mounted in other locations in the shop, all tools are used on this bench. Shelves above the bench provide storage for the tools when not in use. There is also room for some of the author's varied supply of brass and reference books.

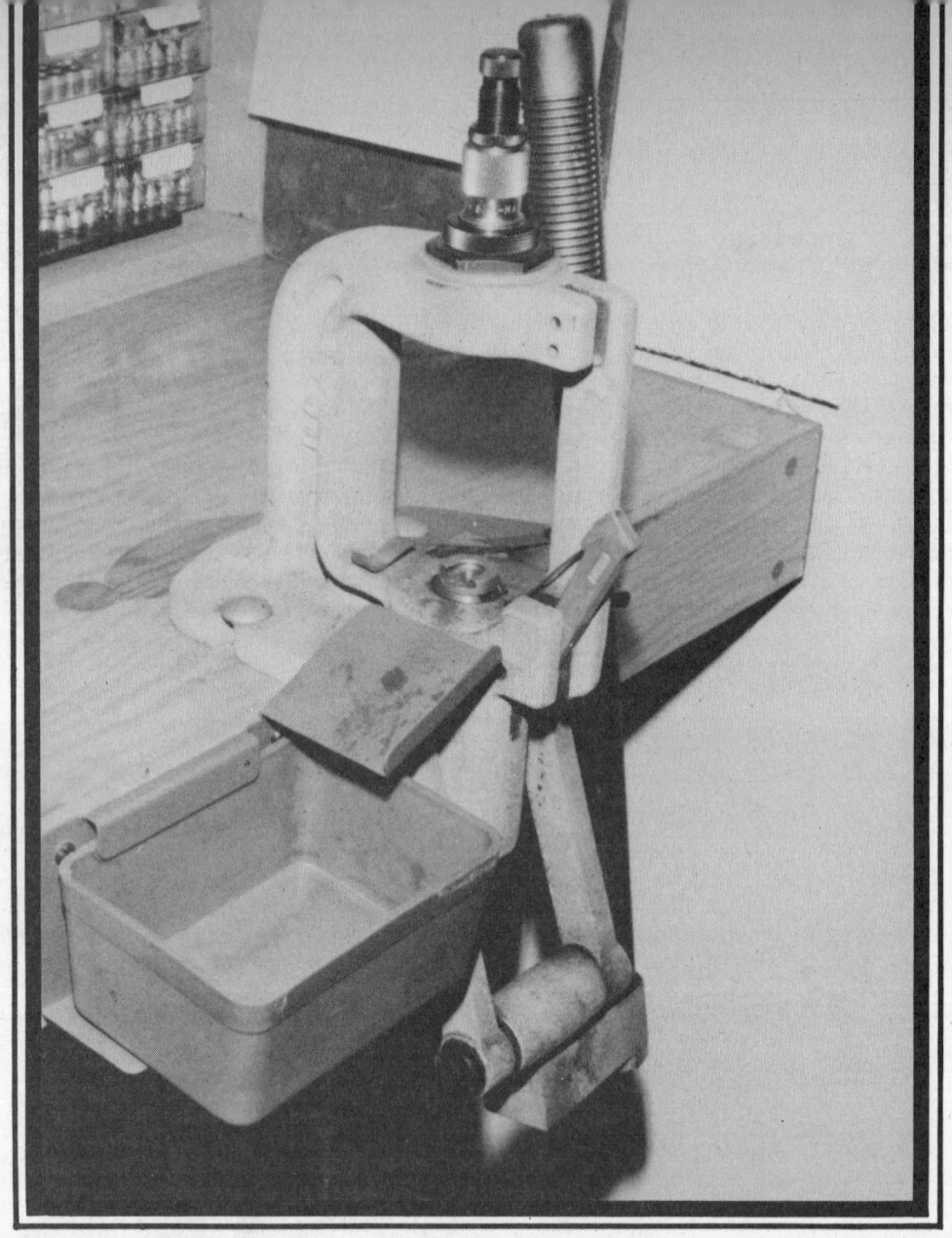

The heart of any loading bench is the press and you'd have to put up a battle to get this worn Rockchucker away from Clapp. For right-handed loaders, the press should be to the right side of the bench as shown. The Rockchucker is mounted far enough away from the wall that there is still access to the space on that side.

of it. We'll also spend some time on record-keeping.

It's certainly true that all reloaders need a place to do their thing. While the firearms press often mentions super-compact loading setups that fold up and store under the bed in the master bedroom, the fact remains that more guys probably bolt their press down in some permanent location. Sure, you can put everything on a plank and C-clamp it to a porch railing or to Mom's kitchen table. Lots of good ammo results from such arrangements, but if you are the sort of reloading handgunner who will spend the bucks to buy such a book as this one, you have a loading bench.

Certain things are true of all good loading benches. First, whether they are "stand-up" or "sit-down" benches, they will position everything you need where you can reach it. I cannot personally understand how anyone would want to stand up to reload, but friend Dean Grennell has loaded zillions of rounds more than I have and he prefers to stand up. My loading bench is low and, when I'm seated at the bench, its top is just a couple of inches above normal desktop height. This positions the main tool, a Rockchucker, so I can see everything that's happening in the press frame.

When I built my present loading bench, I was fresh from bitter experiences with another one. I had several complaints with that hastily improvised model and was determined to solve them with a new bench. For one thing, the old bench was too wide. Positioned against a wall in my garage, the back portion of the old bench was far enough away from easy reach as to become a depository for all kinds of accumulated junk. I reasoned — correctly as it turned out — that a narrower bench wouldn't have those out-of-reach areas that couldn't serve any useful purpose. The new bench is a scant twenty inches wide and I couldn't be happier with it. The narrowness of the bench serves to constantly remind you to put tools away when they are not in use and are in the way. In effect, it tends to enforce organization.

I also wanted a loading bench with tools (other than the permanently installed Rockchucker) mounted so they could be put into use quickly, then moved out of the way when no longer needed. Most of all, I wanted the mounting system to be free of bumps, edges and corners that seem to forever leap out at me and cause the spills and drops which induce language that my mother never taught me. The use of case trimmer, priming tool, powder measure stand and other necessary devices mounted on boards which were in turn C-clamped to the bench was not for me.

Finally, I wanted my loading bench to be heavy and rigid. Despite the industry claims about loading press stress being handled within the press itself, there are vi-

A closer view of the right hand side of the bench shows the open-faced storage boxes used to house bullets. The boxes are about six inches deep so there's room for just one row of bullets. Frequently used tools, such as the allen wrenches and screwdrivers, hang from hooks under the lower shelf. A hardware store small parts cabinet is in the center. It's used for shell holders, expander plugs and all the other little fittings that want to stray.

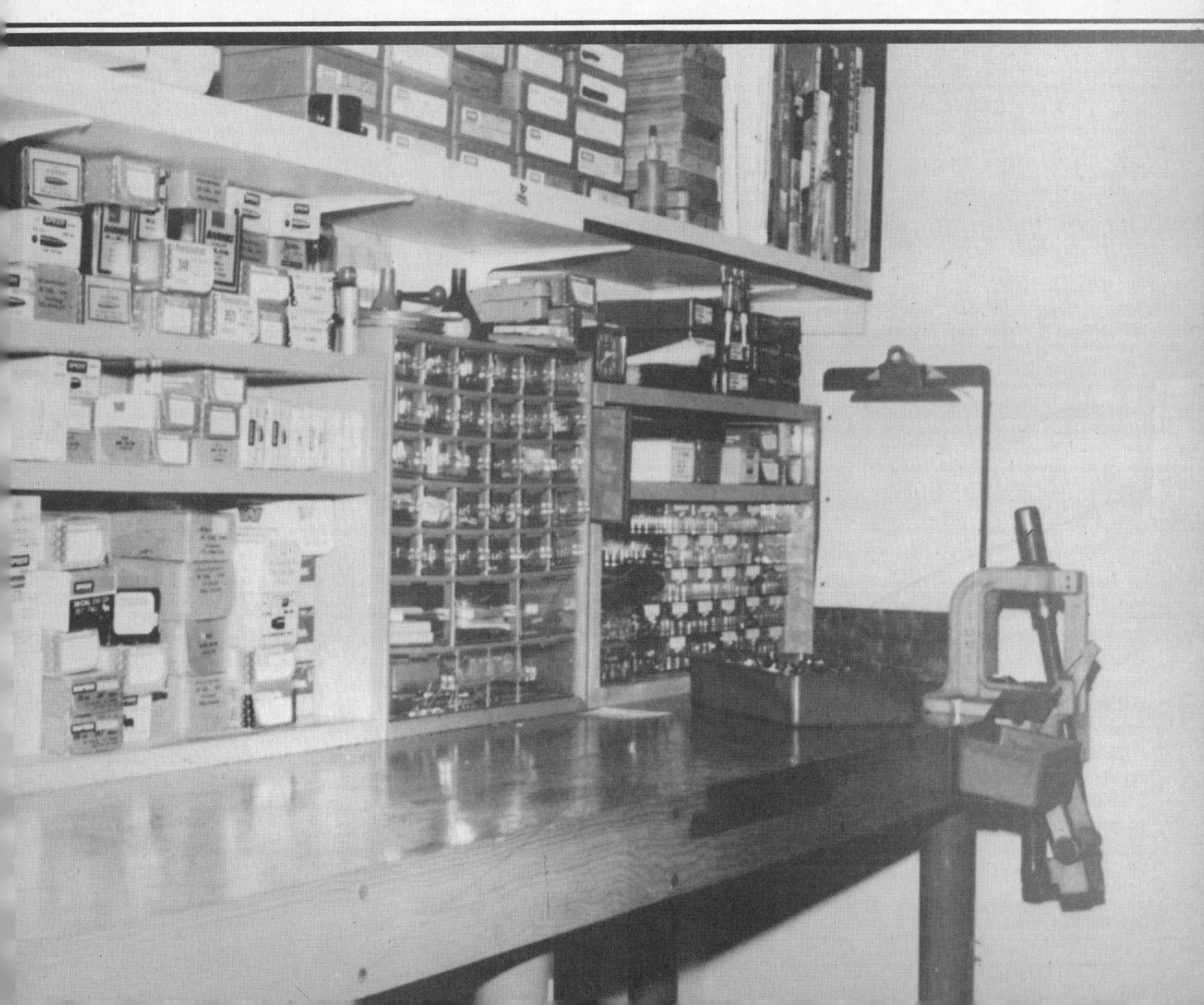

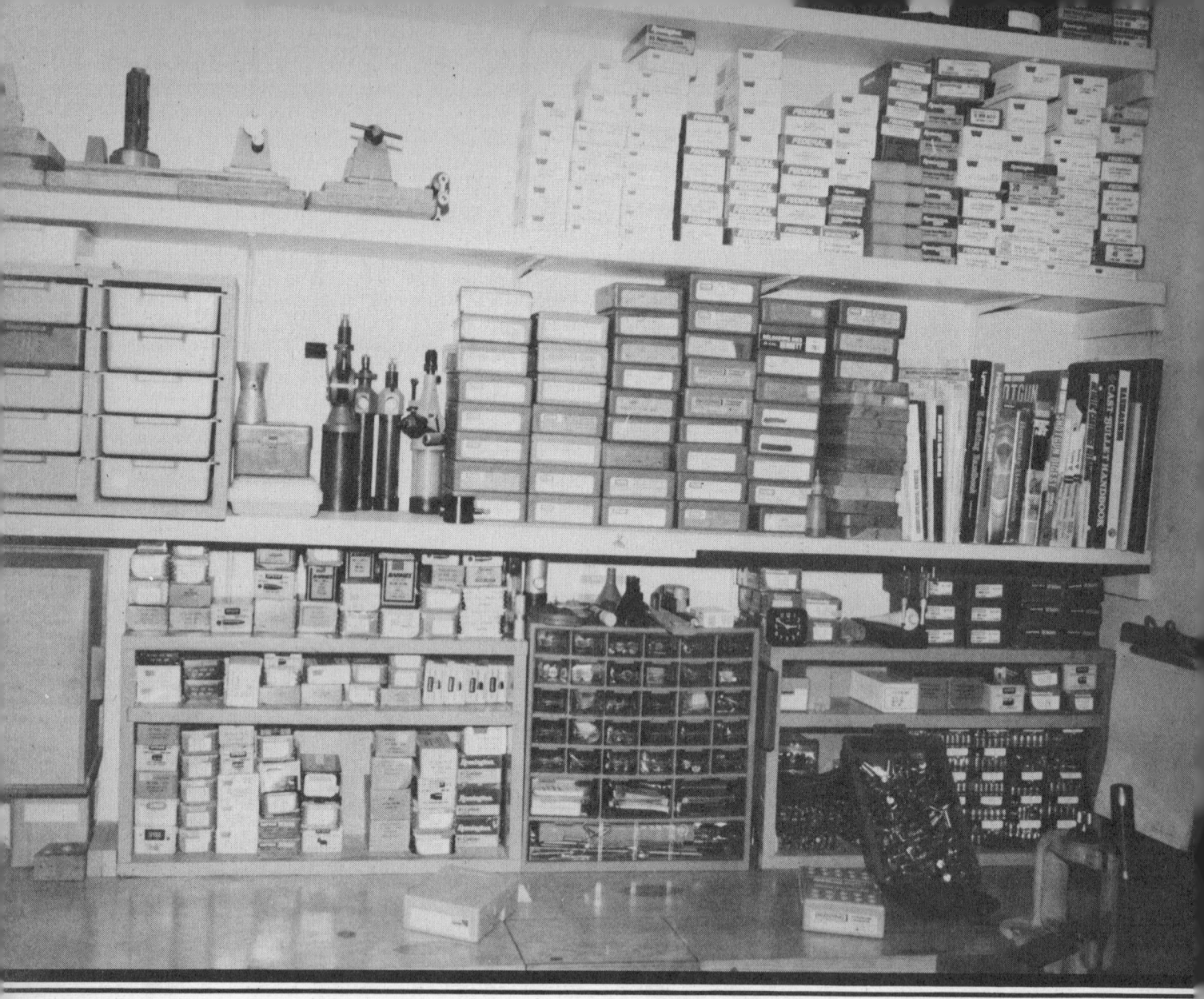

Straight-on view of the bench shows reference books on the right of first shelf and a cabinet of trays on left. The plastic trays are used for brass in various stages of processing. Seen above the trays, on the next shelf, are tools mounted on insert boards. Virgin brass is stored in the original cartons on the right side of the same shelf. The top shelf is barely visible in this view, but surplus film cans are there, holding brass which is load-ready.

brations produced by reloading. If the bench on which the press is mounted isn't appropriately heavy, those vibrations can have adverse effects when transmitted through the bench to other tools. The most obvious of these effects is changes in the reading on powder scales.

After lots of head-scratching and plan-sketching, I came up first with dimensions, then a means of constructing what I hoped would be the ultimate loading bench. I have already mentioned that I prefer to load from the seated position, so desktop height is fine. After experimenting with mock-ups of several widths for the bench, I settled on twenty inches.

The twenty-inch width turned out to be just right. Since the bench goes against the wall of a one-car garage in a California condominium, it doesn't stick out far enough to preclude parking a car. It also is right in the sense that it's deep enough to give up six inches of space on the back edge to the heavy wooden box units I use for housing bullets. On the back edge of the bench, these boxes are out of the way, but I can see at a glance exactly what I have in stock in the way of bullets.

There's also a typical hardware store small-parts cabinet in the center of the back edge of the loading bench. That's for the myriad of tiny doo-dads and thingamajigs that are essential to getting things done. (It's also one of my pet peeves! Doesn't *anyone* make one of these multi-drawed plastic units that will withstand realistic use?)

Twenty inches wide it was, but how long? Eight feet would enable myself and one of my reloading cronies to work at the bench without bumping into one another. It's also the standard dimension of commercially available plywood, but that's just a happy coincidence.

Dimensions thus established, it was time to begin construction. As already mentioned, I wanted to make this bench *heavy* and rigid enough to last for years to come. So I bit the figurative bullet and bought a sheet of the best grade of three-quarter-inch plywood and a lesser-priced sheet of three-quarter-inch particle board. The two sheets were the standard size of four-by-eight feet. Out of these, I proposed to laminate a bench top.

My friend and shooting partner, Hal Hobel, used his venerable old table saw to slice the material into sections for me. Each piece of material howled through the saw, coming out the other end as two pieces twenty inches wide, plus a surplus section about eight inches wide by eight feet long. The surplus sections of both plywood and particle board were put aside for another purpose.

Both the particle board and plywood now are ready for lamination. Since the particle board is not the best material

Detailed view of the bench construction. This photo was taken from just off the floor, looking up at the underside of the right end of the bench. The leg is a section of heavy pipe and there is another like it out of sight to the left. The legs are welded to a flat plate which is then securely bolted to the underside of the bench. The bolts which hold the Rockchucker in place pass through the leg unit.

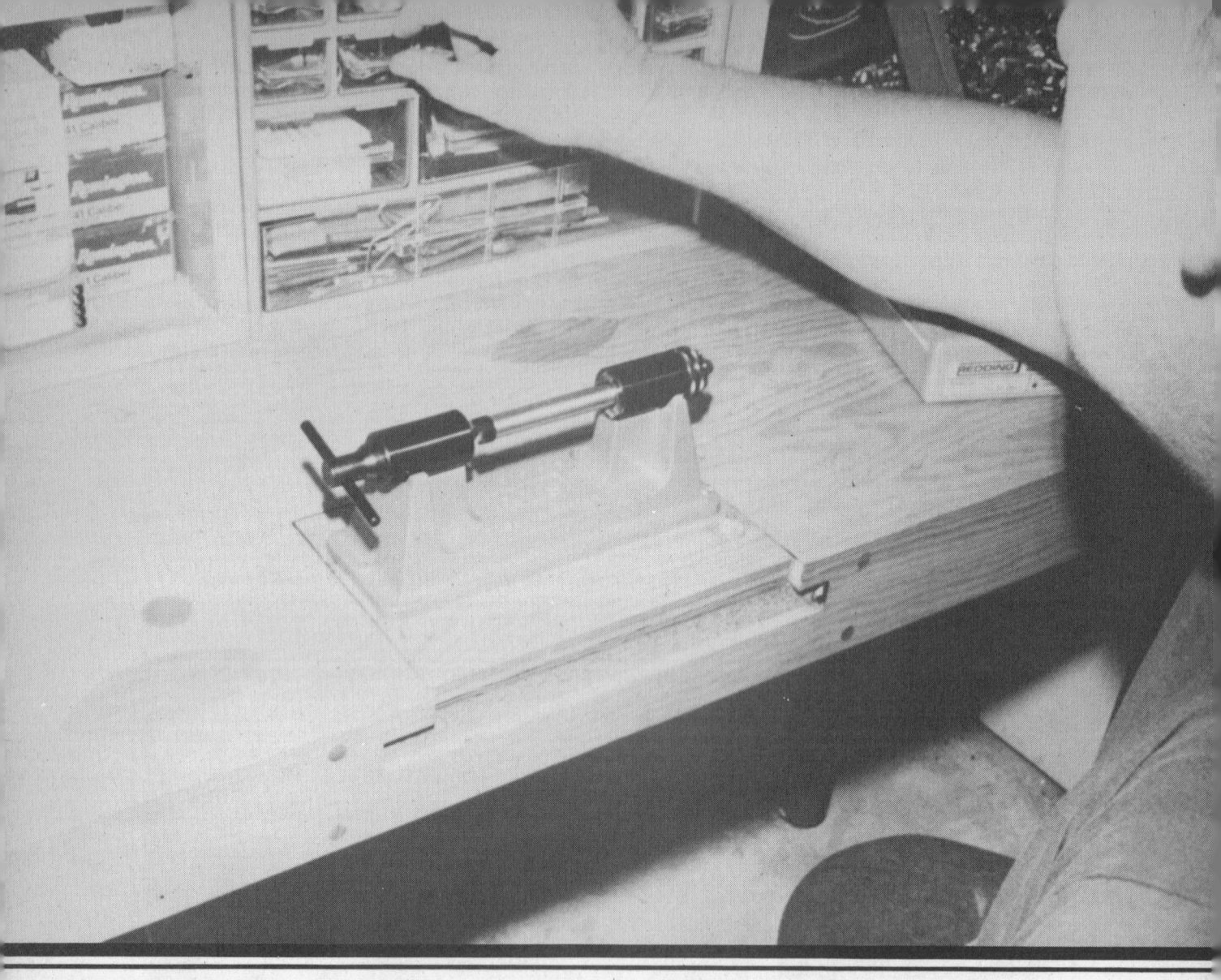

Note the mounting of the RCBS case trimmer. It's even with the top of bench and there is nothing in the way of easy use of tool. Author is ready to set up for case trimming; he's reaching for a bolt in a drawer of the part cabinet. This photo also shows the manner in which the ready supply of bullets are arranged near the main working station of the loading bench. The top of the bench is finished with a good quality varnish.

in the world in terms of tensile strength, it went into the center of the benchtop sandwich. Using lots of yellow aliphatic resin glue, the bench top was sandwiched together. Plywood was the top layer, followed by two thicknesses of particle board and a final lower layer of plywood. The resulting unit was eight feet long, twenty inches wide and three inches thick, composed of four layers of material three-quarters-of-an inch thick. It is heavy as hell.

But it has been in place for the best part of ten years without a single problem, a fact that I'm fond of pointing out to those who criticize my ham-handed over-engineering. The support for the bench comes from welded steel legs.

When the project got underway in the late Seventies, Hal Hobel was fresh from an adult education class on welding. We scrounged up some 2½-inch pipe and some flat steel stock.

Hal welded up three leg units. He put them together using two sections of pipe for each leg unit, welding each leg to a flat section of steel that would be bolted to the underside of the bench top. When completed, the bench is suppported by three pairs of legs and, believe me, it is supported *fully*. The bench has been the birthplace of tens of thousands of rounds of handloaded ammunition and I haven't felt so much as a quiver from it. Similar and just as good legs could be crafted from wood four-by-fours. If you choose, you can buy some professionally made leg units out of the good ol' Sears catalog.

The main press that I use for most reloading is the RCBS Rockchucker. It's mounted to the right side of the bench, with the mounting bolts running through the bench top and the steel of the right-hand leg unit. The only time that the press has moved since originally installed was when I had to unfasten the bolts to mount an RCBS Case Kicker.

The bench decribed so far is a pretty straightforward proposition; nothing more than a whopping heavy slab of wood and glue held level by handmade welded steel legs. But there is more to it than that. You'll remember that I also wanted to include a means by which other tools — ones that are infrequently used — could be set up rapidly for short-term use. Tools in this category are case trimmers, neck turners, powder measure stands and priming tools. Once again, I wanted them to be set up with something other than a lot of extraneous edges and corners that can produce fumbles and foul language.

Every one of these tools comes from the maker with holes which are intended to be used with appropriate screws to hold the tool in place on the work surface. I used the holes to fasten each of them down to a different sort of work surface, one that was level with the surrouding bench, but easily removable when necessary.

At two places on the front edge of the loading bench, I cut sections from the first and second layers of material. This was done *before* laminating the four layers of the bench. The section removed from the top layer measured nine inches wide by seven inches deep. Directly below each of those sections removed from the top plywood layer, another section was cut of the second particle board

This is what makes the whole system work. The board in the author's hands fits the cavity in the face of the loading bench exactly. It is a blank one — there are no tools mounted on it. When it is inserted into the bench cavity, the bench is as flat as the table in your granny's dining room. Details of the bench contruction will be found in the accompanying text. Constructed of several layers of plywood and particle board, it's quite rigid.

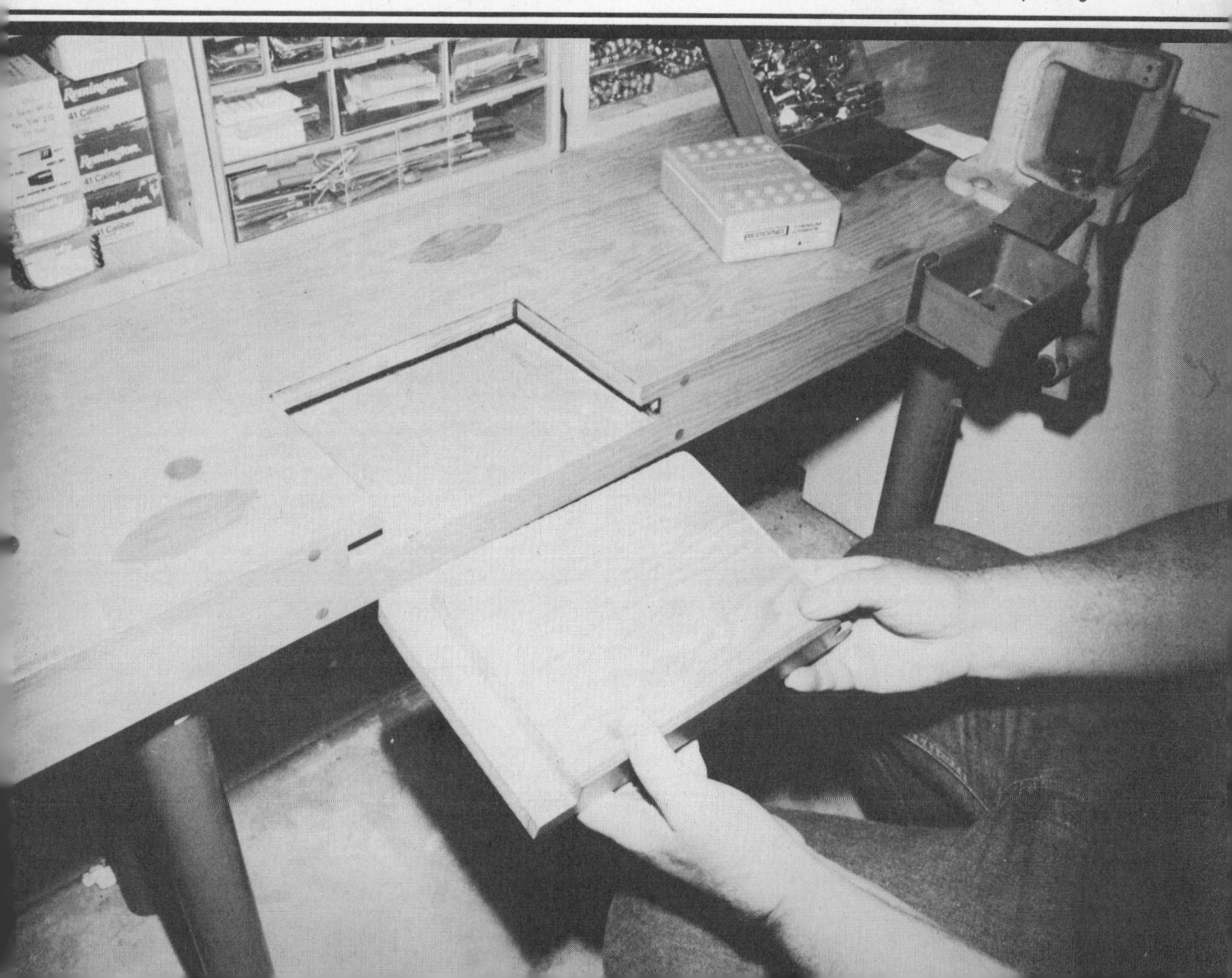

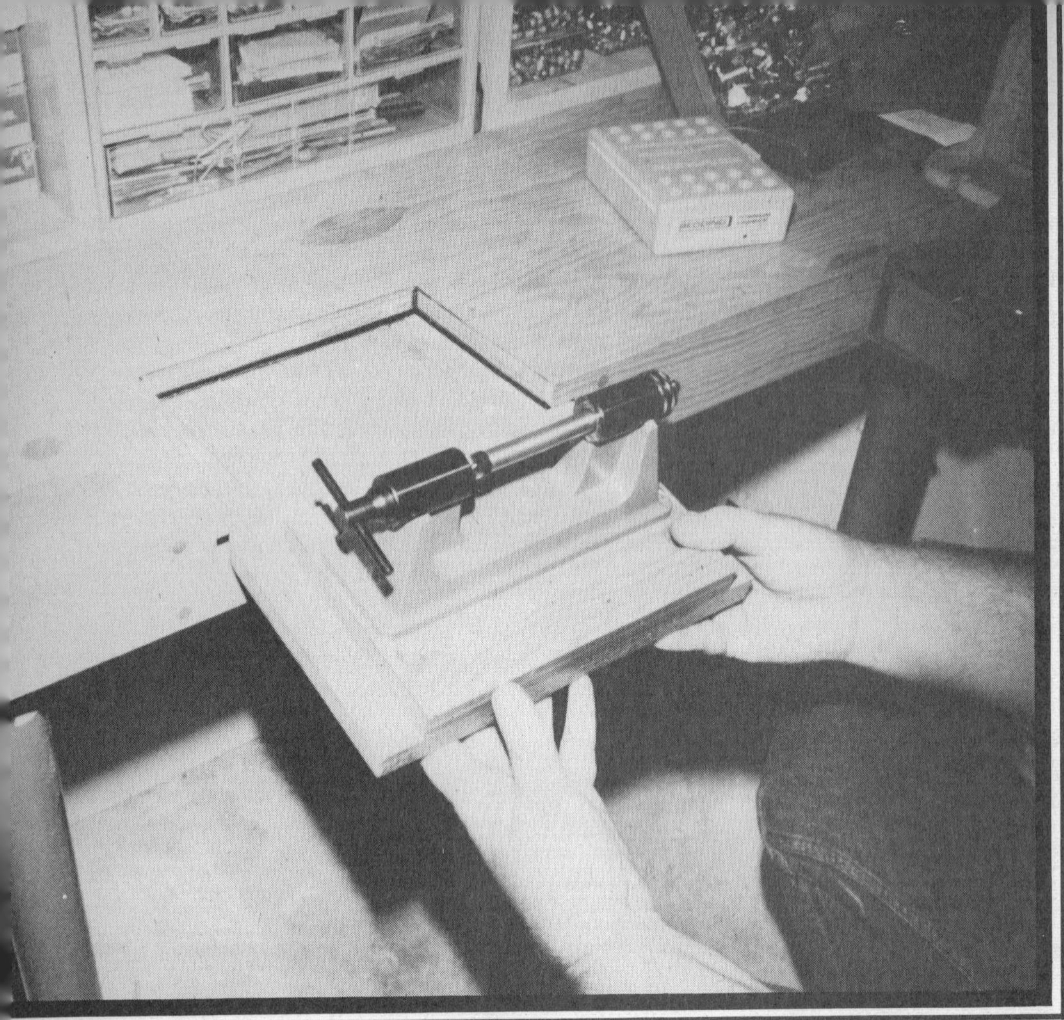

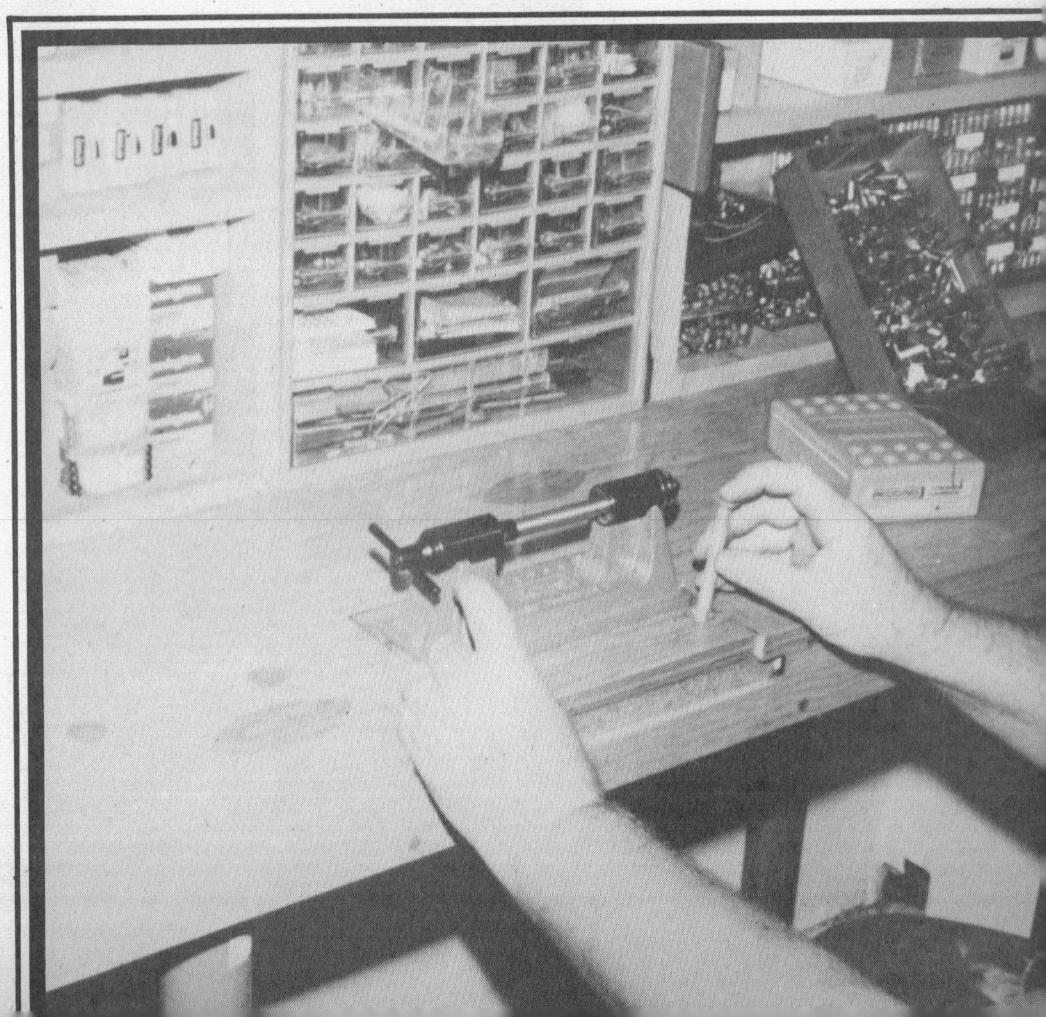

layer. Centered on the cutout above it, that section measured nine by eleven inches. When the four layers of the bench were assembled and glued into place, there were a pair of cavities in the face of the bench.

The two cavities were stepped to be wider on the lower side by an inch in both width and depth. Into these cavities what I came to call insert boards would slide easily. The insert boards were made from that section of material cut from the four-by-eight sheet of the basic material when the construction began. In effect, the positive insert board filled the negative cavity and the dimensions of the bench top were restored to the original.

For normal use, the insert board with no tool mounted thereon was simply slid into place. When it was necessary to use a particular tool — the case trimmer for example — it's easy to slide out the plain insert board and insert another one of the same dimensions. But on that insert board, there's a permanently mounted case trimmer. In order to lock the board in place, there are a pair of countersunk holes drilled in the board. A pair of flat head bolts drop through the holes and matching holes in the bottom of the bench, where they are fastened securely with wing nuts on the bench's underside.

The system works quite nicely. It lets me set up quickly for powder charging, priming, case trimming or neck turning with a minimum of fuss and bother. When tools for doing those reloading chores no longer are needed, they can be removed just as quickly. Slide a plain insert board into place and the bench is restored to plain, flat-topped utility. And best of all, when tools are in use, they are level with the rest of the bench; no extra edges and corners.

The ease of changing the tool setup tends to enforce organization of the loading bench. When a tool is no longer needed, it's in the way, but easy to take down and store on

Facing page, top: Here the insert board with the case trimmer is going into the cavity. The plain board has been removed and set aside. The cavity has been cut to allow an easy slide-in fit. The lower photo is of the author dropping a pair of flat head bolts through holes in the insert board and through the bench. They will be fastened in place with washers and wing nuts on the underside of the bench. Below: When the case trimming operation is concluded, the bolts are removed and the case trimmer withdrawn. Then a plain insert board is slid into place. The system has been in use for several years with complete satisfaction. Can't beat it!

an above-bench shelf. All of this contributes to organized, relaxed reloading. The loading bench built in this or some similar fashion is a pleasant place to work. Most of the other essentials of reloading are stored on shelves above the loading bench and they are readily accessible.

Before leaving the subject of loading bench organization, let's look at one other aspect of the topic. Every reloader needs a scale; this is plainly indispensible. Typically, the scale comes down from a shelf and gets set up on the surface of the bench. That isn't where it belongs, particularly if the reloader is a "stand-up" type with a belt-high bench. The scale must be used carefully and the reloader must read the scale accurately. In order to do this efficiently, the scale should be at eye level.

It also should be close enough that you can adjust it without stretching or bending forward, which is another argument for a relatively narrow loading bench. Either "stand up" or "sit down" reloaders will appreciate a scale mounted at eye level. The one shown here is a Dial-O-Grain from Ohaus and it is resting in a combination box/shelf that positions the beam exactly at eye level for the seated user. When no longer needed, the removable cover for the scale slips back into place, resting in grooves in the frame that make the entire thing almost airtight. The scale is protected from open air, but is ready for instant easy use by lifting off the cover. And you don't have to bend painfully down to read the beam of the scale accurately.

Loading tool manufacturers are in the business to turn a profit. They are in a competitive situation and each would like to have a bigger slice of the pie. Since there are some damned clever heads at work in those design shops, there are plenty of new tools and stuff meant to make the process easier and quicker. Some of them aren't worth a Continental damn; others are priceless. Let's look at a couple of them that are decidely in the latter category.

Any tool that reduces hand movements is going to speed up the reloading process and make it more organized. RCBS makes a nifty little assortment of sheet metal and springs that they call a Case Kicker. For some curious reason, the device — which goes on the Rockchucker,

Have you ever tried to read the graduations on a scale when the scale is well below eye level? That is the way that many reloaders set up their balance. Here's a better way to do it. Instead of awkwardly bending down to see what is going on, set the scale on a shelf so that you are looking straight ahead. The photo also demonstrates the advisability of a narrow bench. In this position, it is easy to reach up and work with the scale controls.

The author's craftsman shooting partner put on his cabinetmaker's hat and turned out this tight-fitting cover which slides into place on the shelf holding the scale. The edges of the cover fit neatly into slots in the shelf. The result is a dust free...

...unit which protects the innards of the Ohaus Dial-O-Grain scale. This scale is a top-of-the-line unit and quite expensive. The extreme ease of using this scale is full justification for the price. It is particularly good for sorting cases by weight.

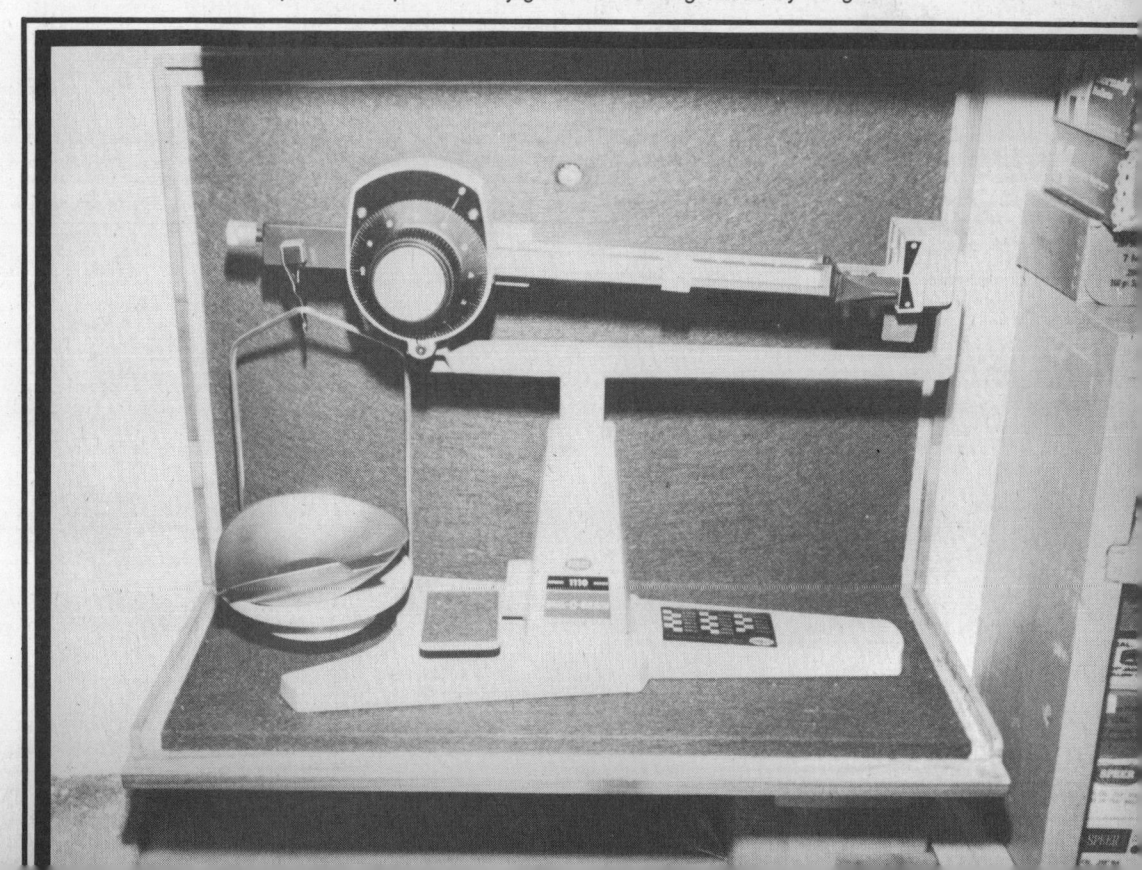

RCBS Jr. and Reloader Special presses — never has received the attention it deserves. The Case Kicker automatically removes a case from the shell holder of a loading press when properly installed and adjusted.

It works like this: The basic structure of the Case Kicker is a pair of sheet steel stampings that screw onto the base of one of the presses mentioned and the bench to which it's mounted. On the right side of the unit, there's a long leaf spring that runs from the base up to and across the top of the shell holder. The tip of the spring curls up and around like the toes of Aladdin's slippers. When the user slips a case into the mouth of the shell holder, he simultaneously raises the ram, which forces the spring back in such a way that it rides against the side of the rising ram.

When the operator lowers the ram, the spring follows it down and near the bottom, pops over the top and against the base of the cartridge in the shell holder. Almost all the way down, the spring tip exerts enough face against the cartridge to push it out of the shell holder and allow it to tumble down an inclined ramp and into a catch box.

The advantage is that when the left hand pushes a case into the shell holder, that's the last time it will need to touch the case in this part of the operation. The right hand never leaves the press handle and the left hand is reaching for another case. Unlike conventional loading press operation, the case is ejected automatically from the shell holder when the ram comes down. I have timed myself informally with one of these units and I believe I can load up to one-third faster with a Case Kicker. That's a whopping big saving in time.

You can speed it up even more by simply applying common sense to the manner in which tools, receptacles and human hands are arranged on the bench. For example, the Case Kicker is a delightful tool to use, but works best with a little organization. I do it like this. Place a plastic pan full of about-to-be-processed cases to the left of the Case Kick-er'ed press. Use a die box or another item under the far end of the box, tilting it to an easy-to-handle angle towards you. Set your left elbow down on the bench so the arm turns a few degrees to the left to pick up a case from the pan and a few degrees to the right to slip it into the shell holder. The right hand just operates the press handle. You can build up

The Rockchucker is great, but it's even better with a Case Kicker. The Case Kicker is a device consisting of a pair of sheet metal fittings, a plastic catch box (seen here to the left of the press on the face of the bench), and a long leaf spring that is visible just to the rear of the shell holder. When you are familiar with this little contraption, you can really speed up your loading cycle. Informal testing indicates a thirty percent time saving.

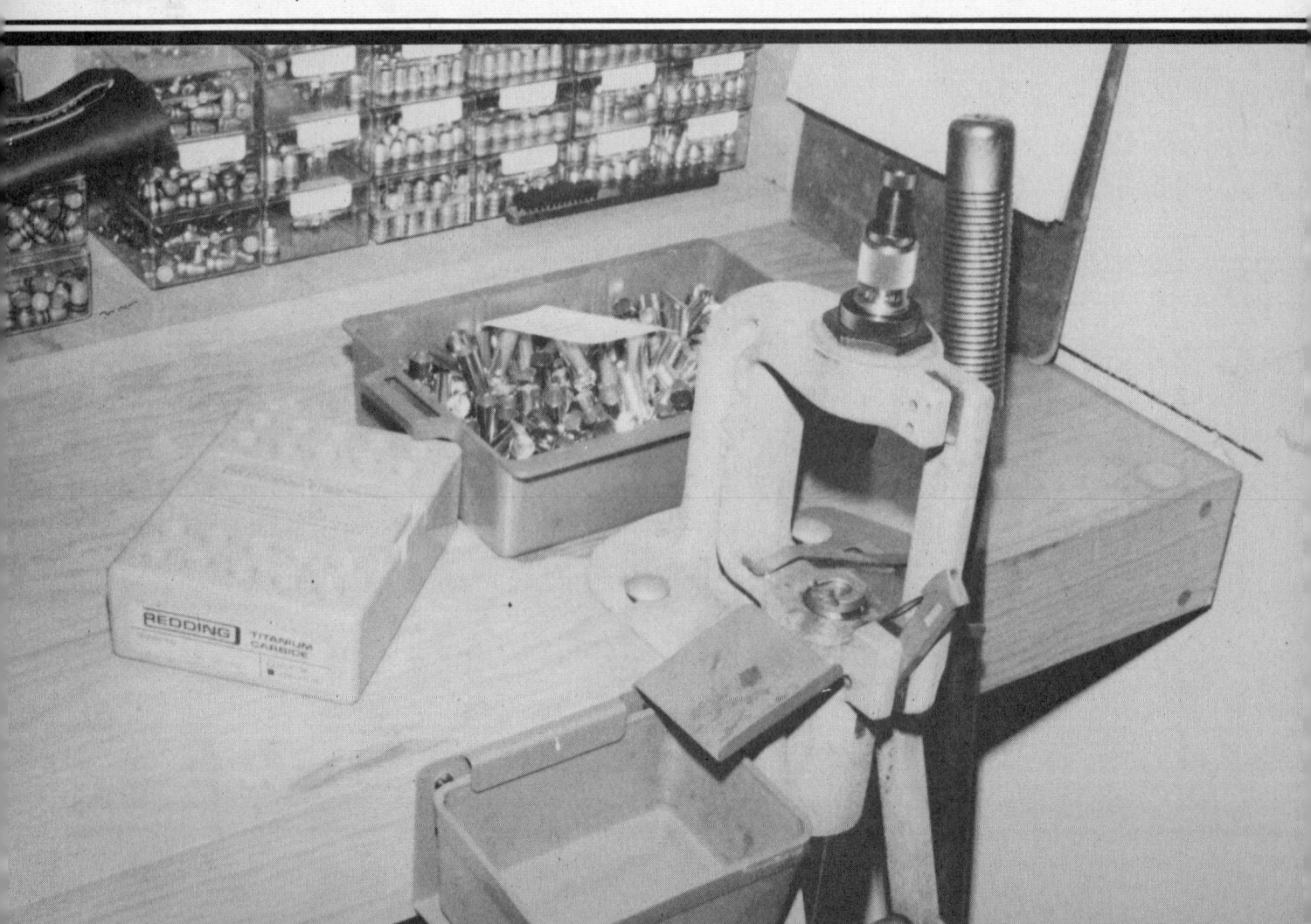

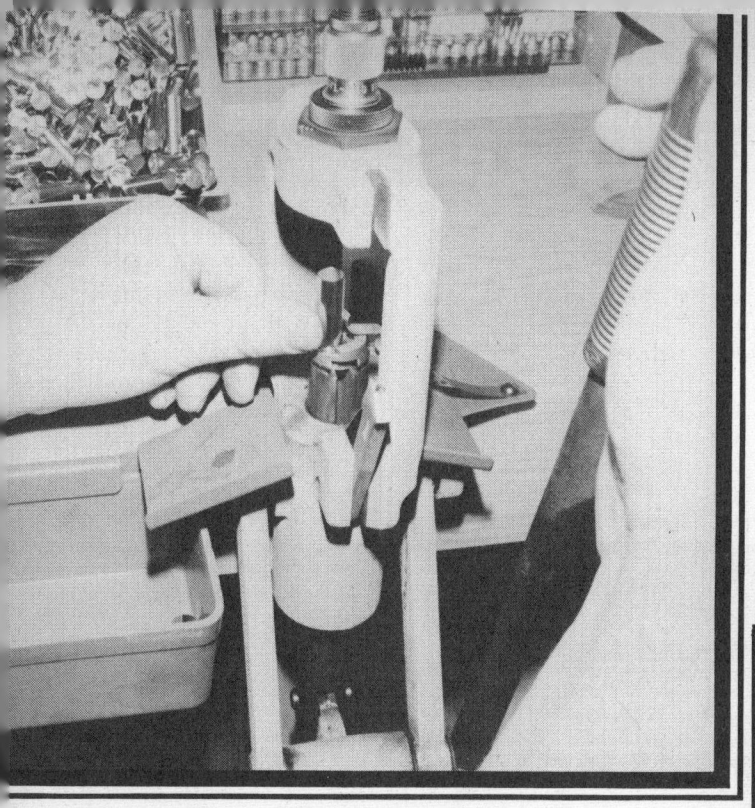

Left: The Case Kicker cycle begins as shown, with the right hand on the press handle and the left hand moving cases to the shell holder. As the left hand pushes a case into the press against the tension of the spring, the...

...right hand starts press ram up. After the case goes into the die for sizing or whatever, the right hand will lower the ram. The Case Kicker spring contacts the side of the case and pushes against its base in such a way...

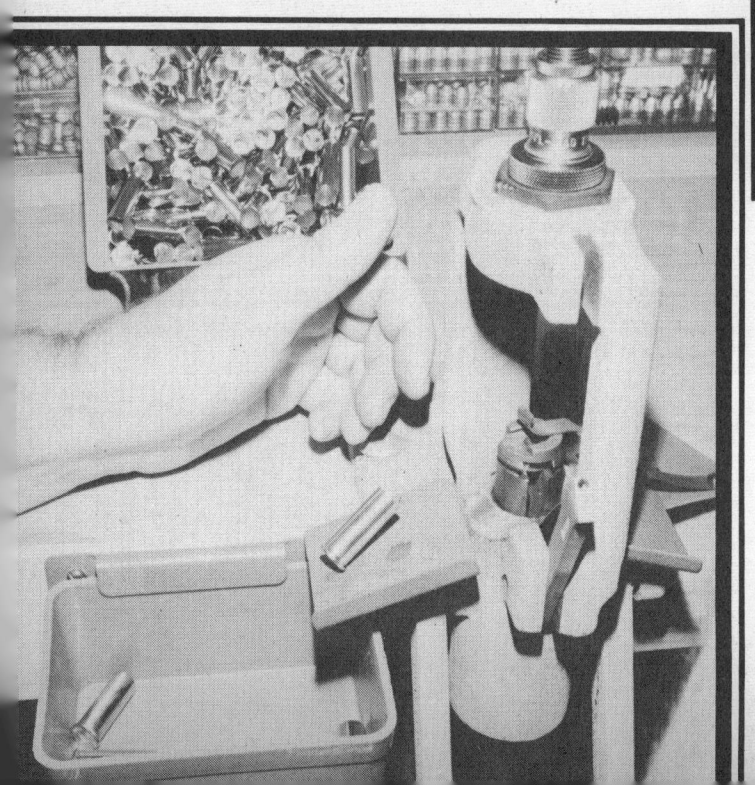

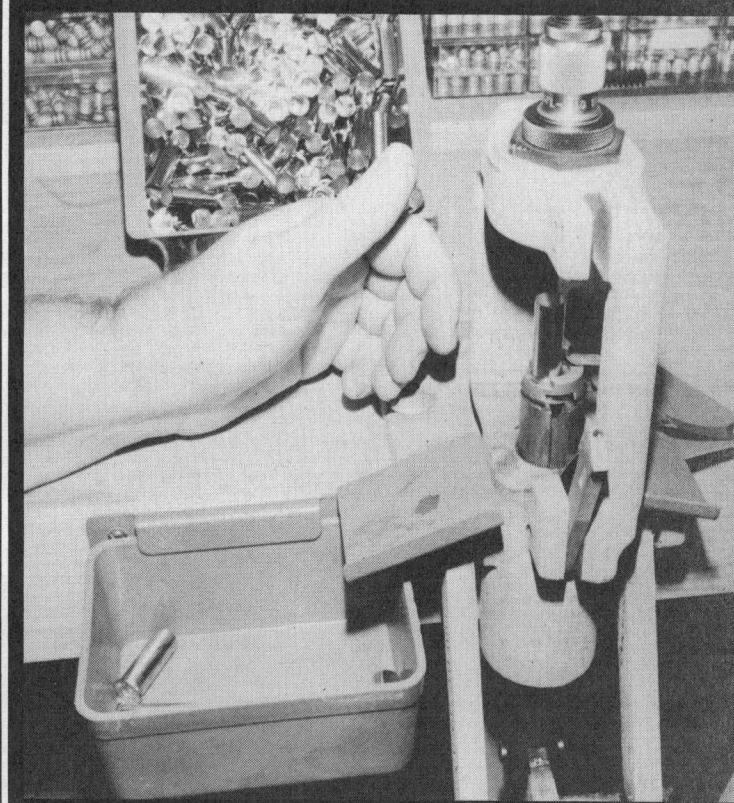

...that the case is forced from the shell holder, tumbling down the ramp into the catch box. The left hand doesn't have to remove the case and can reach for another one. A simple mechanical device that saves lots of effort.

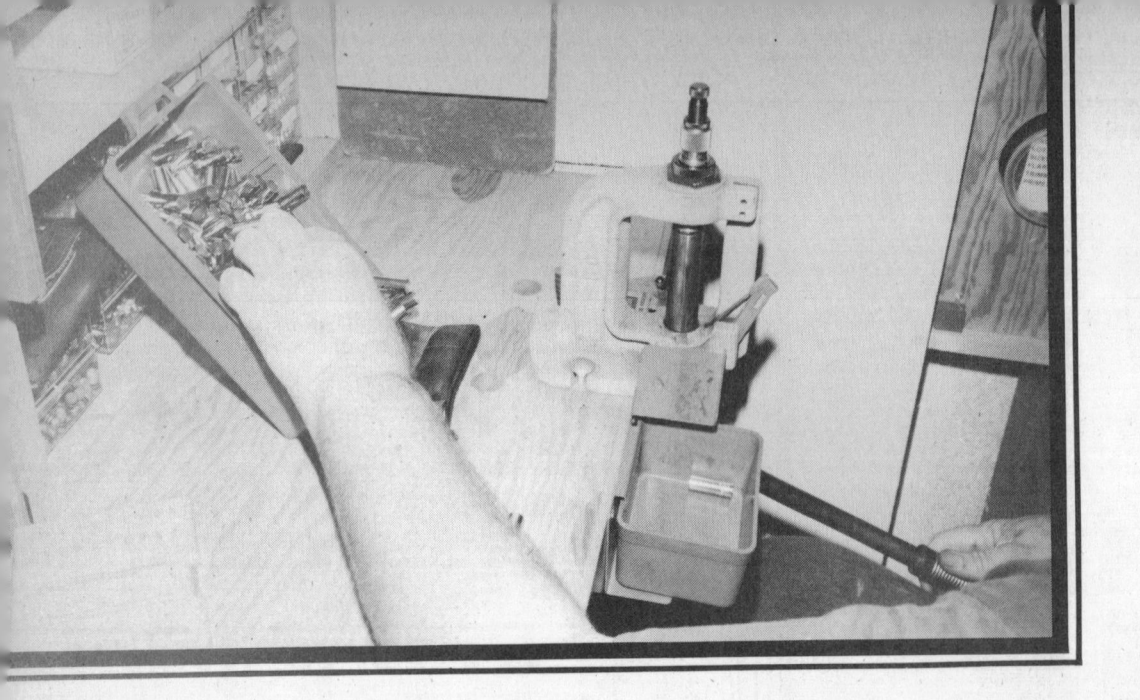

This series of photos illustrates another type of loading bench organization. You can reduce the physical effort and fatigue of reloading by reducing the distance that the hands have to travel. The author's left elbow is on the bench surface, pivoting through a short arc to move cases from the tray to the press. In the top photo, the tray is propped up at an easy-to-reach angle. The left hand picks up a case, moves to the shell holder (center photo) and inserts it. As the right hand raises the ram, the left hand is moving back to get another case (lower photo). If you follow the logic of the sequence, then a clear fault will be obvious: the effort would be even less if the tray of cases were moved closer to the rear of the press! The hand only needs to move a few inches.

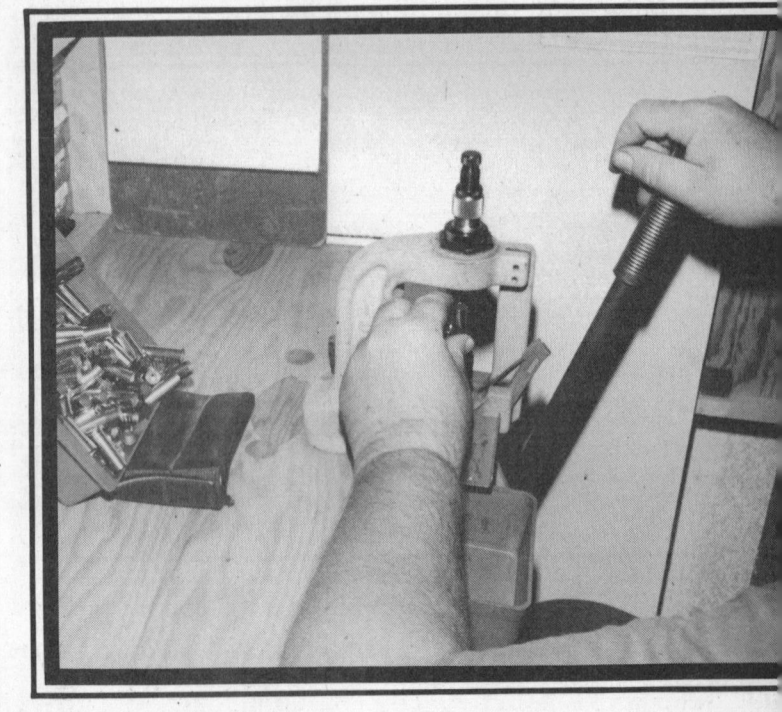

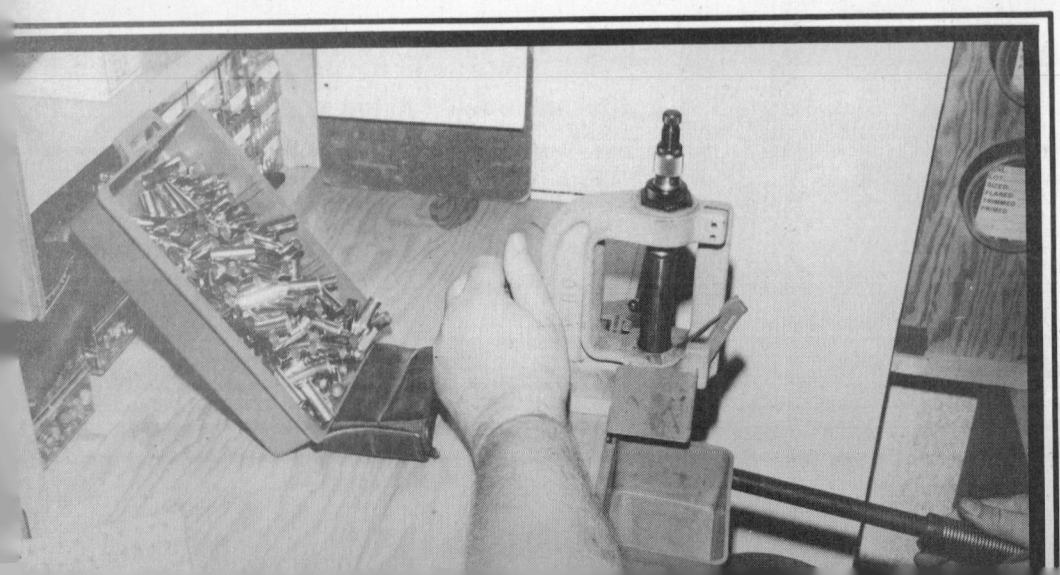

GUN DIGEST BOOK OF HANDGUN RELOADING

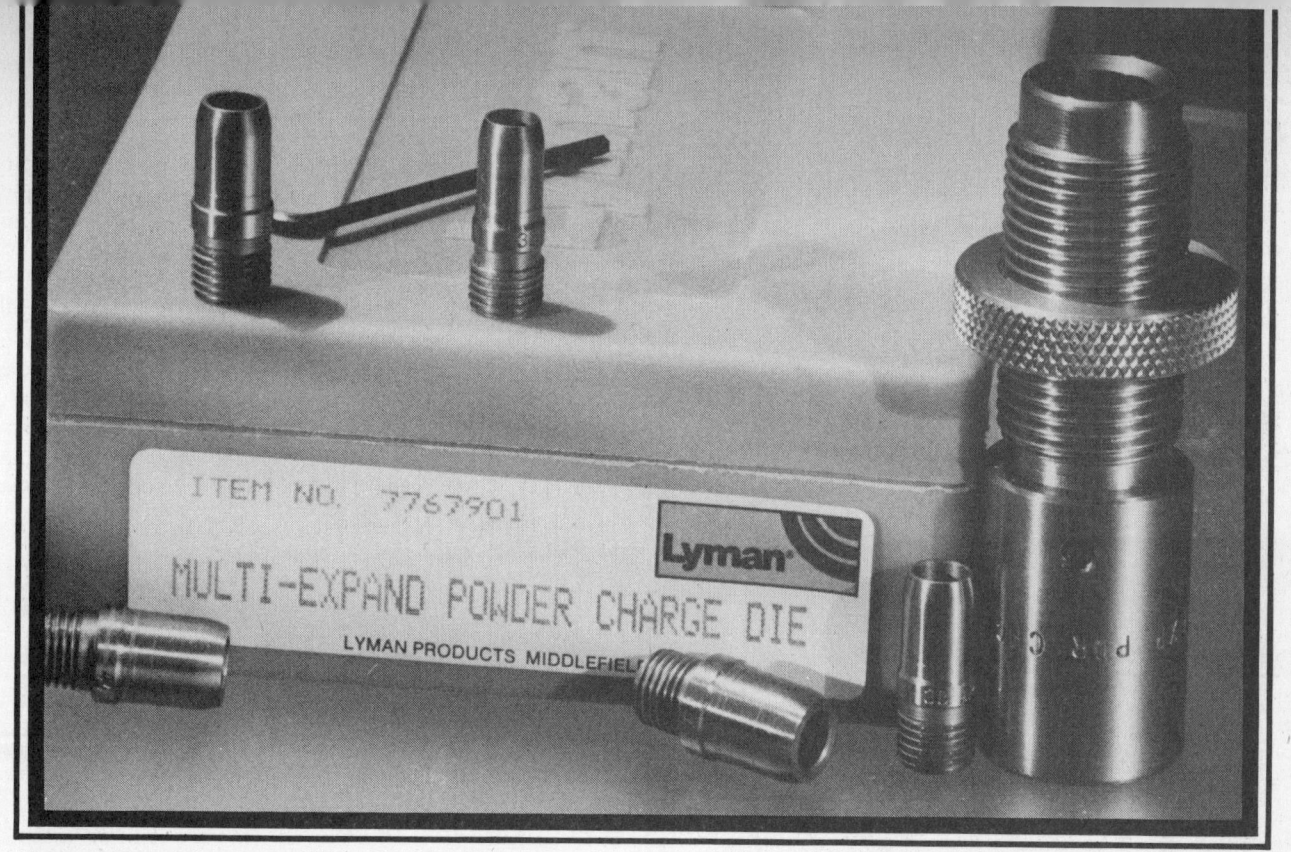

Sometimes an inexpensive tool or special die will help reduce the handloader's work load. Here's one of the best, a Lyman Multi-Expand Powder Charge Die. It uses the "M" type expander plugs and a universal die body. In this die, the reloader can expand any pistol case and also dump the powder charge, since a powder measure mounts on top of the die — one less time the case must be handled.

impressive speed with a little practice. The Case Kicker does some of the work, but your own common sense augments it further.

The placement of items on the bench needs to be done carefully. The results of that care often can result in reduced times and less fatigue. The basic idea is to do no more work than you have to; move your hands and arms as little as possible. And if there is ever some way that you can do two things at once, by all means, do it!

Lyman, the old-line reloading tool maker from Connecticut, has recently introduced a piece of equipment that has great value in not only speeding up the process, but in making better ammunition. The device is called a Multi-Expand/Powder Charge Die. It is an expander die for pistol calibers, but unlike others of that type. The new Lyman die (let's call it the MEPC for short) uses different expander plugs for various calibers, all in the same die. One MEPC will allow the handloader to expand the cases in all popular pistol calibers.

The resulting ammo is likely better, since the expander plugs are made in the Lyman "M" configuration. That means the expander plug has a step that's a few thousandths larger than the body of the plug. The step puts a corresponding step in the mouth of the case, with several attendant benefits. One of them is that the case need not be flared much, if at all. And that translates to longer case life, since there is less working of the case mouth. The case mouth step also is superior in that a bullet inserted in that case starts straighter. Bullets that go into cases straight come out of cases straight and they're more accurate.

MEPC advantages more germane to the subject at hand — organizational efficiency — are significant. Using the MEPC die will save the reloader an extra step. The expander plugs of the MEPC die are hollow and the top of the die is threaded 7/8 by 14. This means that a powder measure can be fitted to the top of the die and powder dumped through the expander die when the case is in place for the expanding operation. Two operations are accomplished at one handling of the case. This is a huge advantage that is well worth the minimal cost of the extra die at under twenty bucks.

While the foregoing pretty well covers proper use of tools and efficiency at the loading bench, it doesn't say much for keeping track of what's in a particular batch of ammunition. That one is pretty simple. I long ago formed the habit of putting all handloaded ammunition into plastic snap-lid boxes of the type popularized by the MTM people. Ammo thus stored will be kept dirt-free and safe. I even went so far as to buy boxes of different color for each of the calibers that I handload: green for .44 magnum,

GUN DIGEST BOOK OF HANDGUN RELOADING

yellow for .45 ACP, brown for .41 magnum, etc.

After a year or so of using ammo box labels from gun stores and from commercial makers, I gave up and started using plain labels from the stationery store. In order to tailor them to my individual preferences, I ordered a rubber stamp for the labels. It has fill-in-the-blank spots for powder charge, primer, bullet and the like. The label, by the way, goes inside the box on the flat inner surface of the lid where there's less chance that it will peel off and be lost.

Ammo produced, identified and stored in this fashion is easy to keep track of. But I needed a system for keeping track of the various-sized batches of brass that were in one

Left: A plywood rack holds one-pound coffee cans full of brass. The clear plastic covers keep the brass clean and a tag showing the contents can be read through the cover. It is a handy way to store brass in ready status.

Below left. Tags are in each brass can with details as to what has been done to the batch, how many times it has been fired and lot #, etc. Down from the top shelf above the loading bench, this heavy plastic film can is used to store seldom-used batches of brass. It helps to keep track of your brass — Handloader's Gold — always.

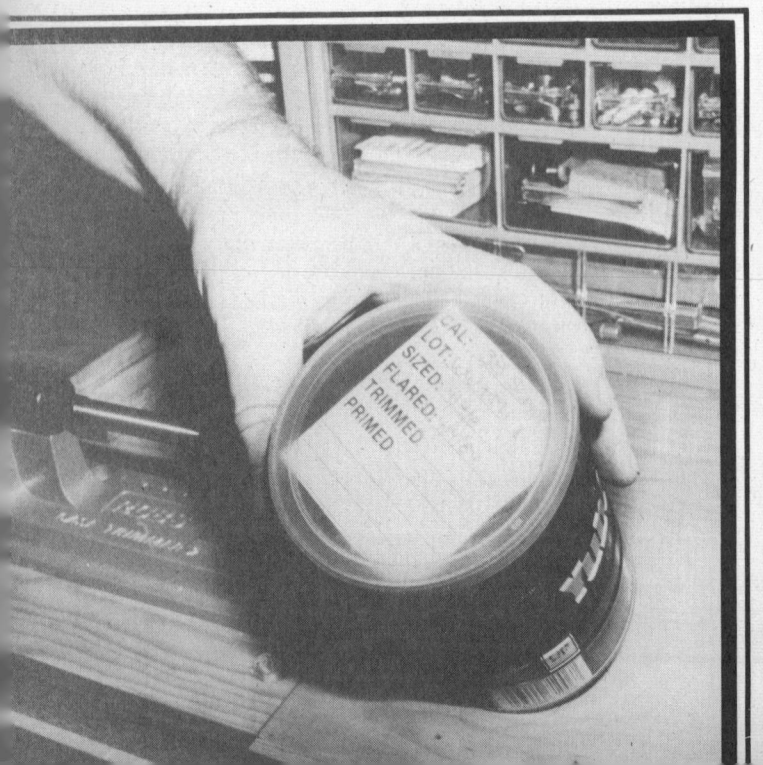

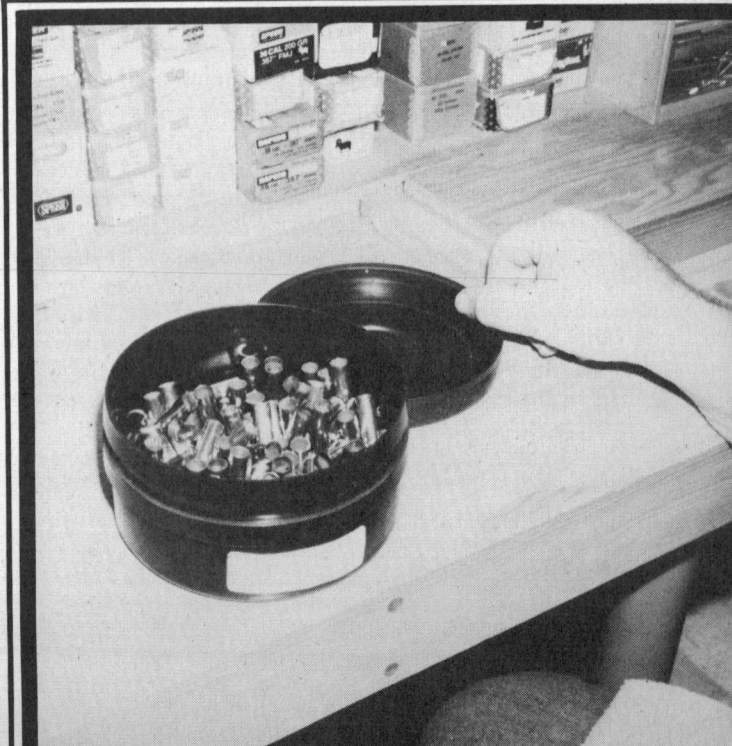

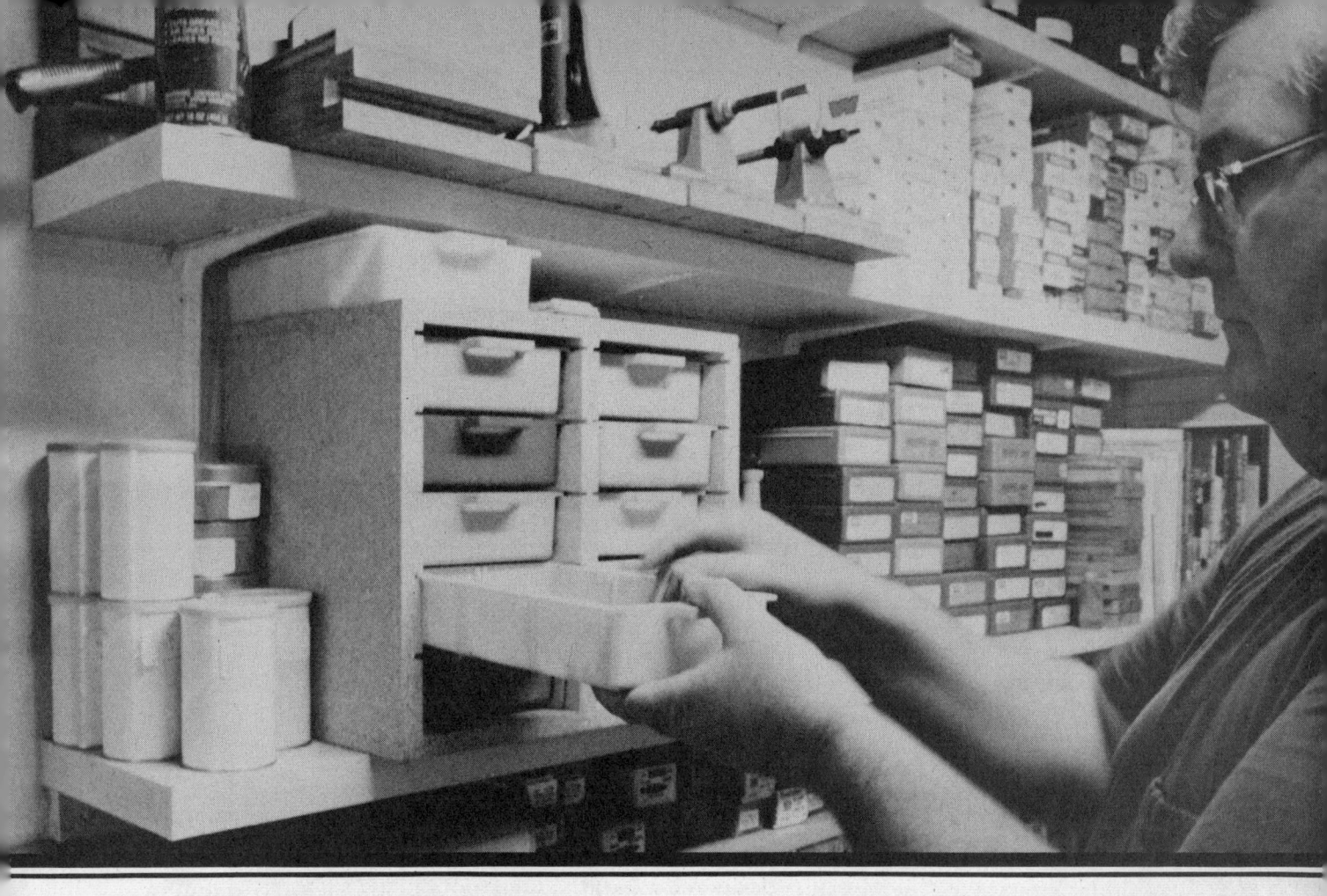

On the shelf above the scale, there's a cabinet holding ten plastic trays. These trays are used to hold brass while it is in various stages of processing. Most often, when the brass is not to be used right away and is ready for more permanent storage, it goes into another type of container. Yes, it is a bit more effort, but it has a lot of advantages. For one thing a bench set up in this way lets a reloader sit down and be loading in minutes.

stage or another of pre-load processing. While developing a system to do this, I was working with a friend who owned a small print shop. He printed up some brass record tags for me that have been easy to use.

When a batch of brass is fired, my first step in processing it for reloading usually is cleaning it in a case tumbler. After that, it goes into containers of various sorts, depending on how long it will be stored and how large the batch is. Without fail, I fill out on of the illustrated tags on each batch of brass and put the tag in whatever container is used to hold it. If the brass is going to be used quickly, it probably will go into one of the open plastic trays I use so much. As each operation is performed on the batch, I check it off the accompanying tag and make whatever comment is necessary. This might be such information as the length to which it was trimmed or what type of primer was used. At any point in time, I can give you a pretty good idea of the history and condition of any batch of brass in my shop.

Dean Grennell shakes his head in wonder at the amount of concern I put into all of this, but I feel the slight amount of time is worth the effort. The system is established and works and I'm not inclined to change it.

There's one area of handloading organization with which I'm not completely satisfied. That is the format for keeping records on the firing of ammunition — chronographed velocities and the like. I have used the MTM forms and the "ditto data" type advocated by John (two Ts) Wootters. Currently, I'm working with the two-tone computer paper with columns and I'm not completely satisfied with it.

You might not be personally satisifed, either, with any of the material I have gone over in this chapter. You may want a different bench and a different record procedure. You might even go off in a huff and do everything differently, that's your business.

But damned if I didn't get you thinking, didn't I? — *Wiley Clapp*

CHAPTER 5

THE SHORT, VIOLENT LIFE OF A LOAD

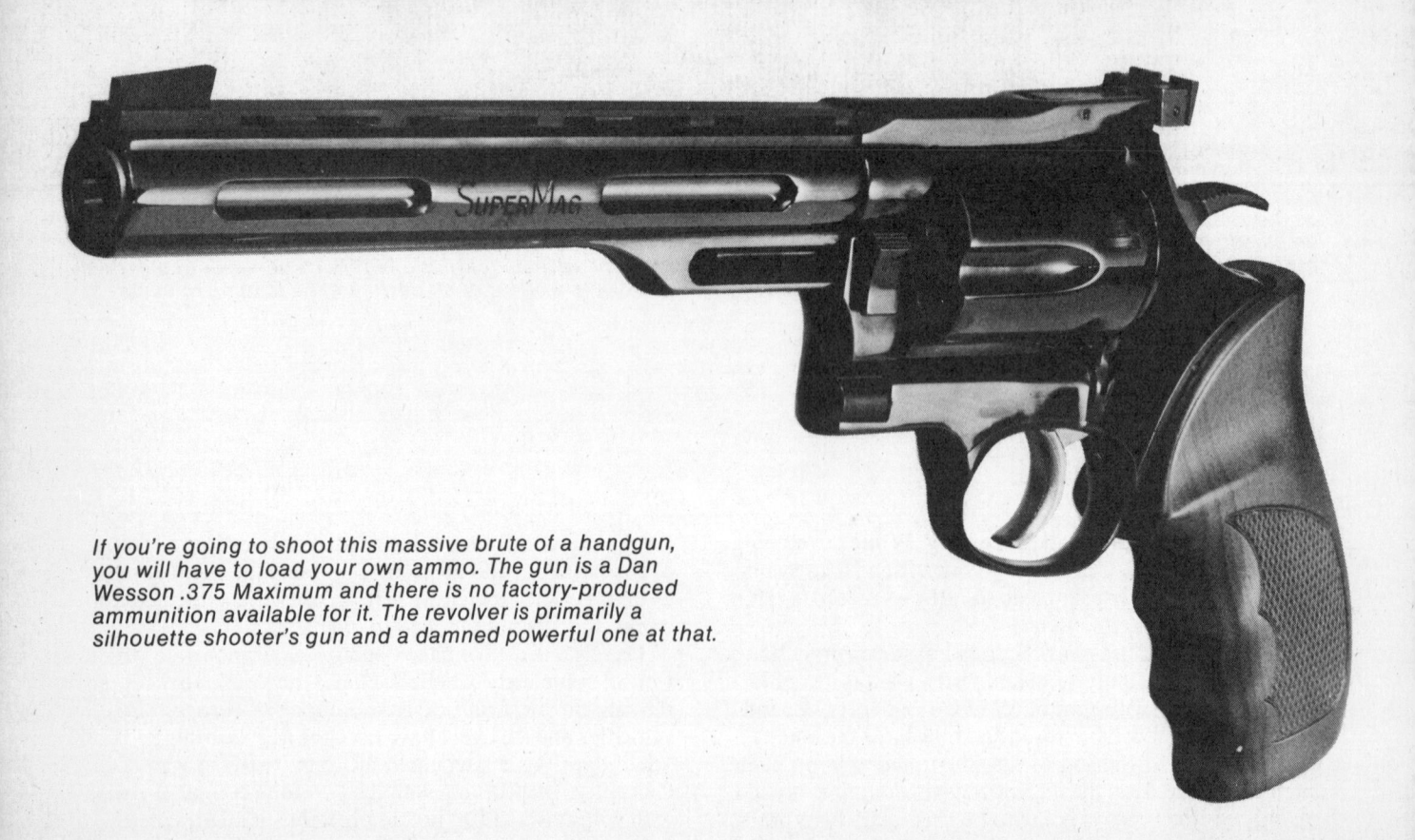

If you're going to shoot this massive brute of a handgun, you will have to load your own ammo. The gun is a Dan Wesson .375 Maximum and there is no factory-produced ammunition available for it. The revolver is primarily a silhouette shooter's gun and a damned powerful one at that.

The Complexities Of Handgun Ballistics, Exactly What Happens When You Drop Your Handgun's Hammer

This tiny Browning .25 auto is a gun for which there is factory ammo, but one that may also be handloaded. Ammo is ammo and the ballistic principles involved are the same as for the Dan Wesson across the page. Despite the power, the principles — internal, external, terminal — don't vary.

ON TV cop shows, "ballistics" refers to a white-robed lab technician who tells the cops whether a suspect's gun is the murder weapon they've been looking for. To a rifleman, the ballistics of a load means its performance statistics — its muzzle velocity and energy, midrange trajectory, bullet drop at target distances and remaining velocities and energies downrange. Then what does ballistics mean to a handgunner?

Ballistics applies equally to rifles, handguns, shotguns, artillery and other means of launching and tracking missiles. It's simply the study of projectiles in motion — the energies that propel them, how they behave and their effects on whatever they strike. Small-arms ballistics is not just one category but several. Some overlap.

Interior ballistics covers what happens inside the gun, from the firing pin's impact to the bullet's exit from the muzzle. Exterior ballistics covers the bullet's travel between muzzle and target. Terminal ballistics covers not only the bullet's travel and behavior in another medium, but also impact, penetration and their effects on both the bullet and the medium.

Interior ballistics covers a very short time — only a few ten-thousandths of a second in a handgun. But a lot goes on in that moment before the bullet comes out the muzzle. When the firing pin falls, its impact drives the entire round forward a tiny fraction of an inch until its headspace device — case rim, mouth or shoulder cone — meets the resistance of its contact surface in or on the breech. This slight forward movement of the cartridge, called drive-in, leaves a tiny space between the head of the case (also the face of the primer) and the face of the breech.

"Headspace" in shooting lingo has come to refer to the way a cartridge is positioned in the chamber, ready for fir-

INTERIOR   EXTERIOR

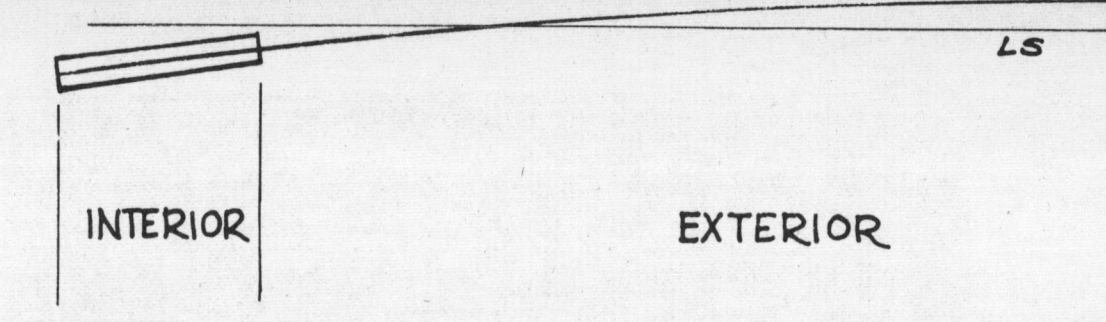

ing (as in "a rimmed case headspaces on the rim"). It has also come to mean — in the definition adopted by the Sporting Arms and Ammunition Manufacturer's Institute, Inc. (SAAMI) — the distance from the face of the closed breech of a firearm to the surface in the chamber on which the cartridge case stops.

SAAMI uses another term — "head clearance" — for the distance between the head of a fully seated cartridge or shell and the face of the breech bolt when the action is locked. Originally, headspace — named after the space between, say, the beans or tomatoes and the underside of the lid in a jar of food canned at home — referred to this tiny space left between the face of the breech and the head of the fully forward case. Call it "headspace" or "head clearance" — but remember this thin space between the case and the breech face.

When the forward movement of the case stops, the tip of the firing pin indents the base of the primer cup. The indentation of the thin primer brass probably begins when the firing pin strikes the face of the cup, but it's soft or incomplete while the case is moving forward and riding with the punch of the firing pin.

Inside the primer cup, the small bulge forced forward by the firing pin crushes the frangible, sensitive primer pellet against the apex of the primer anvil. The legs of the anvil rest firmly on the floor of the primer pocket, so the anvil resists the thrust of the firing pin and the inner bulge that it forms in the primer cup. The pellet of priming compound crumbles between the anvil and the bulge.

Crushing the pellet causes the priming compound to react chemically — and violently. It produces a furious spurt of flame and flame-hot particles that spew through the flash hole to stir, heat and ignite some of the powder kernels inside the body of the case. But the flash hole is only a small opening in the septum or web of brass between the primer pocket and the powder cavity. The powerful primer force pushes equally hard against the inside of the primer cup and the floor of the primer pocket.

During the instant of the primer's glory, some of its fire spurts through the flash hole into the powder cavity. If the powder charge is near the optimum load density (the ratio of the volume of the powder charge to the volume of the powder cavity), there's just enough space among the powder kernels to let the primer flame and fire-hot particles

The three most common systems for headspacing handgun cartridges — and three typical cartridges using them — are seen on the facing page. On the left, the semi-rimmed .38 Super which allegedly headspaces on a tiny lip on the barrel extension. Middle, the rimmed .38 Special, which headspaces solidly on the rear of the chamber wall. The right sketch is of the 9mmP, headspacing the case mouth on a lip in the chamber. The much-maligned .38 Super is now being produced by Colt with case-mouth headspacing.

68   GUN DIGEST BOOK OF HANDGUN RELOADING

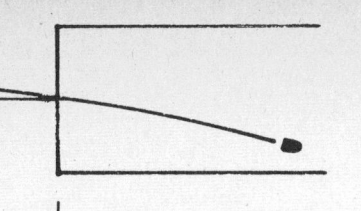

TERMINAL

Internal ballistics (far left) covers everything that takes place from the strike of the firing pin to the bullet's exit from the muzzle. Exterior ballistics (middle) is the study of what the bullet does in flight from the muzzle to the downrange impact on something. Terminal ballistics (right) considers how the bullet and target substance inter-react.

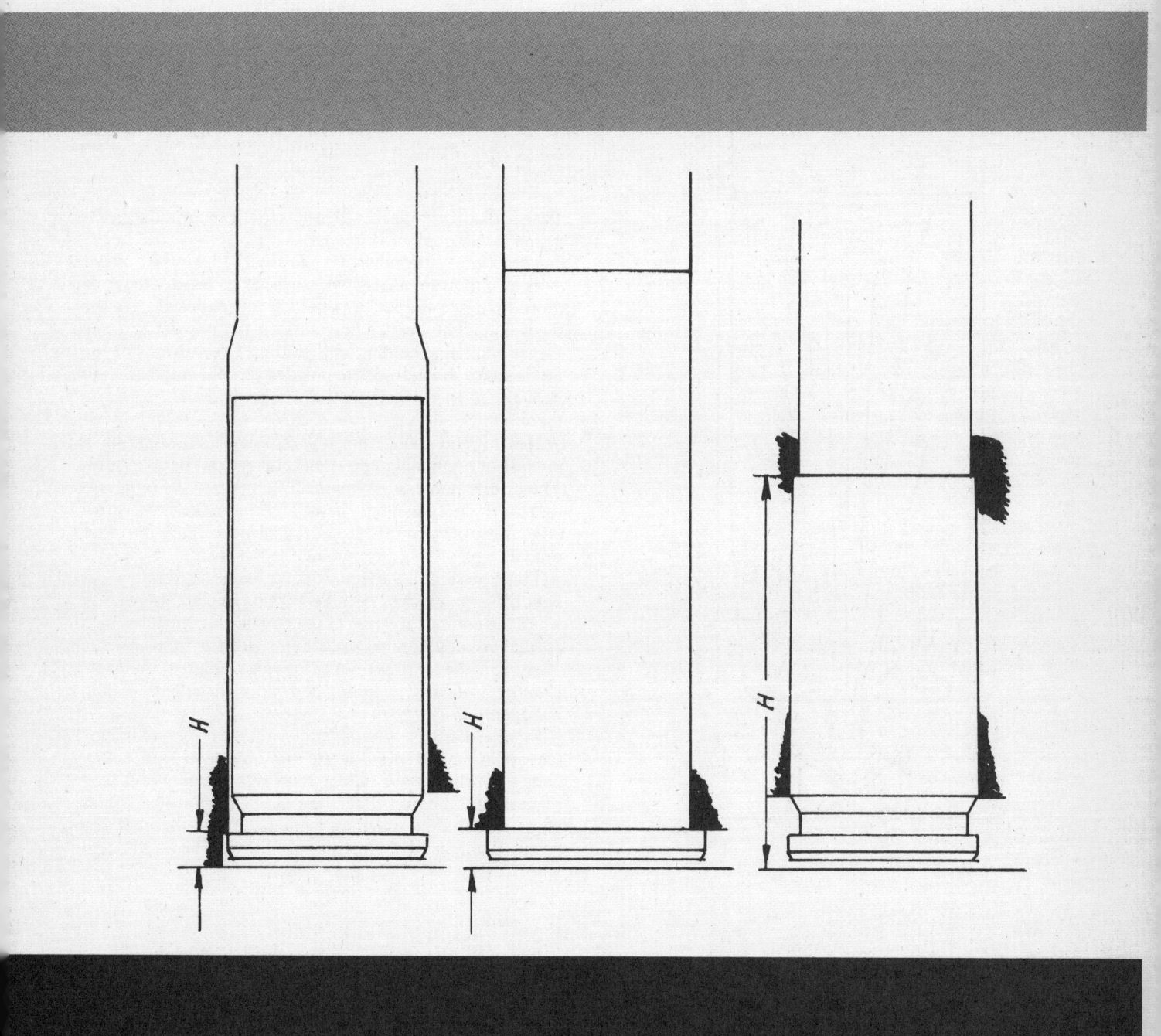

GUN DIGEST BOOK OF HANDGUN RELOADING

The primer doesn't just light off the powder on the other side of the case web separating powder and primer. It's also required to stir the entire powder charge vigorously so that it can light off nearly every granule of powder at about the same time A good primer must be both vigorous enough and hot enough to both stir and light all powder.

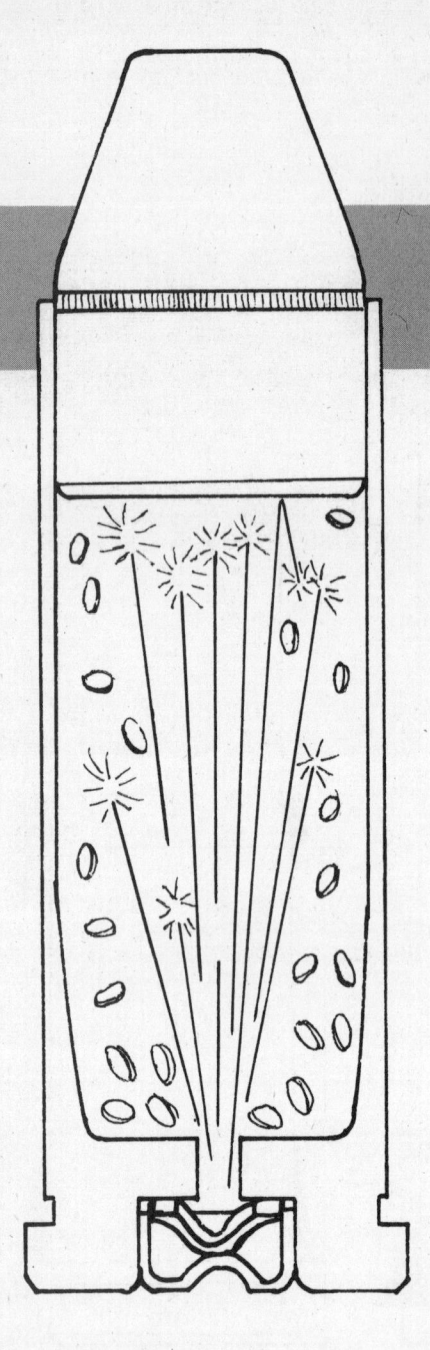

flash forward among these kernels to stir them and to ignite a significant number of them.

The primer's turbulence — its ability to stir the powder — is as important as its heat, since it's necessary to ignite more than just the few powder kernels exposed near the flash hole. The primer thus acts on the powder in three ways — stirring, heating and igniting. The stirring lets the flame reach most or all the powder kernels in the case, to heat them up to their ignition point and then to ignite them.

Heating and ignition are separate steps in the lighting of the powder. Unheated powder kernels can't light or burn. The primer flame must first heat them up to the temperature at which it can ignite them. Thus larger kernels, with a larger ratio of volume to surface, take more flame and more time as well as more turbulence to stir them.

This is essentially the reason for magnum primers — to light off large charges of large-kernel powders, especially when the combustion rate of the powder is controlled by a special coating applied to delay or slow ignition. This is also one of the reasons for making rifle primers significantly more vigorous — hotter and more turbulent — than handgun primers.

Smaller kernels, with a smaller ratio of volume to surface, heat up faster and ignite more readily. This is why handgun primers are milder than rifle primers and why regular or standard primers are at least theoretically less tempestuous than magnum primers. (I say theoretically because some makers' regular primers can be as wild and wooly as others' magnums.

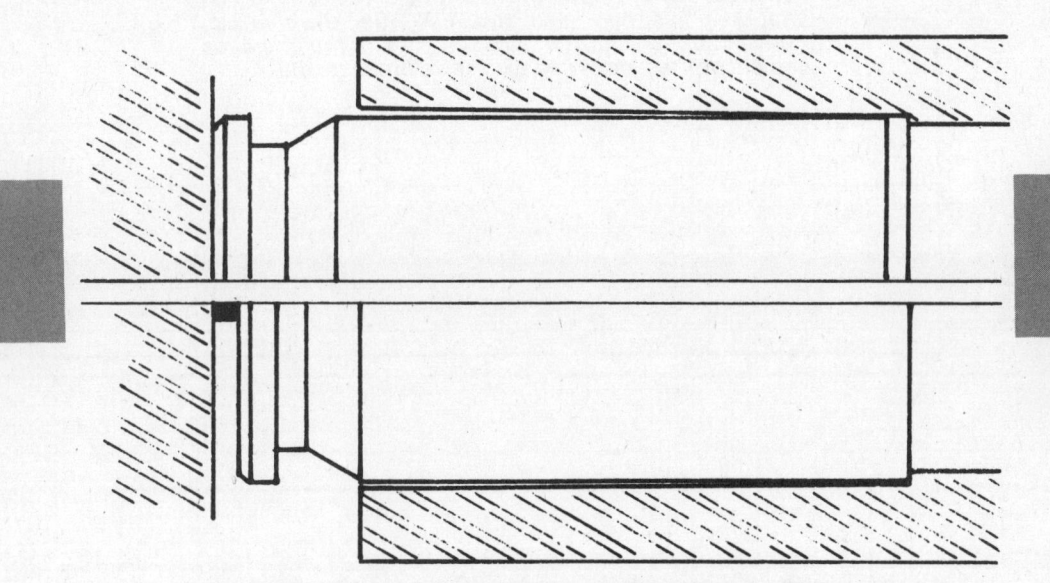

There's always a bit of extra space between the face of the breech and the chambered cartridge. The impact of the firing pin drives the entire round forward, until it headspaces firmly, in what is called "drive-in." In this drawing, the upper part of a sectioned .45 ACP case is driven forward. When the cartridge fires, internal gas pressure will force the case back hard against the face of the breech. This is "drive-back," in the lower section.

Powder burns hot. The lab guys say "with a high flame temperature," but hot is what they mean. When the first powder kernels burn, they in turn raise the amount (British thermal units, Btu) and the quality (temperature, °F or °C) of the heat inside the case. This heat warms and ignites any powder that the primer hasn't already set ablaze and all this burning powder produces still more heat. The powder thus burns slow at first but then faster and faster.

Burning powder produces a composite gas; and it's hot, too — over 3,000°F. The hotter any gas becomes, the more room it wants, so this hot gas expands madly — vastly and rapidly increasing the pressure inside the bullet-stoppered cartridge. The increase in pressure makes the powder burn faster, producing more hot gas, which further increases pressure, making the powder burn still faster and turn out still more hot gas, which....

The process just keeps on going. With heat and pressure both increasing each other over and over in such a short time and a tiny space, something has to give. But before the bullet moves, this extreme internal pressure does a few things to the case and the primer. The growing pressure expands the case until the walls of the chamber stop its expansion.

The pressure continues to build, pushing the head of the case rearward to press it hard against the face of the breech, and pushing the primer hard against breech face and the tip of the firing pin. This rearward movement of the case is called drive-back — the opposite of the earlier drive-in produced by the impact of the firing pin.

As pressure continues to rise toward its peak — thousands

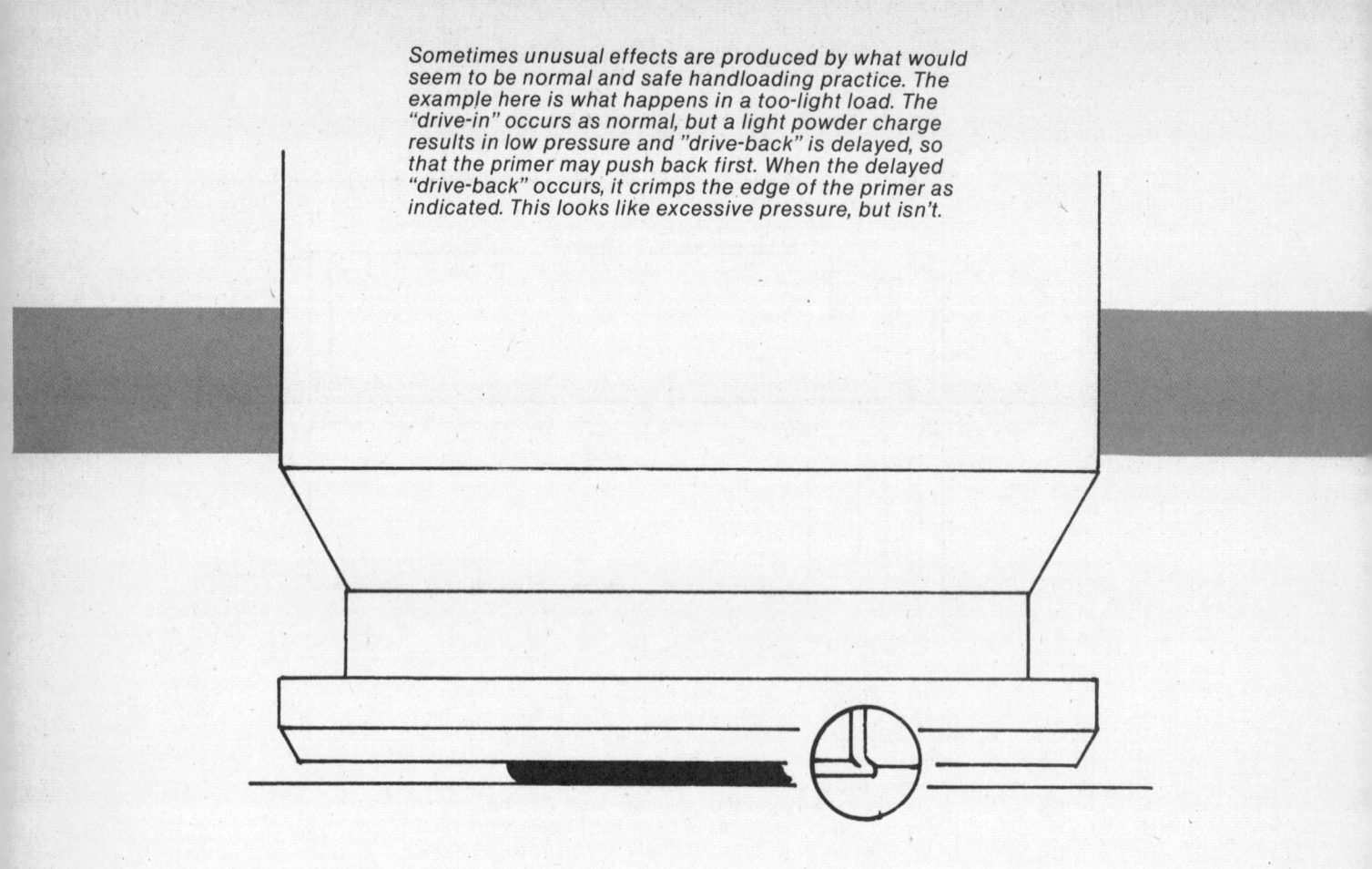

Sometimes unusual effects are produced by what would seem to be normal and safe handloading practice. The example here is what happens in a too-light load. The "drive-in" occurs as normal, but a light powder charge results in low pressure and "drive-back" is delayed, so that the primer may push back first. When the delayed "drive-back" occurs, it crimps the edge of the primer as indicated. This looks like excessive pressure, but isn't.

of pounds per square inch — it presses the comparatively soft brass of the primer cup and the case against the much harder steel of the firing pin, breech face and chamber wall. This pressure can be great enough to imprint on the primer cup and case the tiny marks left by the tools that cut and reamed those steel surfaces.

These are the "tools marks" — the imprints of the tool marks, really, as distinctive as fingerprints or footprints — studied by the firearms examiner or forensic ballistician who compares them as evidence. Extremely high pressure leaves pronounced imprints of these tool marks on the primer cup and the case.

And still the pressure increases. But eventually (if it's legitimate to say eventually about a time lag of about a ten-thousandth of a second), it overcomes the inertia of the bullet. The bullet starts to move, slowly at first but then faster and still faster. The first slow budging of the bullet must overcome the inertia of its unmoving mass, the grip of the cartridge neck — called bullet pull — then an instant later, the resistance of the rifling.

The first movement of the bullet also increases the space in the case behind it. As this space continues to increase — all the while the bullet is traveling down the barrel — it will relieve much of the powder-gas pressure behind the bullet. But not yet. The first bullet movement and the increase in the enclosed space are too slow to reduce the pressure. Pressure continues to rise, though not quite so fast as before.

In a pistol, the bullet begins to enter the rifling after a relatively short travel, sometimes before its base leaves the mouth of the case. In a revolver, it exits the case and almost exits the cylinder before it meets the resistance of the rifling.

The bullet moves forward too slowly to let propulsion

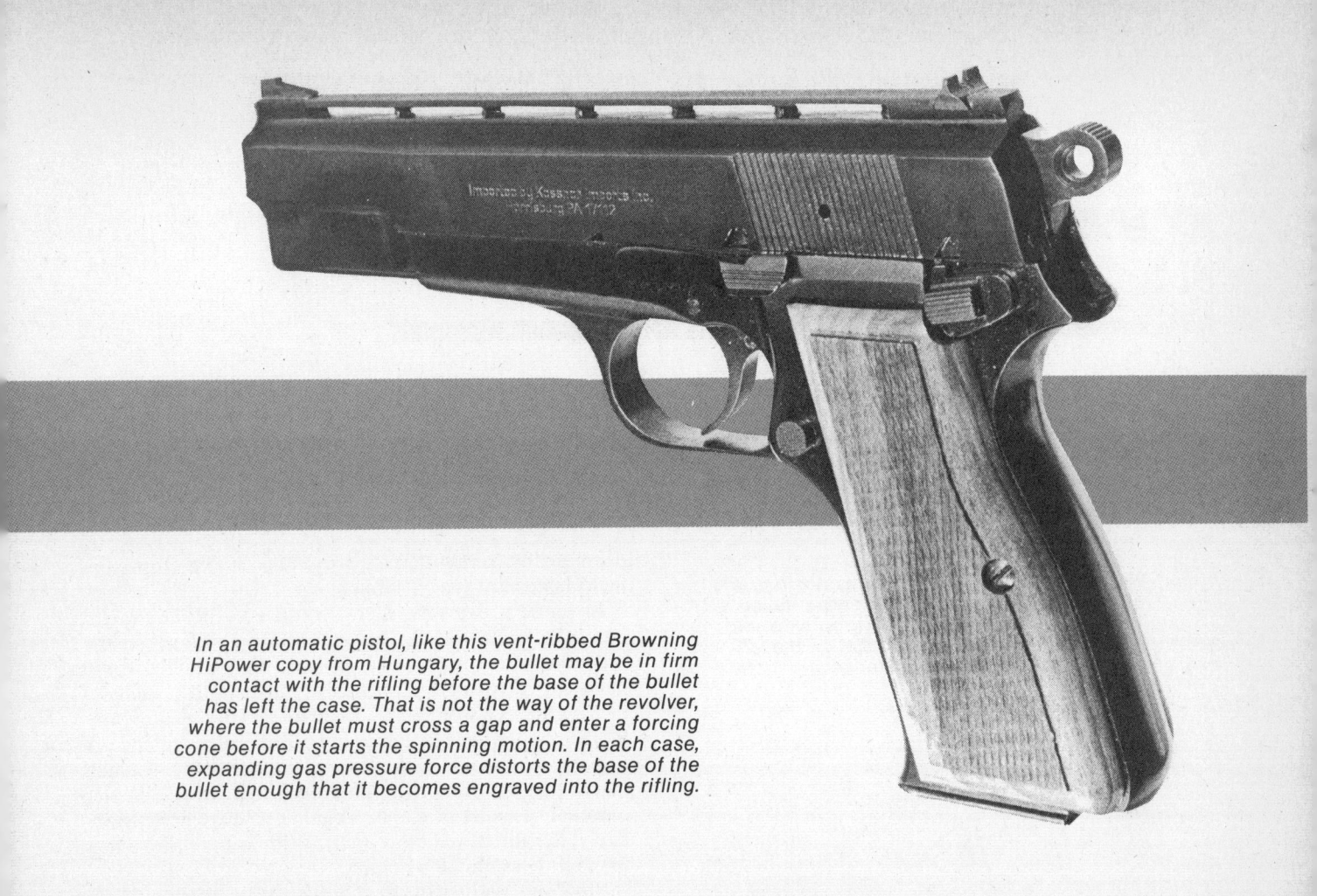

In an automatic pistol, like this vent-ribbed Browning HiPower copy from Hungary, the bullet may be in firm contact with the rifling before the base of the bullet has left the case. That is not the way of the revolver, where the bullet must cross a gap and enter a forcing cone before it starts the spinning motion. In each case, expanding gas pressure force distorts the base of the bullet enough that it becomes engraved into the rifling.

gas expand as fast as it could if the bullet didn't impede it. The base of the bullet therefore swells, squeezed between the pushing gas and its own resisting mass. The revolver bullet leaps forward into the leade, or throat, meets the added resistance of the rifling and loses much of its momentum. The powder gas still thrusts against its swollen base, expanding it further.

The thrust of the propelling gas must swage the expanded base into the throat, then into the rifling. The bullet doesn't stop moving forward, but its forward motion slows dramatically until the gas swages it into the bore and accelerates it down the barrel. In either a pistol or a revolver, pressure continues to rise until the bullet enters the rifling and travels far enough down the barrel for the increasing space behind it to begin reducing pressure.

As the bullet accelerates, it leaves behind more and more expansion room for the propelling gas. Pressure begins to drop, then drops more and more rapidly. The gas continues to expand. It exerts less and less force on the base of the bullet — but less and less force is necessary as the increasing momentum of the bullet requires less force to accelerate the bullet.

By the time the action of an auto begins to open, or the base of the revolver bullet clears the face of the cylinder, the gas has done most of its job. It has started the bullet on its way, has overcome most of the resistance to the bullet's forward travel and must now simply keep nudging it along a bit faster.

How rapidly the speed of the bullet can increase depends now on the friction between the bullet and the surfaces of the bore — also on the twist of the rifling. To understand the effect of the twist, consider the relative effects of two ridiculous extremes — a bore "rifled" with straight lands parallel to the axis of the bore and a bore with

GUN DIGEST BOOK OF HANDGUN RELOADING

The expansion ratio of a cartridge and its bullet is an important ballistic relationship. Expansion ratio is the number of times the barrel travel of the bullet (from A through B to C, at the muzzle) increases the volume of space behind bullet. Volume at A versus volume at C.

Rifling twist affects the internal gas pressure of the cartridge by resisting and therefore slowing the bullet movement. If the barrel had no rifling (right) it would allow the bullet to go straight. The bullet on the left faces a too-fast twist, with more distance and friction.

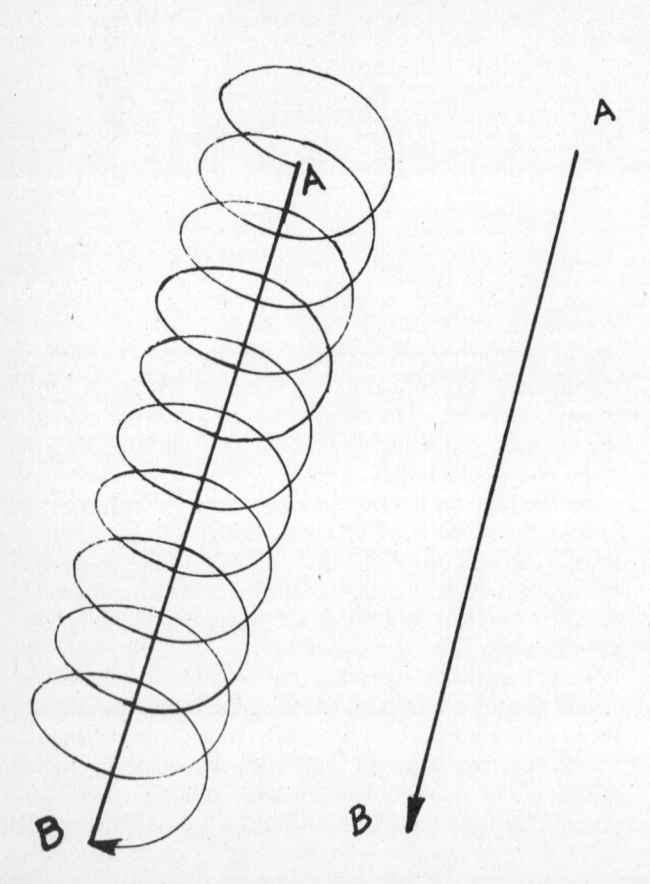

deep, coarse screw threads (say, two or three turns per inch) instead of typical rifling.

In the first extreme, there's no twist, only a straight course of travel for the bullet's bearing surface. In the second extreme, the twist is so great that a bullet would have to revolve two or three times to go forward an inch. Its bearing surface would have to travel two or three circumferences of the bore for each inch of forward movement. A screw thread would make the bullet a bore plug — much like the screw that plugs the breech of a muzzle-loader barrel. The bullet probably wouldn't go anywhere. More likely, the gun would blow apart. With its bore plugged at the rear, it would be a bomb.

Slow and fast twists affect bullet acceleration and friction in the same ways as these extreme bores, but with far less range of variation. In a very slow twist, the radial travel of the bullet is slight, making the net barrel travel of the bullet surface not much longer than the distance from leade to muzzle.

In a very fast twist, the bullet's bearing surface travels farther. Some of the powder's propulsion energy is therefore absorbed in imparting rotation to the bullet, with that much less energy imparting directly forward motion. Also, this increase in the travel of the bearing surface means that friction resists movement a bit longer.

As the base of the bullet leaves the muzzle, the propelling gas is at last free to expand as rapidly as its chemistry and physics demand. Also, it encounters free air — a new supply of oxygen — and its continuing combustion is no longer limited to the much smaller supply of oxygen released from within its own chemical composition.

Its muzzle pressure is now much lower than its peak pressure was, just after the bullet started moving — but much higher than the atmospheric pressure of the outside air. The freed gas finally completes both its expansion and its combustion in an instant, producing that thunderclap of

Bullets from even the best loads shake, rattle and roll like this when acceleration shoves them out against the resistance of the air. For best results, the crown must be dead even and the bullet base must be absolutely flat.

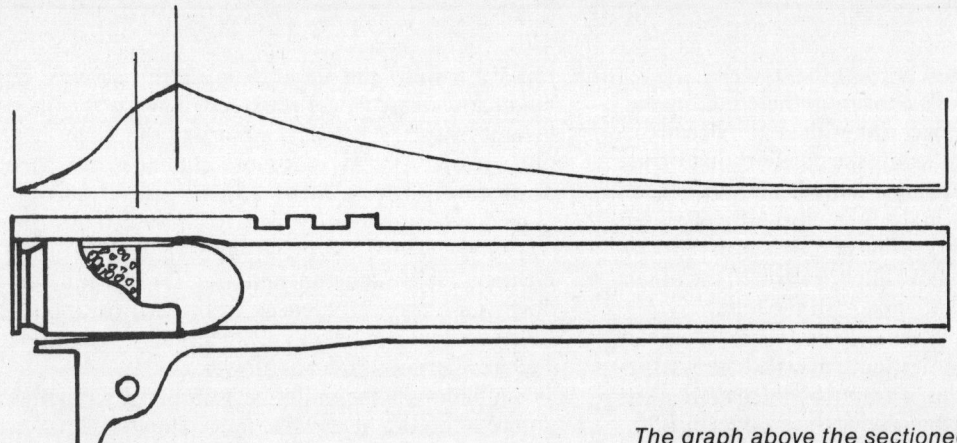

The graph above the sectioned .45 ACP barrel roughly relates internal gas pressure and bullet travel down the barrel. Pressure rises sharply at first, then less sharply as the bullet begins to move. It peaks when the bullet is still near the chamber and drops rapidly thereafter.

sound called muzzle blast. Its outward rush at this instant is also its fastest and weakest movement so far. It exerts a little forward push on the base of the bullet — but not much.

Still, if the base of the bullet is uneven, then its exit from the muzzle is uneven, letting the released propelling gas push more on one side of the bullet base, tipping the bullet to one side. Every bullet tips somewhat and yaws more or less violently for some distance in its early air travel, then settles down to relatively even flight. Too much tipping and yawing ruin the evenness of the bullet's flight. Tipping and yawing aren't consistent or controllable, so too much of this erratic flight increases drag and causes inconsistent trajectories.

The bullet is now as fast as it'll ever be, going at muzzle velocity for only an instant, slowing rapidly as the density of the air resists its travel. The rate of slowing is greater here, since air's resistance, called drag, is directly proportional to the square of the bullet's speed (later, if the bullet flies so far that it's going only half as fast, the drag on it will be only a quarter as great).

Muzzle velocity is a useful reference figure, but it exists for only an instant, at the generally least useful point on the bullet's trajectory. Muzzle velocity is a peak speed — where the speed increase inside the barrel stops and the speed decrease outside the barrel begins. The actual speed of the bullet at any point beyond the muzzle is lower. The bullet's kinetic energy — product of its mass and velocity

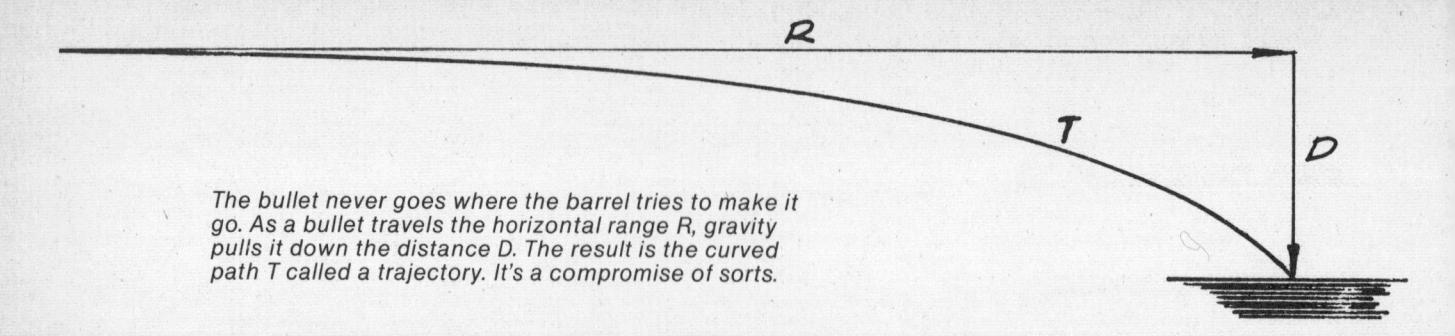

The bullet never goes where the barrel tries to make it go. As a bullet travels the horizontal range R, gravity pulls it down the distance D. The result is the curved path T called a trajectory. It's a compromise of sorts.

— is also lower at any point beyond the muzzle.

The practical efficiency of the cartridge therefore depends on the bullet's speed at impact rather than at its peak just beyond the muzzle. Theoretically, the powders that we use in handloading metallic cartridges contain about 175 foot-pounds of energy (fpe) per grain. The most efficient loads actually deliver far less energy. The percentage of the powder's theoretical energy that a bullet imparts to the object that it strikes is the actual efficiency of the load.

But it's generally handier to compare cartridges' efficiency using their muzzle energies. To figure muzzle energy, multiply the weight of the bullet in grains by the square of its velocity in feet per second, then divide the product by 450,-240 (this number combines the figures used to convert the bullet's weight in grains to a fraction of a pound and its weight to mass).

To figure the efficiency of a load, multiply the number of grains in the powder charge by 175 to get the theoretical energy potential of that load. Divide this figure into muzzle energy. The result — always less than 1.00 and probably below 0.25 — is the efficiency of that load. Convert the decimal figure to percent and express efficiency as this percentage. Derive the efficiency percentages of any two loads in the same way to compare them.

For example, ten grains of powder would theoretically produce about 1750 fpe. But behind a 240-grain in the .44 magnum, it's more likely to produce only 700 fpe — about 40% efficiency. Also, 23.0 grains of 4227 would theoretically produce about 4000 fpe. But with the same bullet, it's more likely to give you no more than about 770 fpe — about 19% efficiency.

Before its exit form the muzzle, the bullet has been supported from below by the barrel. Now, although it's coasting rapidly toward the target, the earth's gravity pulls the bullet downward from the extended axis of the barrel at the instant of exit. At first, the mass of the bullet resists this pull, then gives way to it more and more and accelerates downward.

Its downward velocity increases from 0.0 feet per second (fps) at the rate of about 30 fps per second. In other words, if the bullet remains unsupported for one full second, its downward travel increases from zero to about 30 fps. (Gravity accelerates falling objects 32.16 fps in a vacuum. The acceleration in air is slower.)

But of course the bullet isn't dropping straight down. If you're shooting at a point approximately level with the muzzle, the barrel has to be tilted slightly muzzle-high to compensate for the downward pull of gravity. Also, the bullet travels only slightly upward while it zips along its more obviously horizontal trajectory. Its path levels off quickly as gravity slows its rise, then it drops more and more rapidly as gravity accelerates its fall.

During its slight rise and increasing fall, the bullet goes directly away from the muzzle much more rapidly. It goes up an inch or two, then down several inches in the short time it's going hundreds of yards more or less horizontally. We all joke about the "rainbow trajectories" of slow, heavy, blunt bullets; but the curve of any bullet's trajectory is really almost flat.

The bullet is in the air for only a very short time — so gravity doesn't have much time to pull it downward. Therefore, the effect of gravity is so slight in comparison with the mainly horizontal travel of the bullet that it seems insignificant. But it isn't insignificant, as any long-range rifleman knows.

The trajectory of any bullet — its air path from muzzle to

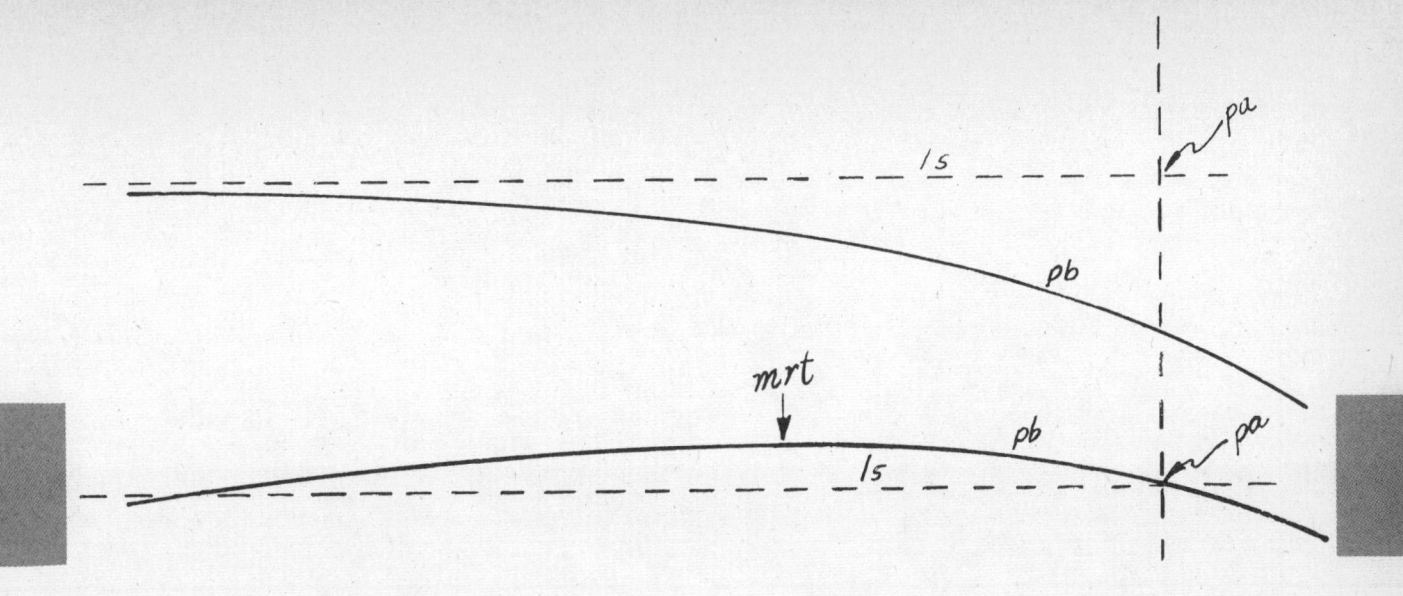

impact — is not an arc of a circle. It's a smooth curve that's more nearly straight at first and curves much more tightly down toward the end, if the bullet travels far enough to lose most of its velocity. The shape of this curve is almost exactly like a long, whippy fishing rod with a can of sand hung from the tip — straighter near the butt, more curved along the middle and more sharply curved downward at the tip.

The general shape of this curve results from the increasing pull of gravity and the decreasing velocity of the bullet. The exact shape of any trajectory is the result of many influences — the muzzle velocity, the mass and the shape of the bullet, the density of the air, the slope of the barrel axis when the bullet comes out of the muzzle and the distance from the muzzle to the point of impact with another substance that changes the speed or the direction of the bullet (or both).

The highest point in the upwardly curved trajectory is always more than halfway to the target. "Midrange trajectory" is not halfway between muzzle and target but much closer to the target.

A bullet with a slow start needs more "arch" or "loft" to go as far as one with a fast start. A cylindrical wadcutter needs a higher launch angle to go as far as a more streamlined bullet. A long wadcutter holds its velocity better than a short wadcutter of the same caliber. The farther a bullet has to go to reach a point essentially level with the muzzle, the higher its launch angle must be.

Cold air is denser than warmer air and the air at sea level and the lower elevations is denser than the air at the higher elevations. Wind is something else. In still air, the bullet's trajectory is a fishing-rod curve from the side and a straight line from above. When the air is moving in exactly the same or exactly the opposite direction, the trajectory curve is slightly longer or shorter. When the air is moving in any other direction, it moves the bullet slightly away sideward from the direction of aim.

A slight breeze at a slight angle can have less effect than a typical aiming error. A stiff wind dead square to the line of aim can blow the bullet noticeably to the side. With a slow bullet, a long range or both, the wind effect is most extreme.

Another phenomenon also makes the bullet travel slightly to the side, even in still air, but its effect is negligible at low velocities and short ranges. The twist of the rifling spins the bullet on its long axis and this spin causes a sideward drift. Only at long ranges is spin drift likely to be greater than a lateral aiming error.

Once the bullet strikes or enters anything besides air, the ballistician's attention turns to its terminal ballistics — the effects of the bullet and the struck medium on each other. Now everything changes — a little or a lot. Even a paper target slows the bullet's velocity a little and any change in its velocity also changes its energy and trajectory. The bullet designer is interested in how the bullet and the target substance affect each other. The shooter is interested in the bullet designer's findings and designs.

The target shooter's first interest in the bullet's terminal ballistics beyond the target is the assurance that an ade-

To strike a point of aim (pa) on roughly the same level as the shooter's eye, the bullet path (pb) must rise above line of sight to compensate for drop. The trajectories are the same in each case shown, but what a difference in effect! It's also an argument for adjustable sights.

While the beginning shooter starts with a .22 rimfire, like the S&W 422 shown here, she/he quickly learns to appreciate the complexities of trajectories and use of adjustable sights. Handloaded ammo in the centerfires will make practice more economical. Right: Winchester makes high quality .38 Super ammo as the Ransom Rest proves, but careful handloaders equal it — sometimes!

Winchester 125 Grain Silvertip .38 Super 1.592" 10 shots

quate backstop stops the bullet without ricochets or bouncebacks. The hunter and defensive handgunner are concerned about the effect of the bullet on flesh — and they want to be sure that after a miss or a shoot-through, the bullet won't be likely to go on to do further, unwanted damage or injury.

Whether the bullet stops within the medium, ricochets off in a new direction or shoots through to travel a secondary trajectory, it enters a new stage in its travel. The one thing that seems to change the least, upon and after impact, is the bullet's rotation. Down to the end of its trip, the bullet spins essentially as fast as it did when it came out the muzzle. In tests handgun bullets stopped by foot-wide cubes of clay leave definite swirl marks in the inside surfaces of the cavities they make.

The bullet's speed and angle of impact help to decide its course beyond the point of impact. A bullet that easily penetrates a car door when it strikes it squarely, for exam-

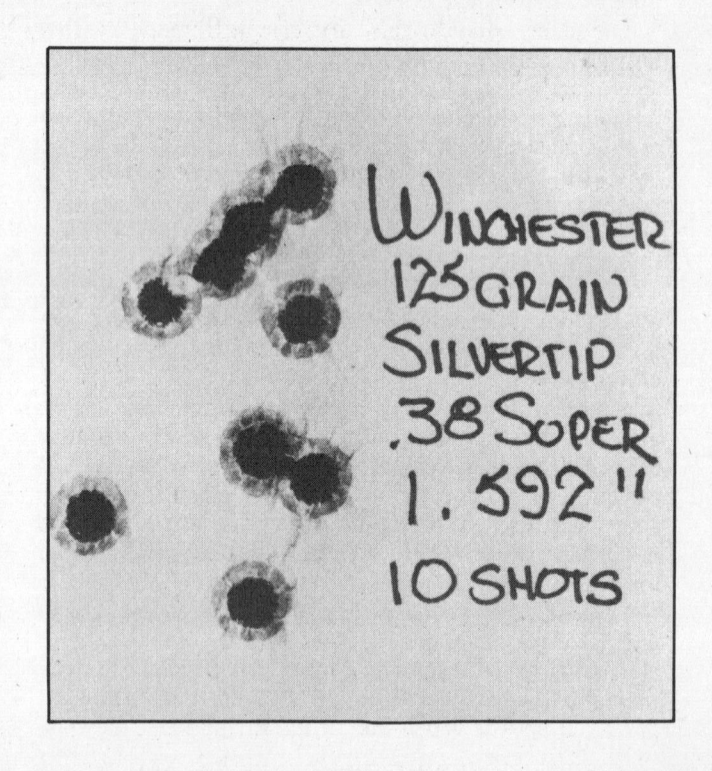

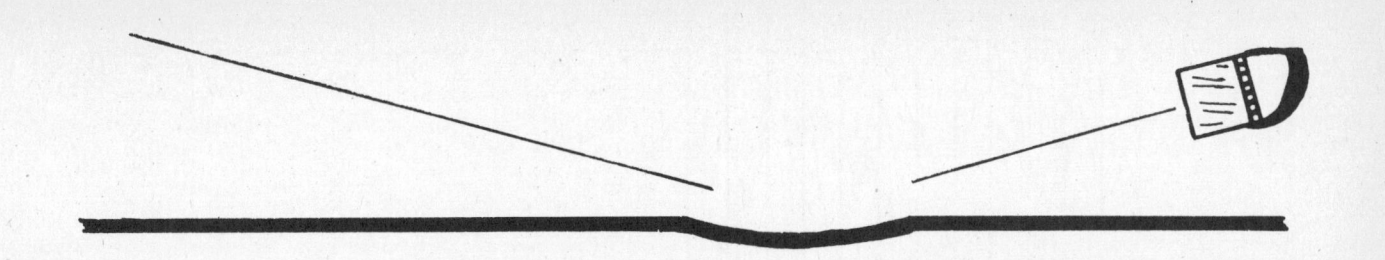

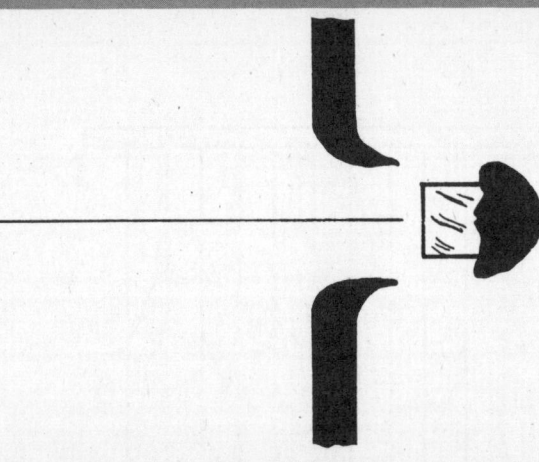

In considering terminal ballistics, the angle of a bullet's strike can make a world of difference in its behavior. Above: The bullet strikes a sheet of thin steel at an acute angle and bounces off, with uneven deformation of the bullet nose. Left: The same bullet tackles steel much thicker, but at a perfect right angle. The result is not just penetration, but also symmetrical deformation. The principle applies to other terminal ballistic media.

ple, just as easily ricochets off the thin steel when the angle of impact is very acute.

A slower bullet tends to follow the grain or direction of varied densities within the struck medium. An old Montana friend, Charles M. O'Neil, once heard a bullet zip past him after he fired at a target tacked to a fence post. He found the bullet's entrance and exit holes on the near side of the post, so he split the post to see the track of the bullet inside. Sure enough, the grain and densities of a knot inside the post had "returned his serve" exactly the way a jai alai player's *cesta* returns the speeding *pelota*.

Contrasting densities within a block or body make the terminal behavior and effects of a game bullet — for rifle or handgun — unpredictable. To see the general effects of contrasting densities, shoot first into a horizontal stack of a dozen or so vertical inch-thick boards with no space between them. Then tack strips to them to hold them an inch apart and shoot into them again — with the same load, from the same range. The spaces between them let the bullet penetrate more boards.

In other materials, differences in their densities affect bullets generally the same as the alternating layers of boards and air. After this simple experiment, you can imagine the effects of wood grain, flesh and bone, even the contrasting densities of flexed and relaxed muscles, on the upsetting and penetration of a bullet — when it hits them square-on and when it hits them at acute and obtuse angles. Even skin and bone can divert a bullet if the impact angle is slight.

The shape, hardness and sectional density of the bullet, as well as the density of the medium, determine the upsetting (change in shape) and penetration of the bullet. The change in shape also determines the nature and extent of the damage inside the medium. An extremely soft, blunt or broadly rounded bullet can damage a lot of tissue over a relatively short stopping distance in any penetrable substance. An extremely hard, sharp bullet can cam tissues aside, leaving a tiny hole that closes behind it, to do comparatively little damage over a very long penetration distance.

The thickness of the medium — from front to back, as the bullet goes — determines how long the bullet and the medium can act on each other. A single thickness of target paper seems to have no effect on a bullet. But line up a great

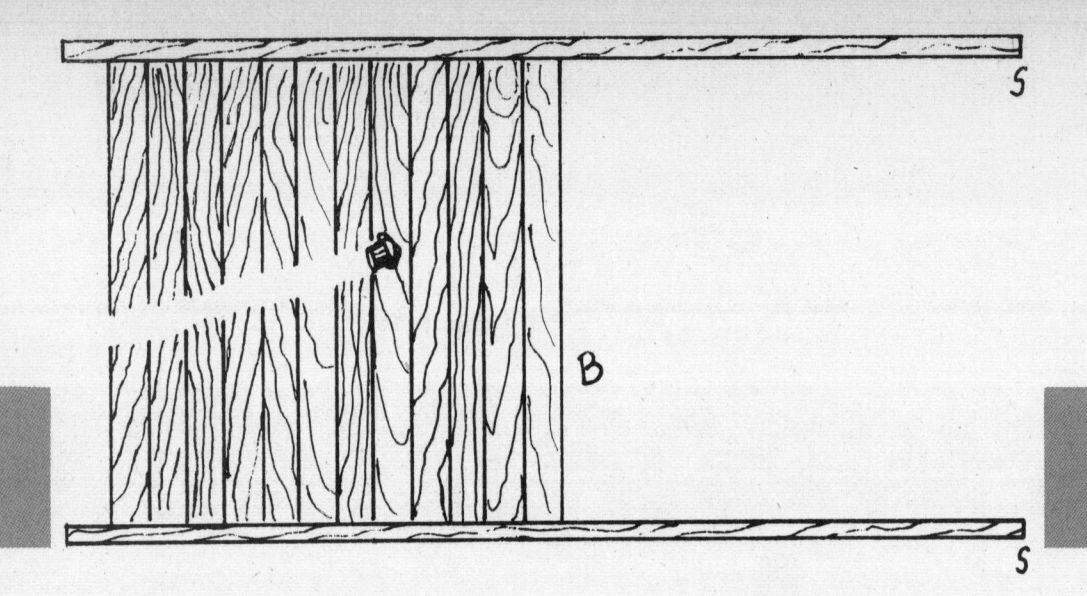

Nail a dozen inch-thick boards (B) between the pairs of slats (S), first tightly together, then separated evenly by about an inch of space. Shoot into them straight-on and a variety of angles — and you'll learn a lot about the effect of variations in the density of the ballistic test substance and about penetration and bullet trajectory.

number of targets and they'll stop the bullet eventually if you've lined up enough targets.

Bind the targets into a thick bundle and you'll need a lot fewer targets than if you'd lined them up a foot apart along the bullet's path. Or consider an even greater extreme — a bullet fired into soft, fluffy snow comes to rest without a mark on it if the snow extends far enough along the path of the bullet.

The composition of the medium — simple or complex — is always a factor to consider in terminal ballistics. In a living body, for example, a wound in one part of the body can transmit shock to organs in another part of the body — so that "just a flesh wound" can mysteriously incapacitate or kill by shock transmitted without an obvious trace. A bullet striking a certain area in a man's lower abdomen, for example, can burst or paralyze a valve near his heart.

Whether you handload for a b-b-b-big p-p-p-pistol like the Dan Wesson at the start of this chapter or a teeny one like the Browning .25, the principles of ballistics don't change. Even the low-pressure loads that will be necessary for this lovely old Merwin & Hulbert — a 19th century revolver — work on the basis of principles that Ken Howell discusses in this chapter. Internal, external and terminal ballistics are a form of applied physics.

Interior ballistics is a subject for technicians to study with labs full of ultra-precise, terribly expensive measuring and recording instruments. But their findings, and the principles basic to their work, are not only interesting to some of us but also vital to all handloaders. The data derived in studies of interior ballistics are the limits that we handloaders have to stay within.

Exterior ballistics is a subject that any of us can study in the field — in a limited way — with good machine rests, low-cost electronic chronographs and simple bullet-drop experiments. Indeed, it's the one category of ballistics that every shooter needs both gut and field knowledge of, if he's to reach the top of his shooting skill and utilize the potential of his gun and its loads.

Terminal ballistics, easily the most complex category in the science of projectile travel and effect, means different things to different people. Many factors in this field of ballistics are predictable in only the broadest general ways, yet study of them is vital to bullet design as well as to investigations for insurance settlements and criminal trials.

So there you have the gist of handgun ballistics in a nutshell — in other words, just a sip from the keg. It's nice that all this happens in a lot less time than it takes to explain any of it. It's also nice that we can learn so much about what occurs between firing pin and target. But ballistics and ballisticians are totally stumped by the performance of one sixgun that I remember.

The jailer in my mother's home town knew that he wasn't an imposing enough figure to awe his lodgers, so he regularly and solemnly described the terrifying ballistics of his six-shooter like this — patting his holster for gentle emphasis:

"I carry a b-b-big p-p-p-p-pistol. It shoots a z-z-z-zig-zag, r-r-ramblin' ball. When you z-zig, it z-z-zigs. When you z-zag, it z-z-zags. C-c-c-c-c-can't miss!"

Nobody ever gave him any lip or worry. But we kids would've given anything to shoot that b-b-big p-p-p-p-pistol! — *Ken Howell*

CHAPTER 6

TESTING

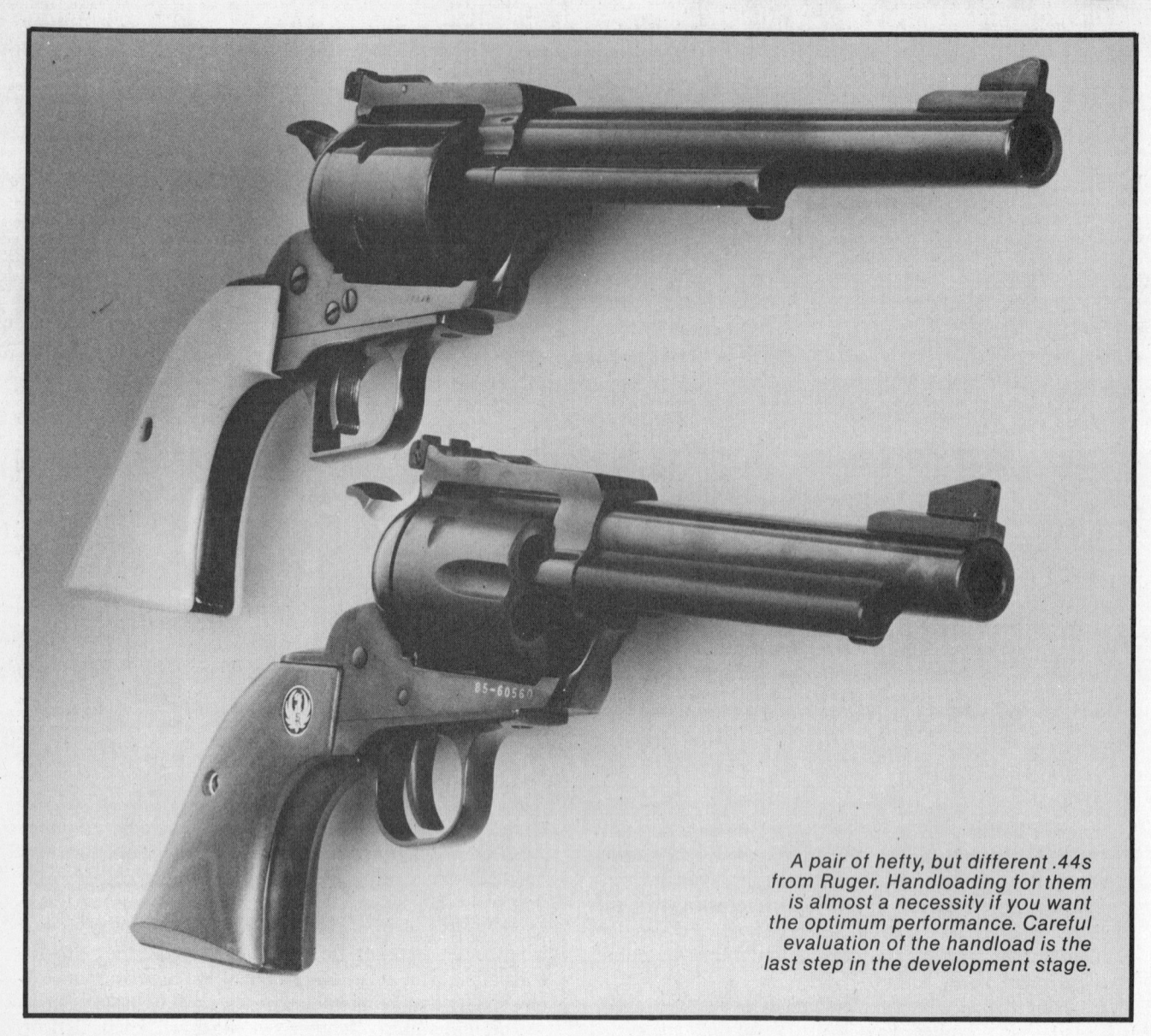

A pair of hefty, but different .44s from Ruger. Handloading for them is almost a necessity if you want the optimum performance. Careful evaluation of the handload is the last step in the development stage.

THE HANDLOAD —

ONCE A handgun handload is prepared, the shooter still must test it to determine whether it is suitable to his particular firearm. Just because a cartridge is quality assembled according to the accepted steps does not mean it will give acceptable performance. There are just too many variables.

As a good example, let us consider the .45 Colt. This old-timer dates from 1873 and, as is the case with many rounds of antique vintage, its specificitions have undergone considerable change in 114 years. Most .45 Colt revolvers of Nineteenth Century vintage were intended for .454-inch bullets. During the 1950s, Colt began using .451-inch for the bore diameters of their .45 Colt sixguns and other manufacturers followed suit. Therefore, almost all correct factory-produced .45 Colt bullets have .451- or .452-inch diameters. However, I have tested more than one .45 Colt revolver that has a .451-inch barrel, but .456-inch chamber mouths. Their accuracy with most .451- or .452-inch

Good Equipment And Common Sense Make Test Results More Meaningful

Left: Author Venturino's test setup includes this bench made of heavy plywood. He also makes good use of old shot bags filled with sand. Like many shooting writers, he chooses the respected Oehler Model 33 chronograph.

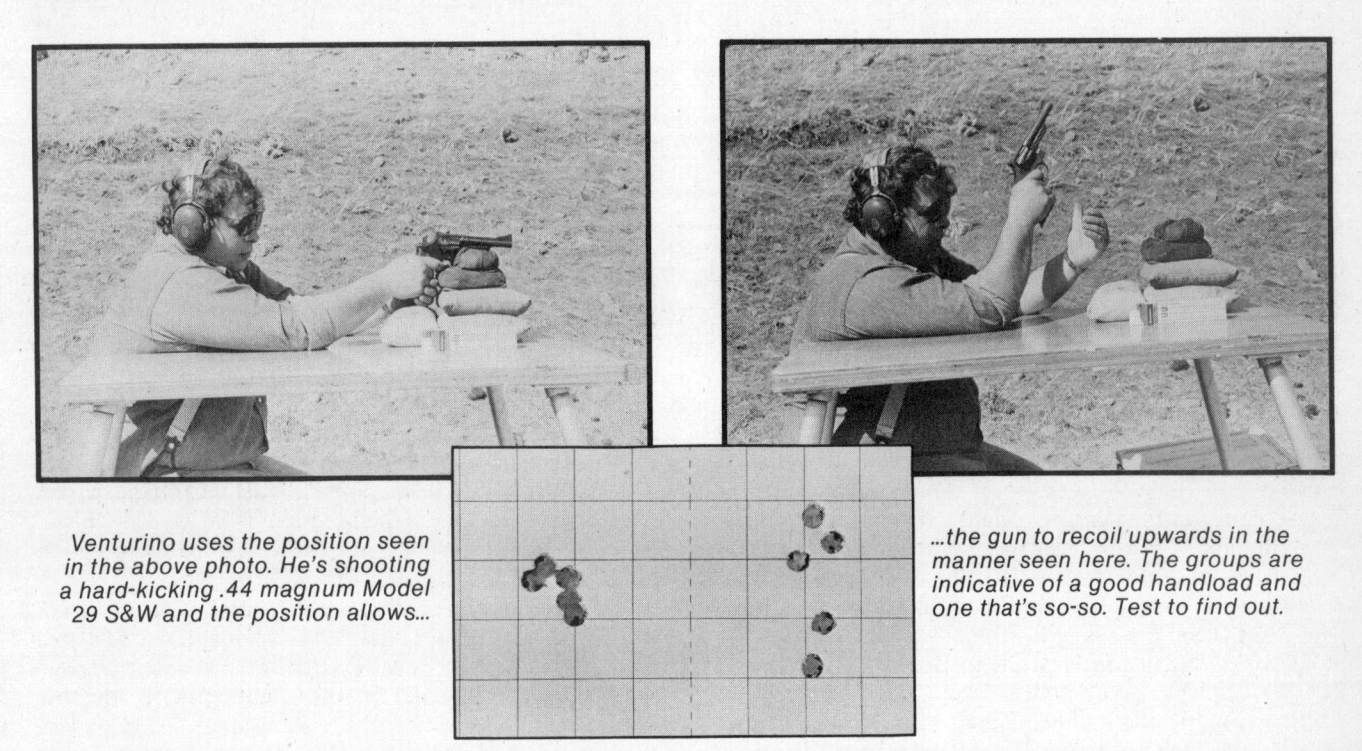

Venturino uses the position seen in the above photo. He's shooting a hard-kicking .44 magnum Model 29 S&W and the position allows...

...the gun to recoil upwards in the manner seen here. The groups are indicative of a good handload and one that's so-so. Test to find out.

factory bullets has been unacceptable. On the other hand, they will shoot nicely with .45-inch cast slugs. Therefore, even if a handloader put the utmost care in assembling .45 Colt handloads with .451-inch bullets and fired them from a .45 Colt sixgun with .456 chamber mouths, he could end up with poor ammunition.

For this reason, many handgun reloaders end up being handgun experimenters. Testing ammunition in individual guns is educational and captivating. Variables are near endless and many will have as dramatic effect on ammo performance as similiar variables will for rifle shooters.

Checking your handgun ammunition performance can be a simple thing. The minimum required is a sturdy bench and some solid sandbags. By resting the gun solidly over the bags and using the best sight alignment, breath control and trigger squeeze possible, the shooter can discover his handgun and ammunition potential. The range should be whatever distance at which the tester feels comfortable with his sights. For some, this is fifty yards, but for handguns with service-type sights, this is often fifty feet. By far, most shooters use twenty-five yards. Many modern shooters involved with the so-called handguns using scopes and

Probably the most definitive device made by which to determine the accuracy of particular ammo is the Ransom Rest. Invented and still marketed by Chuck Ransom, the rest uses inserts faced with a rubbery material and pre-shaped to the contours of various guns. The gun and inserts are held on an arm that is aligned in the same exact spot for each shot. Test results are therefore highly reliable.

chambered for high-velocity bottleneck cartridges feel that one hundred yards is a better testing distance for their devices.

Actually, a scope can be a great aid in testing handgun ammunition. With iron sights, the shooter's eyes must cope with three reference points: the rear sight, the front sight and the target. Eye strain and fuzziness become a definite factor after a lengthy testing session. Conversely, a scope relieves the strain on the eyes, since all one must do is align crosshairs on a bullseye.

Shooter fatigue is another factor to consider, especially with many of today's powerful handgun calibers. The muzzle blast and recoil delivered by a .41 or .44 magnum revolver will cause even an experienced shooter's skill to deteriorate rapidly in a testing session. I probably have much more experience than the average handgun shooter, but freely admit to being able to only get off only a few groups of magnum-velocity ammunition before shooter fatigue sets in and flinch-inspired misses occur.

On the other hand, with such calibers as the .32 S&W Long or .38 Special, test sessions can continue with reliable results until blurring of the eyes with iron sights intervenes.

For such reasons, many handgun experimenters of a more serious nature have come to rely on machine rest devices. Such an apparatus will eliminate or minimize the human element and allow lengthy test sessions.

Currently there are two popular machines on the market for this purpose. One is the Mequon (formerly Lee) Pistol Machine Rest from the Mequon Reloading Corporation (P.O. Box 253, Mequon, WI 53092).

This simple device consists of a base plate with adjustments, a rocker assembly and grip adaptors made for a large variety of handguns. To use it, the shooter removes his handgun's grips and fits the aluminum grip adaptors around the gun's grip frame. For semi-autos, the grip adaptors consist of magazine inserts and the guns must be fired single-shot.

The handgun is fitted to the rocker which then lies on the base. The shooter grips the gun just as he would normally and pulls the trigger. Recoil causes the entire rocker assembly including handgun to pull away from the base plate. For the next shot, the rocker is put on the plate again and aligned with its stops.

The beauty of the simple Mequon rest is that it requires no heavy-duty mounting. Since the handgun on the rocker arm moves away from the base plate during recoil, there is no heavy stress on the shooting bench. In fact, the shooting bench need not even be anchored into the ground.

On the minus side is the fact that the shooter's hand

The Ransom Rest in use with a S&W .44. Bolted to the bench, the gun is ready to be fired by means of a lever device in the trigger guard. When the gun is fired, the entire rocker arm...

...pivots upwards in recoil as seen. Then the arm is pushed back to the lower position and the gun is cocked for firing the next shot. The football-shaped knobs on the base are part of the windage base.

absorbs all the recoil. With milder cartridges, this has no consequence. However, with the big boomers it does, because long sessions with your own mitt wrapped around those bulky aluminum grip adaptors will leave a sore or even blistered shooting hand. During some severe testing I even got to the point that I flinched badly enough with the Mequon Pistol Machine Rest that my jerking of the trigger was pulling shots from the groups. An old glove with the trigger finger cut away will help some with this problem.

Still, in fifteen years of use, I've found the Mequon Pistol Machine Rest a great help and feel it has been a major aid in learning about handgun accuracy. Its simplicity is its greatest boon.

The famous Ransom Rest is a more intricate test device from Ransom International Corporation, (P.O. Box 38-45, Prescott, AZ 86302.)

In the Ransom Rest, the handgun is held between two clamp-type grip adaptors. Again the handgun must shed its grips for testing, but semi-autos can be used with full magazines. It should be noted that some modern revolvers do not have a grip frame as such. They have a metal strip coming from the frame that the grips fit over. Examples are

With a custom .45 pin gun bolted solidly into the inserts, a shooter fires with the lever shown here. A reliable device, the Ransom Rest can be operated inconsistently and the test results will suffer. Most common mistake is to fail...

...to use the little cast ledge that the manufacturer provides. If the shooter pushes against the gun to return the rest to battery, it will change the fit of the gun in those inserts. When that happens, tests are flawed and must be questioned.

the Dan Wesson handguns and the new Ruger GP-100. Such designs may be the wave of the future, but they sadly forget about handgun experimenters using machine rests. As of yet, there is no way to mount such guns in conventional machine rests. They must be tested over a sandbag rest.

The handguns do not recoil free from the Ransom Rest. The recoil is soaked up by the device and somewhat by the bench to which it is attached. That means its mounting must be sturdy; a portable bench often will shift with the recoil of heavy handgun reloads.

On the bonus side is the fact that the shooter does not actually touch the handgun during testing. There is a trigger bar for firing and a return lever to seat the handgun for the next shot. This means thats much longer test sessions can be conducted without human element becoming a factor. In fact, in several years of use with the Ransom Rest, the only fatigue I've experienced has been my feet. That comes from walking fifty yards round trip every time a target must be changed!

With the machine rests one need not use expensive printed targets. In fact, no formal target is necessary. For my machine rest testing any sort of plain, sturdy paper has been adequate. Just be sure it's heavy enough that it does not tear in removing it from the target backing. For machine-rest testing, I get sighted in on one spot, then shoot every group at that point, changing the target every five or ten shots as required.

Doing so will kill two birds with one stone. By not bouncing about with the machine rest's adjustments for

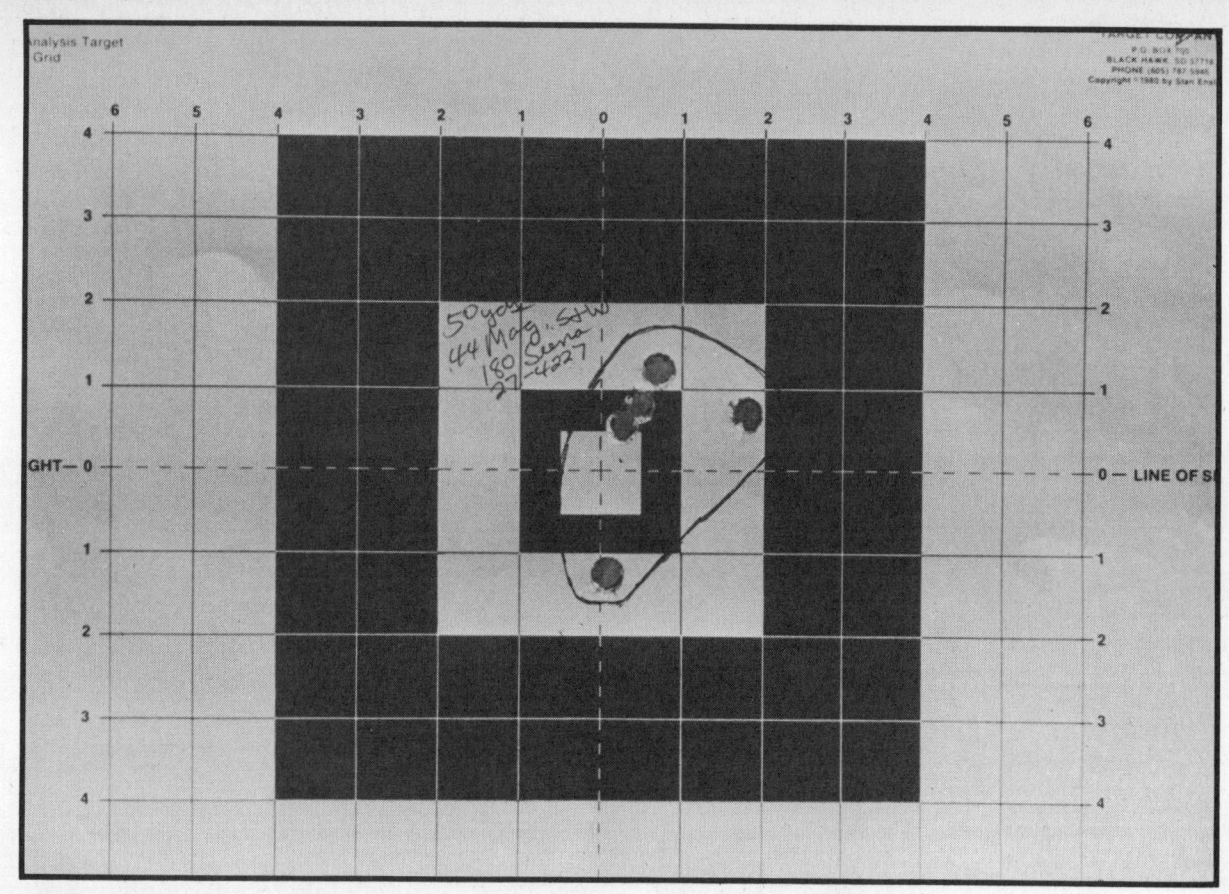

With a solid bench and a good position, a careful shooter can produce some acceptable results. One of the major difficulties is the target at which the gun is aimed. This is one of Venturino's favorites from Rocky Mountain Target Co. The grid is a one-incher and the aiming point is as shown. Good group.

every group, your target backing board will last longer. Also, by leaving the adjustments alone, you learn exactly how various handloads group at different points of impact with the same sight setting. Those changes should be noted for future reference.

The most common handgun target is the standard round bullseye and it is never a poor choice. However, some modern target companies such as Data-Targ, (P.O. Box 700, Black Hawk, SD 57718) are producing handgun testing targets with hollow square aiming points. With modern Patridge sights of square profile, these are fine targets, as the target square can be aligned precisely above the handgun's sights. Another benefit of these targets is their data section wherein pertinent data for that group or groups can be recorded. Such is handy, as you can realize, if you have ever tried to match targets and notes recorded years ago and make sense of it. When doing sandbag testing at either twenty-five or fifty yards, these are the best targets available, in my opinion.

A treatise on handgun handloading experimentation would not be complete without mention of chronographs. Some years ago, only the most serious and dedicated of handloaders had such instruments. With the advances in electronics, however, the cost of hobby-type chronographs has been reduced the point that most any shoooter can own one.

Basically there are two questions to be answered concerning chronographs. "Should I have one?" and "Which one?" The first question is answered by another, "How serious are you?"

If the handgun handloader is only interested in working up quality handloads, he can simply test them for accuracy. If they meet his standards, he can go on with his target shooting, hunting or whatever. The less-then-serious experimenter can interpolate data from the various manuals and come close to knowing his handloads' speeds.

The chronograph comes in handy if handloads are giving one problems. If velocity variation is excessive, the handload will seldom be as accurate as one with less variation. If bullets are leaving the muzzle with a difference of several hundred feet per second, it's a safe bet they are not grouping closely on the paper. If that is the case, the experimenter knows to look to his powder selection, primer type, bullet fit or even the crimp. All will effect velocity variation. If the chronograph shows little velocity variation, but the bullet holes are still spread generously on target, the shooter needs to examine his bullet quality, bullet fit to chamber mouths or bore. The chronograph saves time there and, if you are just plain curious, the chronograph can show you what velocities you're getting for all the time spent at the reloading bench. As a gunwriter, I would not be without a good chronograph. As a fun shooter, it often goes unused for long periods of time.

In the dark ages of twenty years ago, the relatively cheap

GUN DIGEST BOOK OF HANDGUN RELOADING

Left: The Oehler 33 control box offers an immediate readout of velocity. On demand, the unit will display the fastest and slowest shots in a series, then the extreme spread, average velocity, standard deviation.

The bullet travels downrange, crossing the first of a pair of readout screens. First screen starts the electronic clock as the bullet passes and second one stops it. The control unit has a computer which calculates speed.

chronographs affordable to handloaders had some sort of breaking screen. After each shot, the screen — or screens — were replaced. That was both costly and time-consuming. However, in the Seventies, chronographs with electric-eye screens became commonly available. They are the only ones taken seriously by most shooters today.

Most of these chronographs work in the same fashion. The screens have a predeteremined spacing and the instrument measures the average time it takes the projectile to pass between these two points. For many years, the standard by which most other chronographs have been judged is the Oehler Model 33 with its progressive models of Skyscreen systems. The little black box gives direct readout of each shot fired and, on command, will give the high and low velocities, extreme spread, average and standard deviation. It is truly an amazing machine.

In more recent years, another more simple type of chronograph is making its debut. Theses are the machines that contain the computers and screens all in one box, the entire unit set out in front of the muzzle and fired over. The readout is given on a display screen and the unit resets itself after a few seconds. Two such units are the Quartz-Lok 77 and the Pro-Tach. I have used both and have found their simplicity of operation to be habit forming. You simply set them on a tripod and start shooting. If fact, if you are brave enough, you can set them under your target and get

The Pro-Tach is a new unit on the market which locates all of the electronics downrange. It is quick and easy to set up and use, but you had best shoot well!

downrange velocities. "Bravery" is the key word in using this type of chronograph, because one tends to be a bit tense with a relatively costly piece of equipment out in front of the muzzle. You sure don't want just anyone pointing his gun at it! While testing the Quartz-Lok 7, I had a gas check separate from a cast bullet and wipe out the digital readout screen. The manufacturer now warns buyers not to test gas-check bullets with that device.

A chronograph is an interesting piece of equipment and I wouldn't be without one, but I don't use it to record my each and every shot.

Testing your handloads is the logical conclusion to the ammunition assembling process. Without doing so, the shooter might be as well off firing factory ammunition. For decades handloaders drummed into each others' heads that their homemade ammunition always was better than the factory-produced stuff. That is not always the case. But factory ammo can be excellent and, conversely, handloads can be pretty bad, unless they are investigated sufficiently to determine their worth. Sometimes the test results can be surprising, in either direction!

— *Mike Venturino.*

CHAPTER 7

If the muzzle of this auto seems a bit over-sized, it's because this HiPower has been re-barreled to the new .41 Action. That means loading with .41s.

COMMERCIAL BULLETS FOR HANDLOADERS

HANDGUN HANDLOADERS who rely on store-bought bullets never have had it so good. At present, there are four major producers of jacketed handgun bullets collectively offering over one hundred designs ranging from .25 to .45 caliber.

For many years Hornady Manufacturing Company (Box 1848, Grand Island, NE 68802), Speer (Box 856, Lewiston, ID 83501) and Sierra Bullets (10352 South Painter Ave., Santa Fe Springs, CA 90670) were the premier jacketed bullet manufacturers. Now the respected name of Nosler (Box 688, Beaverton, OR 97075) must be added to that list. Additionally swaged lead handgun bullets are produced by two of the Big Four, Hornady and Speer, and a wider variety of smaller companies. Sole

90   GUN DIGEST BOOK OF HANDGUN RELOADING

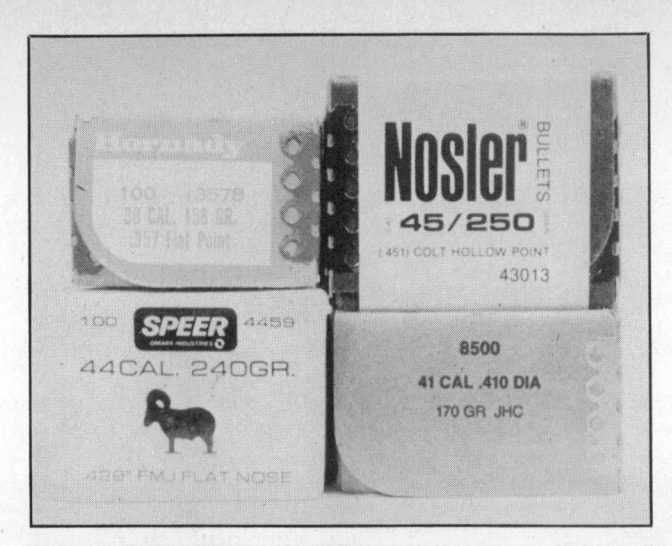

product of the Alberts Corporation (519 E. 19th St., Patterson, NJ 07514) is swaged lead projectiles.

Actually, in today's world of specialized single-shot "handguns" chambered for almost every cartridge imaginable, about any jacketed projectile could be considered a handgun bullet at some time or the other. A good example would be the venerable .35 Remington. For decades, it was considered a fine dense-woods deer-hunting caliber in lever actions; now it often is used in single-shot handguns for all sorts of big game hunting. Are all 200-grain-and-up .35 caliber bullets used in reloading the .35 Remington now considered handgun bullets? That is hardly the case,

Left: Bullets come from the major makers in a bewildering array of sizes and weights as represented here. Lots of handgunners have taken up handloading as factory ammo costs have gone through the roof. Below: Nosler, long a maker of premium rifle bullets, now offers pistol types.

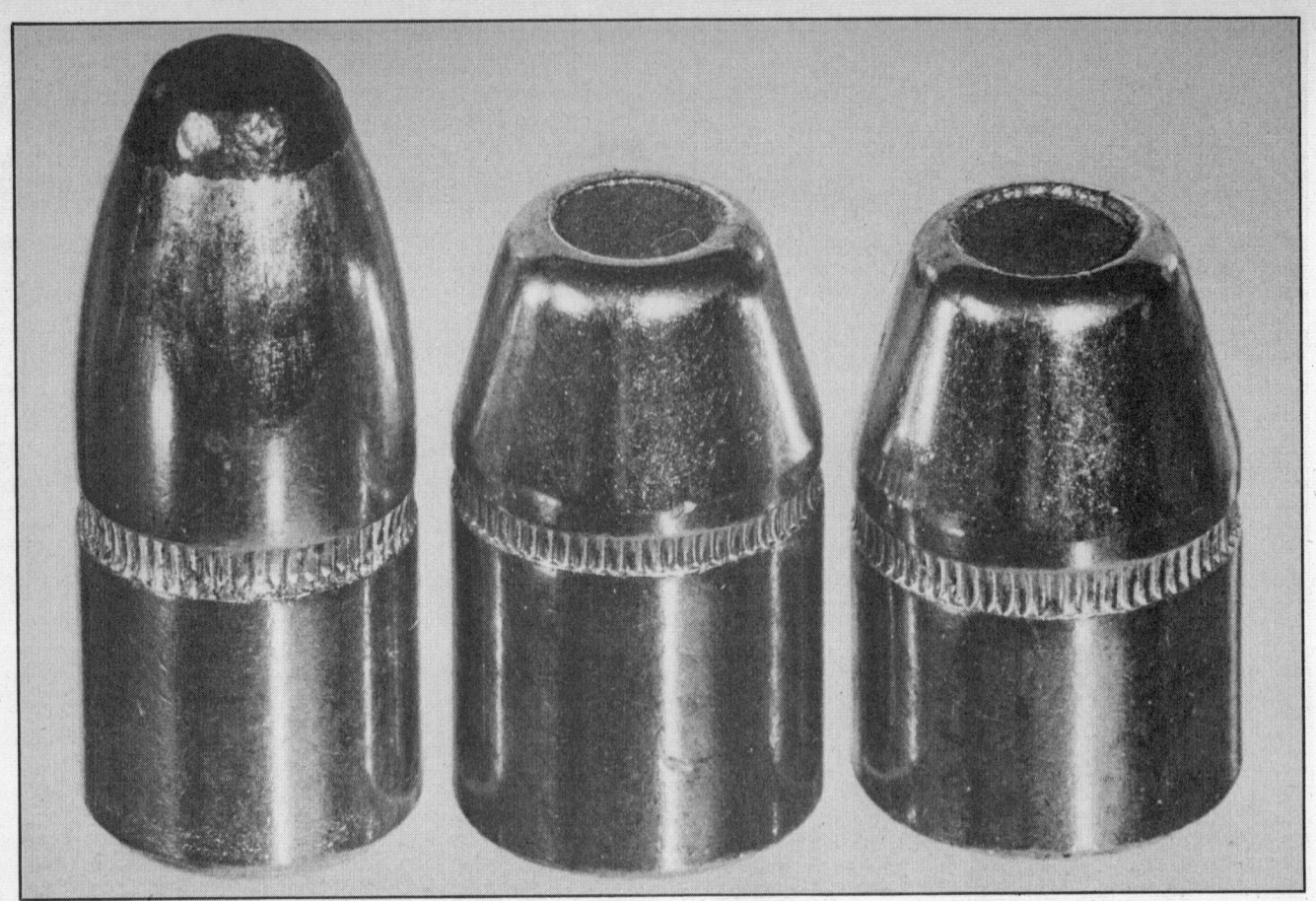

This trio of bullets makes an interesting point. All of them are jacketed — all are the same weight: 220 grains. Diameters are .375, .410 and .429 respectively, all pistol. Below left: The big and little of available pistol bullets.

Above: A modern, well-designed automatic pistol bullet has little if any lead exposed at the tip. They don't feed well that way, so the makers have developed other designs.

and most shooters feel the .35 Remington cartridge and its suitable bullets still belong in the rifle category.

With that behind us, let's consider the offerings of handgun bullets currently commercially available.

With the steady increase in popularity of semi-auto pistols in this country, there has been a corresponding increase in the number and types of suitable bullets in those calibers. At the low end of the spectrum is the .25 ACP. Hornady, Sierra and Speer all make 50-grain .251-inch full-metal-jacketed (FMJ) bullets. On the surface, all three are identical in shape to the traditional round nose of .25 ACP factory manufacture. In all my travels and conversations, I never have encountered a handloader of the diminutive .25 ACP cartridge, but someone must be doing it to create the demand that is evidently there.

Only two of the Big Four bulletmakers offer .32 ACP bullets: near identical 71-grain .311-inch round nose FMJs by Hornady and Sierra. At 9mm bore size, however, the selection goes wild. The 9mm is the handgun wave of the future in this country and those producing shooting com-

Hornady makes a wide variety of 9mm bullets; the 115-grain JHP is a long-time favorite of many handloaders.

The .380 can't use the heavier 9mm bullets, so Hornady produces some light ones in the same .355-inch diameter.

Here's a new bullet on the market, the 200-grain .451-inch TMJ. It is intended for IPSC competition loads in .45 caliber guns. The TMJ terminology means a Totally Metal Jacketed slug. Author Grennell coined the term.

ponents are obviously aware of it. Bullet weights in .355 inch are 88, 95, 100, 115, 124, 125 and 130 grains. Types are full metal jacket, hollow points and soft points. In addition to jacketed bullets, Speer and Hornady list 125- and 124-grain swaged lead round nose bullets respectively. Alberts, whose only bullets are of swaged lead, also has 125-grain 9mm round noses.

The old war-horse .45 ACP suffers a shortage of commercial bullets by all the above makers. For 1987, Speer added 185- and 200-grain full jacketed semi-wadcutters to the 200-grain JHP, 225 JHP and 230 FMJs they already produced. Hornady makes 185- and 200-grain full jacketed SWCs, a 185-grain JHP, and two 230-grain FMJs; one is a round nose, the other a flat point. Sierra offers a

GUN DIGEST BOOK OF HANDGUN RELOADING 93

So far, Norma has the whole market in 10mm auto bullets. That will change as more of the guns come into use and the demand develops. These are 200-grain flat points.

This is another limited use number, but one that shows signs of increasing demand. .32 magnum guns are made in quantity, need .312 bullets.

185-grain JHP, 185- and 200-grain full jacketed slugs and, finally, a 230-grain FMJ round nose. Nosler's .45 ACP bullets are a 185-grain JHP and a 230-grain flat-point FMJ. Lastly, there are .45 ACP lead bullets by Speer, Hornady and Alberts. Hornady's lead .45 is a 200-grain SWC; Speer's are a 200-grain SWC and a 230-grain round nose; Alberts has both of those, plus a 225-grain lead hollow point. All jacketed .45 ACP bullets are .451 inch, while the lead versions are generally .452 inch.

Those are the current semi-auto handgun bullets from the big manufacturers, but how soon will we be seeing 10mm bullets from the same companies? If the new semi-auto caliber catches on, as is predicted, bulletmakers will not leave handloaders in the lurch.

Revolver bullets for the handloader start at .32 caliber, but only a short time ago that would not have been true. It would have been said that .38 caliber was the starting point. The change was caused by the advent of the .32 H&R magnum cartridge. For .32 revolvers, Hornady and Sierra fill the gap with an 85-grain JHP for the former and a 90-grain JHC for the latter. Bullet diameters are .312 inch which, incidentally, make these JHPs also usable in the .32 S&W Long and the ancient .32/20.

In recent testing of a Ruger SSM .32 H&R magnum, both of the above bullets were found to be capable of 1⅛-inch, twenty-five-yard sandbag rest groups over 8.0 grains of 2400 powder. Among the .32 caliber lead bullets, there is a 90-grain hollow-base wadcutter and a 90-grain semi-wadcutter from Hornady. Speer and Alberts produce other 90-grain hollow-base full wadcutters in .32 caliber. Gen-

There are millions of revolvers in use chambered for the .38 Special and .357 magnum cartridges. For that reason, the bullet makers offer an incredible variety of slugs weighing from 110 to 180 gr. Take your pick!

erally speaking, the hollow-base wadcutters are target bullets for the .32 S&W Long, but that does not preclude their use in the newer magnum. Lead alloy .32 bullets range from .312 to .314 inch.

The full range of factory produced bullets for the .38 Special, .357 magnum and .357 Maximum are too numerous to detail fully here; the selection is enormous. Weights range from 110 to 180 grains. Types vary from a frangible 110-grain Blitz by Sierra to totally non-expanding silhouette bullets by all four jacketed bullet producers. A relative newcomer to jacketed handgun bullets, Nosler already has four offerings in .357-inch diameter, while Speer has eleven. By far the most common type of .38/.357 jacketed bullet is the hollow point or, as Sierra refers to them, the hollow cavity. Others rely on only a fairly generous soft lead nose for expansion.

Originally the most popular .38/.357 bullet weight was 158 to 160 grains, but high-velocity influences starting in the mid-1960s caused the 100- and 125-grain weights to become popular, especially in law enforcement circles. Today, it seems as if the two extremes have compromised and .38/.357 bullets of about 140 grains are growing in popularity. Sierra, Speer and Hornady all have jacketed hollow point (or cavity) bullets of 140 grains.

Lead bullets of this diameter cannot be forgotten. Because of the problems of leading, such bullets are best used

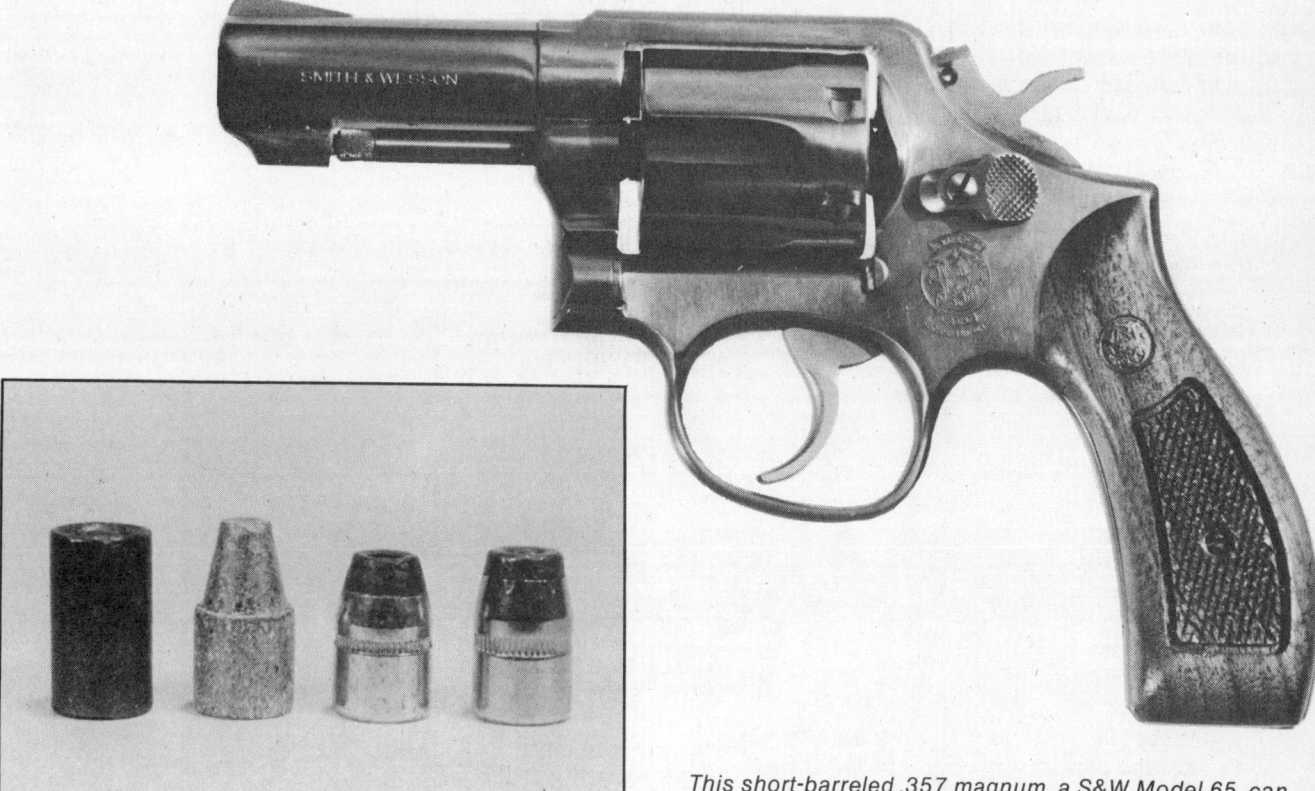

Above: Four types of .38/.357 bullets are, from left, swaged lead wadcutter, swaged semi-wadcutter, hollow point with partial jacket and then jacketed flat point.

This short-barreled .357 magnum, a S&W Model 65, can use bullets of any available weight, but careful choice will enhance the performance of the gun/ammo combo.

GUN DIGEST BOOK OF HANDGUN RELOADING

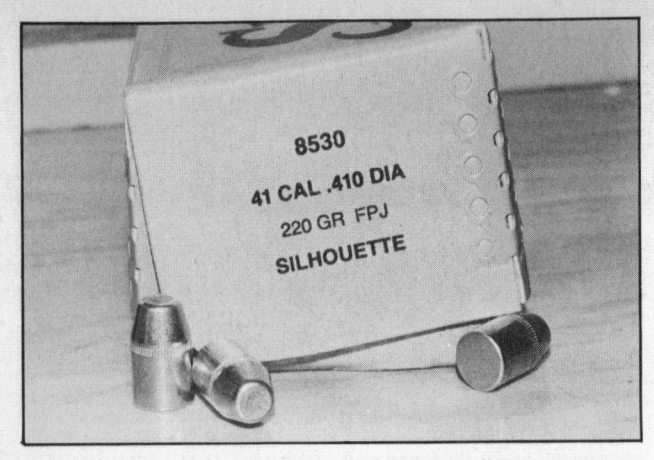

An example of how the makers will respond to demand: this Sierra is designed and sold for silhouette shooting.

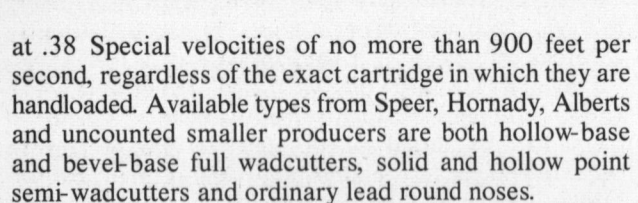

The .41 Hornady JHP is a fine bullet and it used to be their only .41. Now they offer a 210-grain silhouette.

at .38 Special velocities of no more than 900 feet per second, regardless of the exact cartridge in which they are handloaded. Available types from Speer, Hornady, Alberts and uncounted smaller producers are both hollow-base and bevel-base full wadcutters, solid and hollow point semi-wadcutters and ordinary lead round noses.

Although the lead round nose form of bullet in .38 Special is considered worse than archaic by most of today's handgunners, it should not be snubbed totally for less than serious use. It may be a poor man and animal stopper, but it still can be superbly accurate for informal shooting. In my own testing of .38 Special caliber at speeds of about 800 feet per second, the round nose swaged lead bullet was the easiest with which to achieve tight groups of all the swaged lead bullets. Jacketed .38/.357 bullets usually are .357 inch, while the lead alloy ones measure .358 inch.

For the .41 caliber, bullet selection suddenly narrows drastically. No one has ever said that the .41 magnum is a poor cartridge; nothing could be further from the truth. However, it never has become a popular cartridge like the .357 and .44 magnums. Jacketed .410-inch bullets are limited to ones and twos from all four makers. Nosler has a 210-grain JHP, Hornady has a 210-grain JHP and a 210-grain full jacketed flat-point for silhouette shooting. Speer produces a 200- and 220-grain JHP, both with long lead noses. Sierra, the big pusher of .41s, has 170- and 210-grain JHCs and a 220-grain "full profile" silhouette bullet. Swaged lead .41s also are scarce; the only major supplier is Alberts with solid and hollow point semi-wadcutters.

Despite the vast popularity of .44 caliber revolvers, bullet selection is not that large. Some may be surprised at this, but the reason is that .44 caliber revolvers are not particularly versatile, especially the most popular .44 magnum. As big bores, they are meant for big jobs; hence the limited bullet selection.

With the possible exception of swaged lead bullets,

With the advent of the .41 Action, there's a need for more variety in .410-inch bullets. This 170-grain Sierra is the best choice, but it was designed for revolvers.

96

GUN DIGEST BOOK OF HANDGUN RELOADING

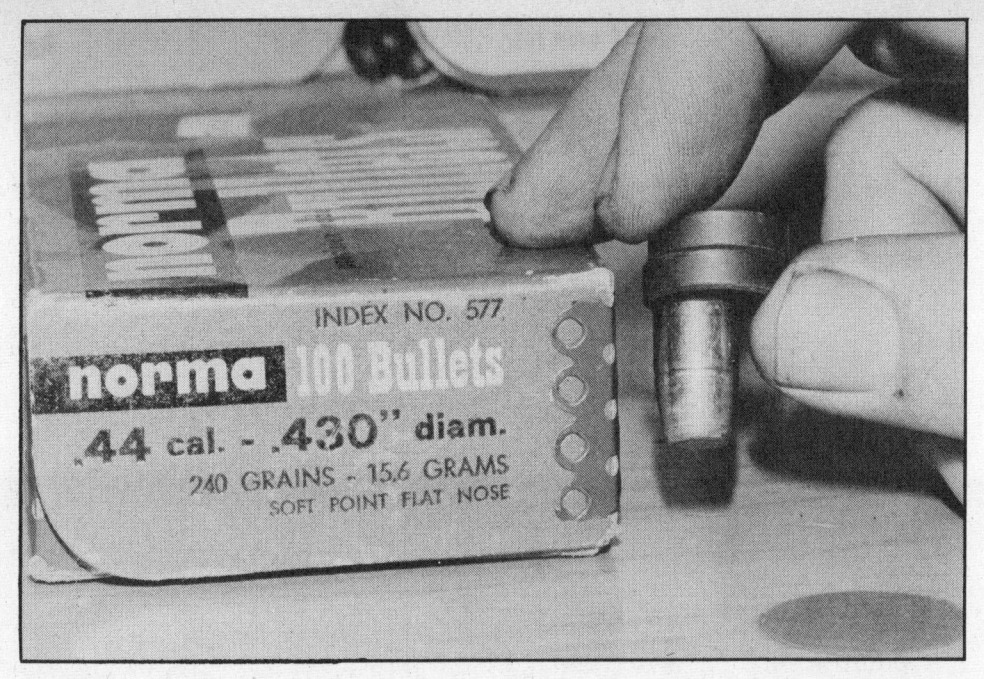

There are some special purpose bullets around that are no longer commonly seen. This is the Norma 240-grain JSP. With a mild steel jacket, it's a great pig bullet.

factory-produced .44 caliber bullets are intended for big-game hunting or long-range silhouette shooting. In other words, they are meant for high-velocity, high-pressure loads.

Speer leads with six .44 caliber jacketed bullets. Three are billed as magnums. Those are 200- and 240-grain hollow points and a 240-grain soft point. All three bullets feature a curved ogive with a metal jacket covering the ogive. Two of Speer's bullets have more of a traditional semi-wadcutter shape and are a 225-grain hollow point and a 240-grain soft point. Lastly, there is a rather pointed 240-grain full metal jacketed bullet for the silhouette shooters.

Hornady keeps their .44 caliber bullet selection more simple with 200- and 240-grain JHPs and a 240-grain flat point for the silhouette boys. Sierra has 180-, 210- and 240-grain JCH hunting bullets and 200- and 250-grain full profile silhouette bullets.

My own most accurate .44 magnum handload with jacketed bullets is the 180-grain Sierra JHC over 27.0 grains of IMR-4227 powder with Norma brass and CCI-350 primers. In an informal test with that load in a Smith & Wesson

Another special purpose bullet that is currently available: This Barnes JSP weighs a whopping 275 grains. For hunting use, Barnes makes one even heavier at 300 gr.

GUN DIGEST BOOK OF HANDGUN RELOADING

When the gun makers produced sturdy new guns in .45 Colt, the bullet makers replied with modern bullets for the old round. You can buy a good bullet in red, yellow or green boxes.

Here's the Hornady slug for the .45 Colt, weighing 250 grains. The Sierra entry is lightest at 240 grains. Below: Swaged lead bullets are constant good-sellers in the product line of Speer and Hornady. Lots of these go into practice ammo for police departments: 158 gr. SWCs.

Model 29 revolver, five machine-rest twenty-five-yard groups averaged a tiny 1.06 inches.

Factory lead bullet selection of .44 caliber bullets is slim. Speer and Hornady both offer 240-grain semi-wadcutters. Alberts also has a 240-grain SWC, plus has a hollow point version at 230 grains. Diameters of .44 caliber bullets vary. Speer uses .429 inch; Hornady's are .430; Sierra splits the difference at .4295 inch; and Nosler lists theirs at .429 inch. Lead alloy .44s by the major suppliers measure .430 inch.

Lastly, we come to .45 Colt bullets. It seems that only a few years ago, .45 Colt handloaders had the option of buying the swaged lead bullets then being offered by Remington or Winchester or casting them. That was true in the days when the .45 Colt revolvers around were mostly the Colt Single Action Army, rechambered .455 Smith & Wesson hand ejectors or the Colt New Service model. At the speeds generated by safe loads in those handguns, lead bullets were more than adequate.

However, that is not the case today. Bill Ruger started it all by offering the .45 Colt in his strong Blackhawk model. Since then, Thompson/Center has introduced the Contender in .45 Colt, Interarms offered the recently discontinued Virginian Dragoon and currently Freedom Arms is making their super-strong, five-shot single-action as a .45 Colt. Such guns as the above created a demand for jacketed .45 Colt bullets.

Sierra, Speer, Hornady and Nosler all offer a single bullet billed as ".45 Colt." Sierra's is a 240-grain JCH, Speer's is a 260-grain JHP and the other two are 250-grain JHPs. As for bullet diameter of these slugs, the Speer and Nosler are .451-inch, the Sierra is .4515 inch and Hornady's is .452 inch. All of the above companies' .45 ACP

Hornady swaged bullets, like the 148-grainers above, are produced with a distincitve type of knurling on the bullet shank. It's sort of "waffled" and it holds a dry waxy lubricant.

bullets measure .451 inch, meaning that they too can be used in the .45 Colt.

A smaller source of .45 caliber handgun bullets is the Freedom Arms Company (Box 1776, Freedom, WY 83120). Besides the super-strong handguns, this outfit also produces the properly constructed .452-inch bullets needed in those revolvers. Their weights are 240, 260 and 300 grains. Both hollow and soft points are available.

Swaged lead .45 Colt bullets are again scarce. Speer has the 250-grain semi-wadcutter and Alberts offers a similar style. Also, I had good results with the Speer 230-grain .422-inch swaged lead round nose .45 ACP bullet in 800 feet-per-second-velocity .45 Colt handloads. Bullet diameter of swaged lead .45 Colt bullets is .452 inch.

Aside from the major jacketed bullet manufacturers, there are many smaller companies across the nation selling their products regionally. Some are into making swaged lead bullets and many others are custom cast bullet producers. Products range from excellent to poor, but must be individually tested to determine.

The nationally distributed companies all have distinct reputations and followings, but that does not mean all their bullets perform smartly in all loads. Handgun reloading is not that simple. Jacketed bullets do not always expand as desired. Velocity and related factors such as barrel length and target range are the regulating factors.

It is the handgun handloader's responsibility to work up loads for whatever gun in question that will cause the bullets to behave as desired. Bullet producers put the quality into their production, but it is our job to put the quality into the assembled handload, then direct it properly to the target, whether it be paper, animal or a simple tin can. — *Mike Venturino*

Above: The Speer swaged 200-grain bullet is intended for use in .45 ACP practice fodder and it serves well in that role. Below: Jacketed hollow point, flat point and commercial cast flat point bullets give the reloader a wide choice in projectiles. Many weights are available.

GUN DIGEST BOOK OF HANDGUN RELOADING

CHAPTER 8

HANDGUN BULLETMAKING

Producing Your Own Projectiles Takes A Big Bite Out Of The Cost Of Reloading!

A RELATIVELY modest investment in equipment can set you up with production facilities for making your own cast bullets. Once you have the hard-rock basics in hand, adding other calibers, weights and designs involves nothing much more than a new set of mould blocks to fit the handles you already have, plus perhaps a new lube/sizer die and nose punch.

If you've been peering at price tags in the local gun stores, you may have noted that ready-made bullets cost a respectable sum. Homemade bullets can cost a whole lot less, provided you do not place a high value upon your personal labor. If you can charge the hours off to recreation and muster some skill at scrounging the raw materials to make up the proper alloy, cast bullets become interestingly economical. Jacketed bullets involve some amount of cash outlay for the jackets, but viewed from any angle, there is much to be said for being able to make your own bullets.

For but one consideration, it's often possible to make bullets that cannot be obtained in any other way. Such bullets may or may not offer attractive advantages and there's only one way to find out, if they do or not: Try it!

I can confide that there is a special glow of satisfaction in getting a good group or making a spectacular shot with a bullet you made with your own hands. That gets down into the area of intangible rewards on which there is no tax, to the present.

CAST BULLETS

This is one of the two basic techniques for bullet production, the other being swaging. Other approaches are possible, of course. For example, bullets can be turned on a lathe, one at a time. The French *Arcane* bullets were turned from electrolytic copper, though probably on an automatic screw machine. That only suggests that there are exceptions to virtually every rule (including this one). Casting and swaging account for the vast bulk of all projectiles produced at home or in the factories, so we'll discuss the two approaches, individually.

To produce cast bullets, you need certain essentials, including a supply of suitable alloy or the ingredients for making same; a means of melting and dispensing the alloy; a set of mould blocks and handles; a suitable tool for knocking over the sprue cutter on the mould; some flux to condition the alloy; some shooting glasses or other suitable eye protection and adequate fresh air or ventilation for the operation.

Other things are handy, such as a pair of leather-palmed welder's gloves for handling hot artifacts or a pair of needle-nose pliers for picking them up. It's nice to have a small propane torch to help the initial melting of the alloy along or to pre-warm the mould blocks, and an ingot mould is handy for mixing your alloy and decanting it into convenient pigs for future use. Several sources make ingot moulds and most of them produce ingots too small to be practical. The finest mould I've ever found is the top half of an old stainless steel GI messkit. That produces two kidney-shaped pigs, weighing approximately six pounds apiece and the pigs can be labeled with a felt-tip marker, once they've cooled. I'm not at all certain how readily accessible GI mess kits remain in today's marketplace. I've had my pair for the past few decades and would not sell them for any price.

A typical working temperature for casting bullets is around 700 degrees Fahrenheit/371 degrees Celsius and it's best to turn it down a bit from that point, once things are warmed up and functioning. Any tin in the alloy — and the alloy *needs* tin — tends to burn off and oxidize at about 700° F or hotter.

In the early Fifties, the dawn of my own reloading endeavors, I cast many thousands of bullets by melting my alloy in a discarded frying pan atop the electric stove in the kitchen. It can be done that way, but I do not recommend the approach. Bullet casting is a rather malodorous operation to carry out in a closed system without ventilation; and

Hensley & Gibbs' #292 mould produces the 230-grain flat point bullets above and the same maker's #938 mould turns out the 175-grain conical point bullets below: an example of the variety in bullet casting.

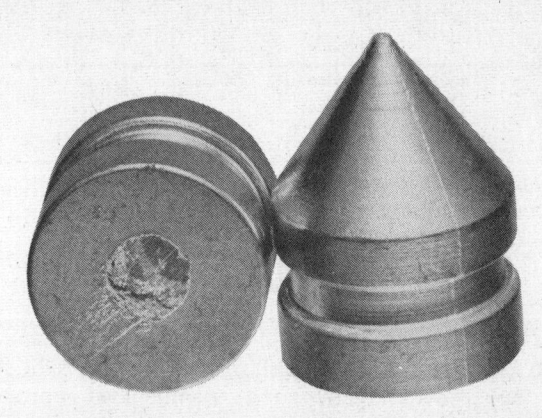

some of the fumes and vapors are impressively toxic, as well. In the process of acquiring my education, I managed to slop molten alloy upon the blue kitchen linoleum and into the side of a bedroom slipper I happened to be wearing at the time; the latter is a remarkably attention-getting maneuver.

If you're melting the alloy in a pot, you have to dip it with some manner of ladle to pour it into the top of the sprue-cutter on the mould. It is possible to put a small pot of alloy atop a gasoline-fired camp stove and produce cast bullets, but

A jacketed version of the Spelunker bullet is shown in the cross-sectional drawing at right. As you'd suppose, it offers remarkable expansion. It also has good accuracy capability in .357 or .429 diameters. Group at left was in .350 Rem mag and, at right, from a .44 magnum. Note the "ticket-punch" holes in paper!

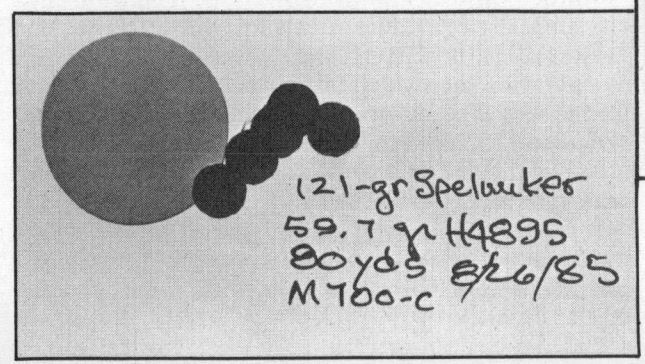

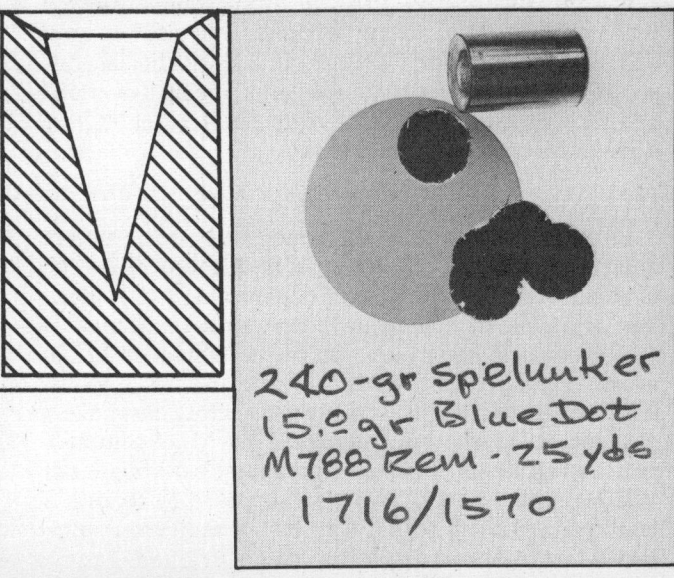

121-gr Spelunker
59.7 gr H4895
80 yds 8/26/85
M700-C

240-gr Spelunker
15.0 gr Blue Dot
M788 Rem - 25 yds
1716/1570

GUN DIGEST BOOK OF HANDGUN RELOADING

Inspect the bases carefully. Irregularities such as seen here preclude any hope for accuracy and are good cause to recycle the rejects back into the pot for another try. Also check for a sharp edge around the bases.

Hensley & Gibbs #333 mould produces these lightweight wadcutters, at typical weights of 66.5 grains or so. In the .38 Special accuracy is good, recoil is zilch and a pound of alloy produces over 100 of them. In the .357 magnum or 9mmP, notable velocities are possible.

The H&G #253 (left) and #291 are for use in the .41 magnum and other cartridges using the .410-inch bullet diameter. My usual procedure is to put lube in the crimping groove, as here, and seat the shoulder of the #291 flush to the case mouth, applying a taper crimp.

all it takes is a light jostle to tip over the whole shebang and send a tidal wave of molten alloy coursing across the floor and unwary feet. Again: *Not* a recommended approach.

Such perilous subterfuges may — arguably — have had some justification in the early Fifties, when reloading was a pursuit for no more than the dedicated few. Up here in the latter Eighties and beyond, reloading has become a respectable activity and other manufacturers offer a lot of good gear at surprisingly moderate prices. Yes, I'm grateful that I stayed around to see the happy situation develop.

Today, you can purchase a thermostatically controlled, bottom-delivery bullet-casting furnace of escalating credentials for somewhere in the ballpark of $50 to $200; perhaps even a trifle less. You also can get electric melting pots from which to dip with a ladle for substantially less, if that's your need and fancy. Whatever you get, I suggest you secure it to a sturdy working surface with at least one solid C-clamp to prevent the specter of tipping it over and all the dire consequences that might entail. After all, dire consequences are the worst kind.

CASTING ALLOY

Bullets for use in centerfire handguns customarily are composed of an alloy consisting of lead, antimony and tin in about that descending order of percentage. If you can line up an understanding and cooperative garageman or tire dealer, willing to part with his accumulation of old wheel weights for a modest amount of cash or trade goods, it is possible to make reasonably satisfactory cast bullets from reclaimed wheel weight metal (WWM). Addition of even a small amount of tin will improve the performance of WWM quite usefully. WWM is supposed to contain a small percentage of tin but that may or may not be present.

When working with WWM, the first step is to melt it down and skim off all the little steel clips and other residue that rises to the surface of the molten mass. It is best to flux the mixture when doing this. Beeswax works for the purpose, after a fashion, but not at all well. Added to the melt, it will smoke furiously and, if the temperature is above its flash point, the smoke will burst into flames, thereby presenting a possible fire hazard which no one really needs nor wants.

The preferable fluxing agent is some stuff called *Marvelux,* obtainable from Brownells, Inc., (210 S. Mill, Montezuma, IA 50171). Get the smallest quantity they offer, because a little bit goes a long way and the silverfish will eat the label into illegibility before you see the bottom of the can. If you don't believe that, look at the label on the can in the nearby photo.

Marvelux is a white, crystalline substance that melts in contact with molten alloy, producing little or no smoke and fumes. For the past few decades, my stirring stick has been the stainless steel handle that parted from the bowl of the soup ladle, about the time the crust of the earth was commencing to cool a little. I also have a grundgy old spoon that I use for skimming the dross and crud off the top of the melt, to discard in an old tin can. Humble artifacts such as these can be surprisingly useful to the bullet caster.

As you stir the Marvelux or other flux into the alloy, it helps bring foreign matter to the surface and, at the same time, it combines the alloy ingredients into a smooth and uniform mixture.

Brownells also can furnish a dial-type thermometer for checking the temperature of the alloy. Merely dip the stem in the alloy and hold it for ten seconds or so, until the needle stops moving, then withdraw the stem and set it aside. The thermometer is not designed for continous, long-term immersion. By use of the Brownell thermometer, I was able to ascertain that the thermostat dial on my RCBS *Pro-Melt* casting furnace is calibrated quite closely.

If you're starting out with a large amount of alloy and want to get it melted as quickly as possible, it is helpful to play the flame of the propane hand torch on the upper surface of the alloy.

I wish I could arrange to have the following warning

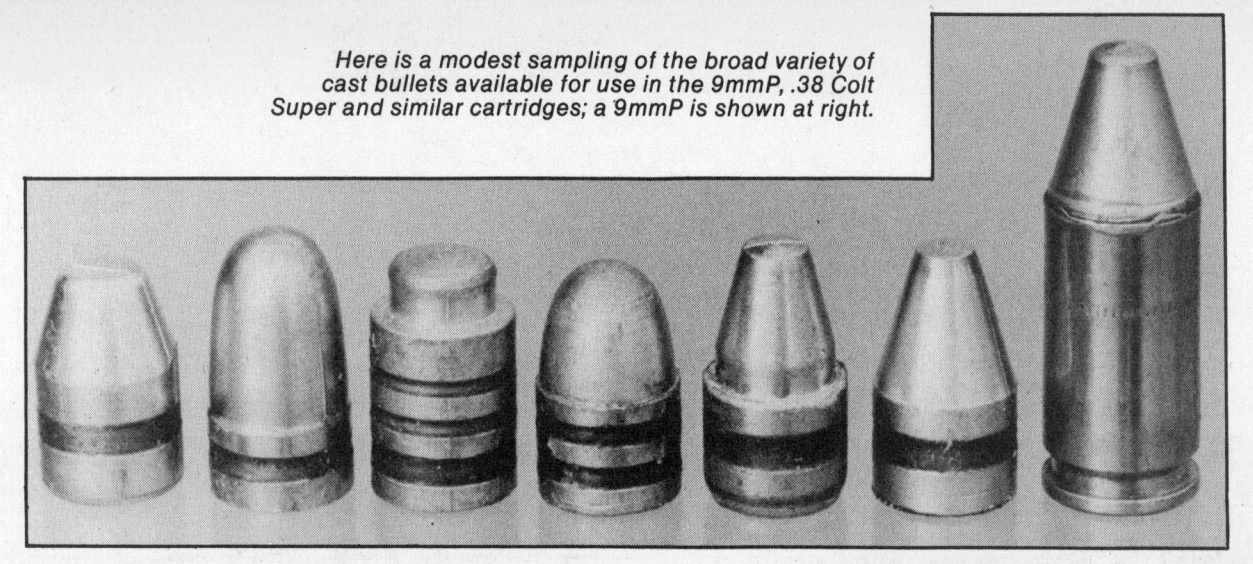

Here is a modest sampling of the broad variety of cast bullets available for use in the 9mmP, .38 Colt Super and similar cartridges; a 9mmP is shown at right.

printed in red ink, but it just isn't possible, so I'll go all-caps for special emphasis: WHEN WORKING WITH MOLTEN ALLOY, ALWAYS WEAR EYE PROTECTION! It is not uncommon for the stuff to pop and snap, throwing random spatters in all directions. You do not want to get one of those in an eye.

Some sources decree leather gloves, long sleeves, buttoned collar and stop just short of a ski mask. As I do most of my casting with the mercury in the high eighties or low nineties, such a costume would be apt to make me collapse on the driveway, perhaps in small pieces of arbitrary size; a victim of what we might be so brash as to term *heat procrustration*. I do my casting barehanded, wearing my usual short-sleeved shirt and jeans over crepe-soled shoes. I can tolerate small spatters of alloy just about anywhere except in the eyes and sometimes have done that.

Linotype alloy makes delightful bullets in its pure state, but I regard that as almost sinful. Lino alloy gets scarcer with each passing year. One part (by weight) of lino alloy with six pounds of cleaned wheel weight metal makes a casting mixture I tend to regard as ideal. If the mould cavities are clean and preheated, it does a beautiful job of filling all the corners to produce bullets as close to perfect as you're ever apt to see.

The details on three main ingredients of casting alloy are:

METAL	DENSITY	MELTING POINT	CHEMICAL SYMBOL
Lead	11.4	621.3F/327.4C	Pb
Antimony	6.62	1166.9F/630.5C	Sb
Tin	7.3	449.4F/231.9C	Sn

Density sometimes is termed specific gravity and refers to the weight of a given volume in comparison to an equal volume of water. Thus, lead is 11.4 times heavier than water and nearly twice as dense as antimony or tin. It is this variance between densities that makes the final weight of a bullet from a given lot of alloy vary considerably from bullets cast of another alloy from the same mould. In general, the lighter the weight of the bullet, the less lead it contains and *vice versa*.

Some bullet mould designs incorporate an area of slightly reduced diameter at the base for installing gas checks, so as to protect the base of the bullet.

You probably are wondering how the antimony gets melted at 1167F, when the pot only heats to 700F or so, right? It is a metallurgical peculiarity that metals in alloys tend to melt at well below their nominal melting point in the pure state. An extreme case of that is an alloy called Wood's metal that melts somewhat below the boiling point of water, despite the fact that all of the ingredients have melting points substantially higher.

At the same time alloys composed entirely of lead and antimony may seem impressively hard in the cool and solid state, but they foul the bores in a most abominable manner. As the alloy cools, crystals of antimony solidify long before the lead freezes. That results in a network of antimony crystals with lead filling the spaces between them. Addition of tin corrects that, as the melting point of tin is even lower than that of lead and it stays in mixture with the lead to the end, hardening it usefully and going far to prevent bore fouling.

As noted, the tin content will commence to oxidize and burn out of the alloy at about 700F. Even if you turn up the thermostat a little to get the mixture melted quickly, you'll

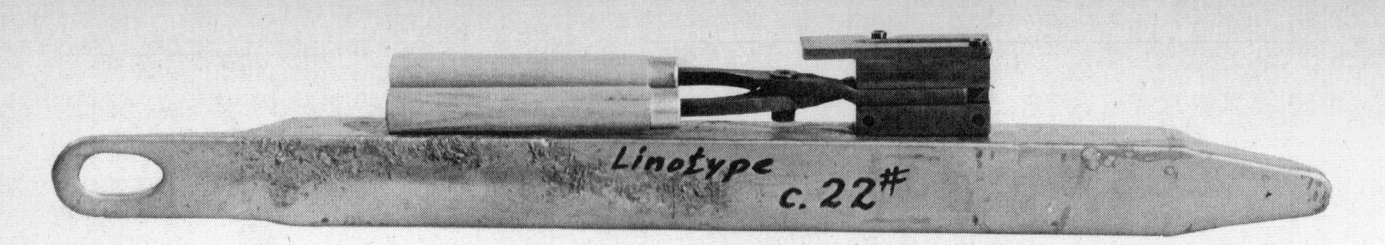

A four-cavity H&G mould reposes atop an ingot of linotype alloy. This is the standard form in which the lino alloy is supplied and it weighs 22 pounds.

H&G can furnish the handle and mould for making the lead-headed casting hammer shown here. It does a fine job of knocking over the sprue cutter and tapping the bullets free of the cavities. The head weighs about one pound and it's a simple project to melt off the old head and cast a fresh one when it becomes overly worn.

want to turn it back down to 650F or so as you commence casting.

MOULD PREPARATION

Mould blocks usually are made of ferrous alloys, meaning they contain some amount of iron; a metal justly notorious for its eagerness to oxidize into rust. Other moulds are made of aluminum alloy, perhaps even brass or stainless steel, thereby avoiding most — if not all — such storage problems.

When it comes to ferrous mould blocks, I can tell you one approach that does *not* work and that is the one of leaving the cavities filled with the last bullet cast. In theory, that keeps air and moisture out of the cavities. In practice, it doesn't work.

I sometimes have allowed the empty mould to cool down to ambient temperatures — read "room temperature," if that makes better sense — and stored them in a gallon-size, heavy-duty Zip-Lock Baggie, with as much air as possible squeezed out of it before sealing the closure. This appears to work reasonably well, provided you do not get a puncture in the plastic during storage.

Otherwise, I apply a liberal coating of rust preventative to the cavities and working surfaces such as the underside of the sprue-cutter. For this, I prefer Break Free to anything else I've tried to the present. Don't forget to shake it before applying.

When it comes time to remove the protective agent, RIG 3 degreaser or Outers Crud-Cutter work extremely well. They dissolve nearly all of the protectant. You need to remove every last lurking molecule of the stuff in order to get decent castings.

Preheating the blocks is a good way to eliminate nearly all the remaining hydrocarbons which, if exposed to hot alloy, generate a puff of gas to produce bullets somewhat resembling silver raisins, quite unfit for ballistic employment.

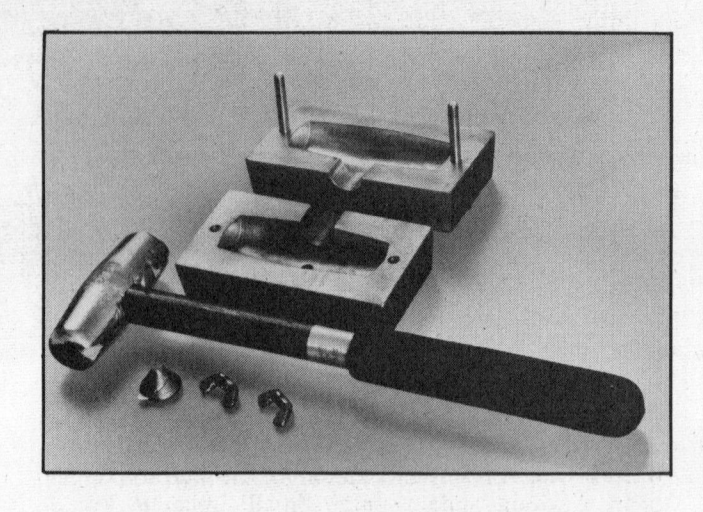

You can rest the mould blocks on the burner of an electric kitchen range for some few minutes, with the dial set about midway between Med Lo and Med Hi, but don't let yourself be carried away in your enthusiasm. If overheated and poured full of alloy, you are in the embarrassing position of having some bullets that refuse to solidify. You will have to stand there for some number of minutes, handles tightly clenched in hand, until the telltale surface of the exposed sprue changes from bright silver to frosty white, indicating a return to solid state.

The second approach is to spritz the cavities and surfaces liberally with RIG 3 or Crud-Cutter, then light the propane torch and play the tip of its flame from cavity to cavity, surface to surface, for a judicious short while. Again, moderation is the key, lest you get moulds that keep the alloy molten.

What you want and need is a mould block scrupulously free of remaining preservatives at a temperature 100-200F or so below the melting point of the alloy. That state of affairs — and only that — will result in usable cast bullets.

If your mould blocks get too hot in the process of filling and emptying them, you will know by the appearance of the bullets. Instead of a sleek silver surface, they will commence to look distinctively frosty. That's a sign you need to cool the blocks down a bit. That can be done in any of several ways. You can rest the lower surface of the blocks on a *heat sink,* to bleed off heat by means of conduction. The only source I know of for readymade heat sinks is Peter Kirker, 402 Rancho Vista Road, Vista, California 92803; phone 619/724-9611. He has them in two sizes and I'd get the larger one, although either works extremely well. You'll have to inquire as to the current price, as such things change as time goes along.

You also can cool the mould by a directed airstream, using a fan of some sort. That also helps to solidify the sprue and that's definitely to the good. The sprue is the

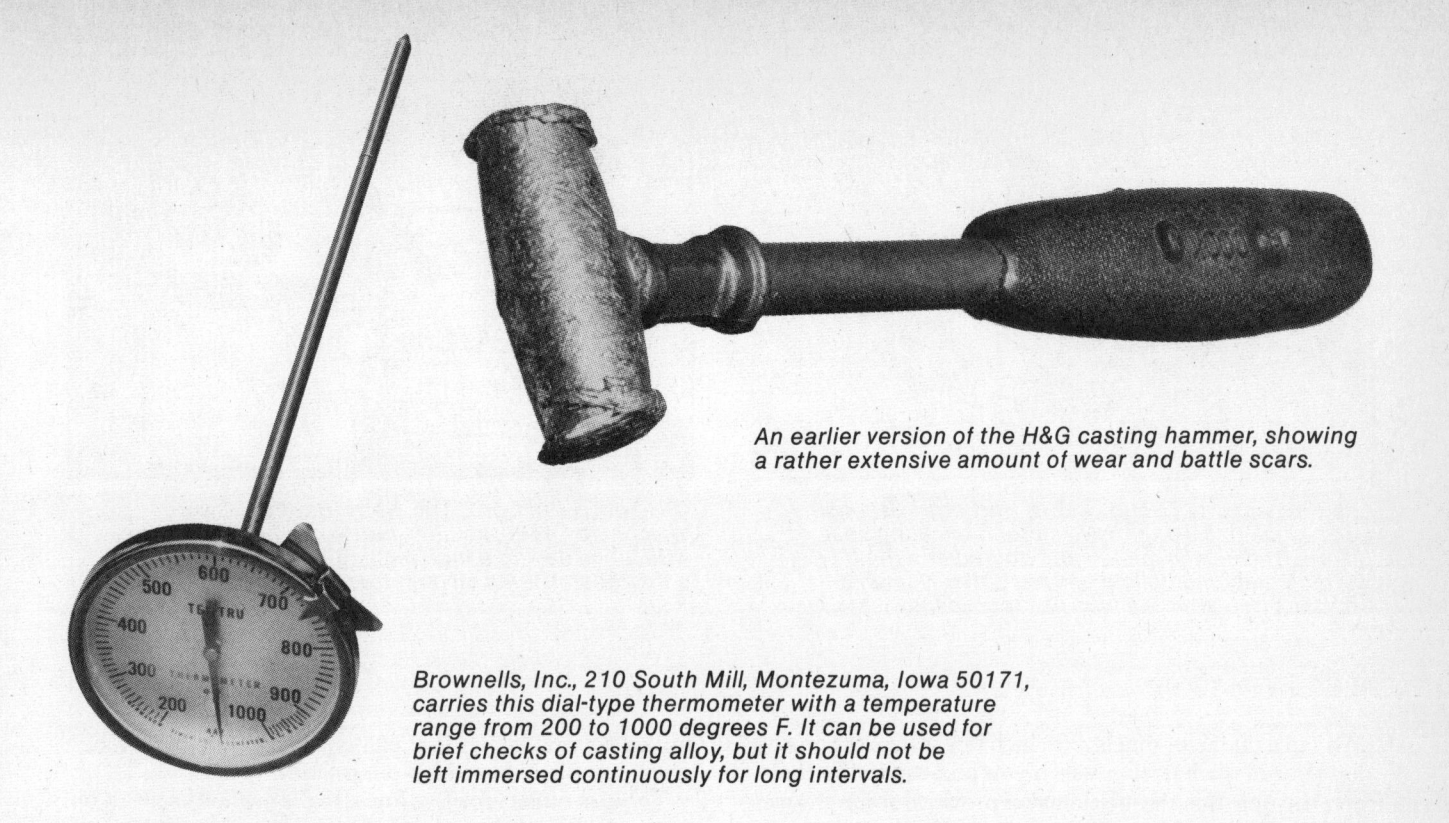

An earlier version of the H&G casting hammer, showing a rather extensive amount of wear and battle scars.

Brownells, Inc., 210 South Mill, Montezuma, Iowa 50171, carries this dial-type thermometer with a temperature range from 200 to 1000 degrees F. It can be used for brief checks of casting alloy, but it should not be left immersed continuously for long intervals.

exposed excess alloy in the knock-off plate of the mould. When freshly filled, it has a bright, silvery appearance. After some brief interval, it will — or should — change in appearance to a crystalline white color, indicating a reasonable degree of solidification has taken place.

If you get impatient and knock over the sprue-cutter too soon, the alloy will only be partially solidified, if that, and you will get curved silver streaks on the upper surface of the mould blocks. If the deposit gets too thick, it will prevent close contact between the sprue-cutter and the blocks, resulting in over-diameter fins at the base of the bullet. Such deposits can be removed by means of a single-edge razor blade wielded with delicate care.

CASTING TECHNIQUES

The sprue-cutter must be knocked over with a nonmarring type of mallet or similar device. Do not use a steel hammer or you will burr the cutter into unsightly ruin. For many years, I used the plastic handle of a large screwdriver. One time, it slipped from my grasp, dropping into some pretty hot alloy and errupted into spectacular fireworks. For the next several years, I used a length of scrap maple flooring, about eight inches long. A friend in Piqua, Ohio, gifted me with some incredible mallets, having heads of highly sophisticated polyurethane plastic. This material was so hard and tough it could drive a small finishing nail into hardwood without leaving a mark on the plastic, but it would not mar metal; not even soft aluminum. Had he marketed those mallets, he could have made a fortune, but he never did.

The best mould-whacker currently available to the general public comes as a kit from the old mouldmaking firm of Hensley & Gibbs, who get their mail out of Box 10, up in Murphy, Oregon 97533. Direct your inquiries to Wayne Gibbs. It consists of a handle coated with rubber at the grabbing end and a mould for casting a head of lead alloy, similar to the kind used for casting bullets. After extended use, the head gets to looking battered and amorphous. That's no problem at all. Just plunge the head down into the cauldron of molten alloy and melt off the old head, then clamp the handle back into the mould and cast a new head onto the end of the handle, good for whacking off a great many thousand sprues. When that one gets to looking tacky, just melt off its head and cast a fresh one. It is, I think, an unbeatable system and Gibbs can quote you the current market price for it.

Freshly cast bullets need to be rapped free onto a nonmarring surface. I use the shallow cardboard boxes used for packing Eastman enlarging paper in the eight-by-teninch size. If you don't happen to purchase that a hundred sheets at a time, any similar cardboard tray probably will serve about as well. Knock the sprues into a second tray and recycle those back into the melting pot from time to time. Scan the accumulating bullets and, if you spot a defective specimen, use the pair of needlenose pliers to move it over with the sprues for recycling. Be callous and cold-blooded. Remind yourself that a defective bullet could result in a miss when you really need a hit. Watch for rounded lower edges and air-holes in the center of the base or cavities in the base between the center and outer perimeter. Anything that could affect the gyroscopic stability of the finished bullet adversely is justified cause for condemnation and a return to the fluid state for a further chance at utter perfection. Remind yourself: You might shoot that one at something you'd rather not miss!

LUBE/SIZING CAST BULLETS

Bullets, as cast, are fairly close to the desired diameter, but a trifle oversized and not necessarily quite as symmetrical as one might wish. Ideally, you want to bring them

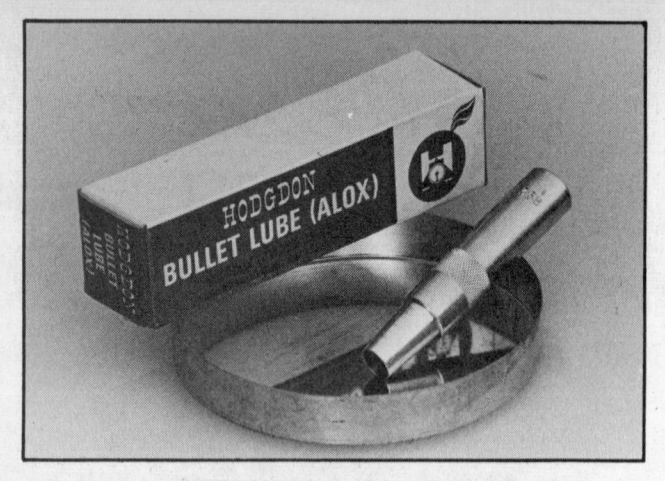

Lee Precision, 4275 Highway U, Hartford, Wisconsin 53027, has this inexpensive kit for lubricating cast bullets. The lube is melted and poured into the tray with the standing bullets and the cutter is used to cut them free after the lube has resolidified.

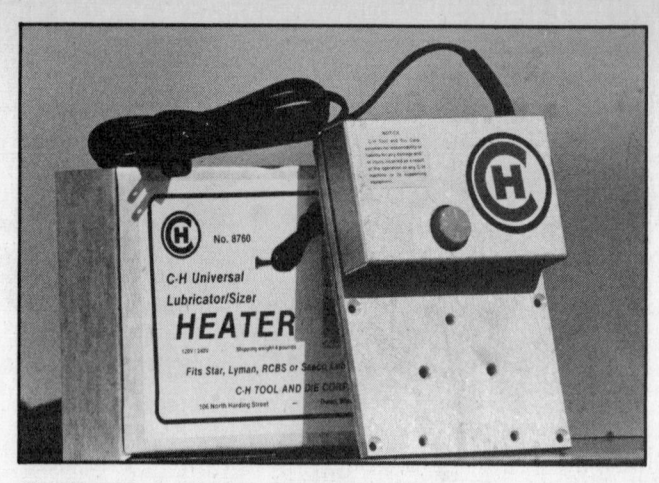

C-H Tool & Die Corp., 106 North Harding, Owen, Wisconsin 54460, has this lube/sizer heater for use with lubes having a high melting point. Controlled by a thermostat, it fits all popular lube/sizers.

down to a diameter that is .001 inch larger than the groove diameter of the barrel in which you propose to fire them. You will note that the mouldmaker provided some grooves in the full-diameter portion of the bullet base and it is highly desirable that such grooves be snugly filled with a suitable lubricant to ease their passage down the bore. Hereinafter, lubricant will be referred to as lube, pronounced "loob."

Some bullet moulds are produced with a reduced-diameter step in the base of the bullet to accept a shallow protective cup known as a *gas check*. The intent of such designs is to prevent the erosion or deformation of a lead-alloy bullet base by high-temperature powder gas. Some of the powders of relatively coarse granulation sometimes deform the bullet base from impact of the unconsumed granules at the instant of firing and gas checks reduce the harmful effect of that to a useful extent.

Lube/sizers are made by firms such as Redding/Saeco, RCBS and Lyman. To lube/size to a desired diameter, such as .358 inch for use in .38 or .357 guns, buy the sizing die and bottom punch for that diameter, adding the top punch to fit the nose of the bullet you plan to use. As all of the mentioned firms make bullet moulds, it is hardly surprising that each can supply top punches to fit bullets from their own moulds — and there is some crossing of party lines.

Bullet lubes come in many makes, formulations and flavors, each uniquely wonderful, as its makers will earnestly assure you. Most lubes contain a substantial percentage of beeswax, plus a somewhat softer material to make it more plastic at typical room temperature. In my early reloading career, I made my own lube by blending beeswax with common Vaseline petroleum jelly, and varied my formula according to the time of year: more Vaseline in cold weather and less in the hotter part of the year.

Some of the contemporary lubes are so stiff as to require application of added heat to make them flow into the grease grooves. C-H Tool & Die, (106 N. Harding, Owen, Wisconsin 54460) makes and markets a thermostatically controlled electric heater that can be used with typical lube/sizers to utilize these high-temp lubes.

The best bullet lube I've found to the present uses — you guessed it! — beeswax, plus a specific formulation of lithium-base grease. It works nothing but great and that's the good news. It is an absolute devil to make up and that's the bad news. The lithium grease has an absurdly high melting point quite close to the flash temperature of beeswax. Various makers have tried their hands and bowed out of the game.

Many makers of the bullet lube include a percentage of Alox, along with the usual beeswax and perhaps other ingredients, as well. Others manufacturers have made hopeful exploration of lubes that can be applied by dipping, spraying or other techniques, either before, after or without sizing the cast bullets. Some of these work reasonably well, although only at rather modest velocities; up to 850-900 fps velocity or so. As far as I know at time of writing, we still are awaiting the big breakthrough in this particular line of endeavor — a trifle impatiently, I might add.

As reloading equipment goes, lube/sizers are not cheap. It is possible to do the same job with simpler equipment and more time plus elbow-grease. Readers who find their interest channeled along such lines should obtain a current copy of the Lee Precision catalog; $2/per copy, at time of writing. Other approaches are provided for in that source.

SWAGED BULLETS

Swaging — pronounced "sway-jing," not swagging — involves cold-forming the bullet under high pressure by use of specialized dies and presses. Put under sufficient pressure, lead-based alloys can and will flow like toothpaste and that is the phenomenon that makes home production of swaged bullet possible. It may not even be necessary to obtain a specialized press. Some makers offer swaging dies that can be used in a sturdy, well made, conventional reloading press to pretty good effect. Not all reloading presses, it should be noted with all decent haste, are built and

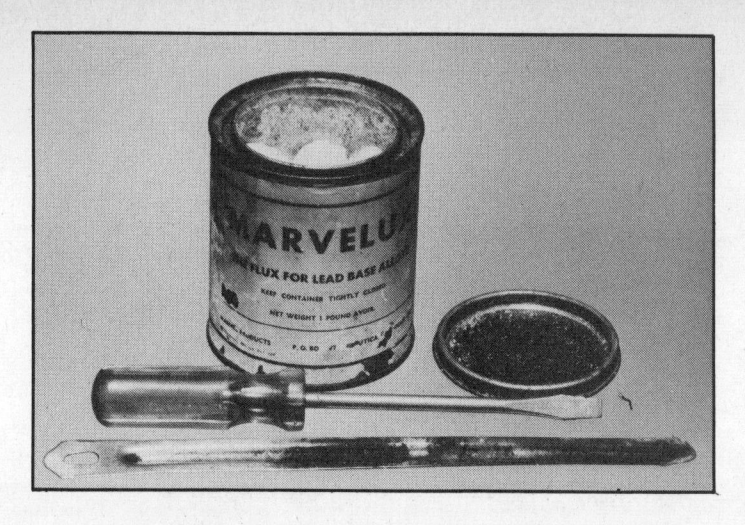

Brownells — address on page 105 — distribute and sell Marvelux flux for use with bullet casting alloys. It helps to remove impurities and mixes the ingredients of the alloy more intimately for finest results. Right, Outers Crud Cutter is a pressurized solvent that works well for degreasing mould block cavities.

geared to handle the stresses entailed in swaging bullets.

C-H tool & Die, whose address was quoted earlier, offers a series of bullet swaging dies in popular calibers for use in a sturdy reloading press and, for good measure, they also manufacture and market one of the best of all possible choices by way of a press for the job. It's listed in their catalog as the C-H Heavyweight CHamp and to the best of my knowledge, it's the last of the all-out, go-for-broke, *standard-scale* reloading presses on the market. To clarify a little, up in the latter Eighties, we are passing through an interregnum in reloading in which nearly anything is possible and a bewildering number of things are rather commonplace.

In a sharp about-face from the fad for caliber .17 and even smaller stuff in the early Seventies, the gun world now has its eyes all set, glazed and agleam on the heavy stuff. There is a growing enthusiasm for firing the caliber .50 Browning cartridge, one shot at a time, braced against your personal shoulder. At a much earlier point in my career, I touched off between a quarter and a half-million rounds of that cartridge; a load I continue to regard with nostalgic respect and admiration.

The guns in which I set off that heavy aggregate of cartridges were soldly fixed to large aircraft — such as the Consolidated B-24 *Liberator* — or to equally durable ground mounts anchored in massive blocks of reinforced concrete.

I have never fired a caliber .50 rifle and I don't plan to do so. Offhand, I can't think of anything I want or need to hit that hard. The .50 Browning is not a handgun cartridge — not so far, at least — but handguns have been made up to fire bullets of half-inch diameter and even a bit larger. What I'm saying is that anything can happen, and probably has. Oversize presses are now being made for reloading .50, even 20mm ammo. That's why I specified standard-scale.

In swaging jacketed bullets, you need the press, dies, swaging lube, jackets and cores. The cores often are cut from lead wire, in which example you need a supply of lead wire in the suitable diameter, plus a core cutter for cutting the wire into lengths that weigh slightly more than the desired final weight of the core.

A further alternative to cutting lead wire for the cores is to use cast bullets as the cores. Some favor the first approach, some the second and others use both. I fall into the last category. When swaging bullets for the .38 or .357, I often use an old Lyman four-cavity mould for their No. 311316 bullet, a gas-checked design intended for use in the .32-20 WCF and similar cartridges. With jackets about .5-inch in length, that particular core usually produces completed bullets that weigh about 140 grains; a weight I tend to regard as about right for that caliber.

When making up cast bullets for use as cores in swaging, work with a moderately soft alloy. While some equipment is capable of swaging fairly hard lead alloys, the resulting bullets do not fill out as well and they are not apt to expand as well as those with somewhat softer cores.

Bullet swaging lube can be obtained from suppliers such as Corbin. A small amount is applied to an uninked rubber stamp pad or similar surface and rubbed in. The jackets are rolled across the pad to deposit a small amount of lube on the sidewalls of the jackets. This is necessary to prevent undue wear on the swaging dies.

It was noted that the cores were produced to weigh slightly more than the desired final weight of the completed core. The excess core metal is removed by one of at least two techniques. Some swaging dies have a small bleed hole at the center of the bullet tip and surplus core metal is

Peter Kirker, 402 Rancho Vista Way, Vista, California 92083, makes and markets these two sizes of heat sinks, used for bringing down the temperature of moulds.

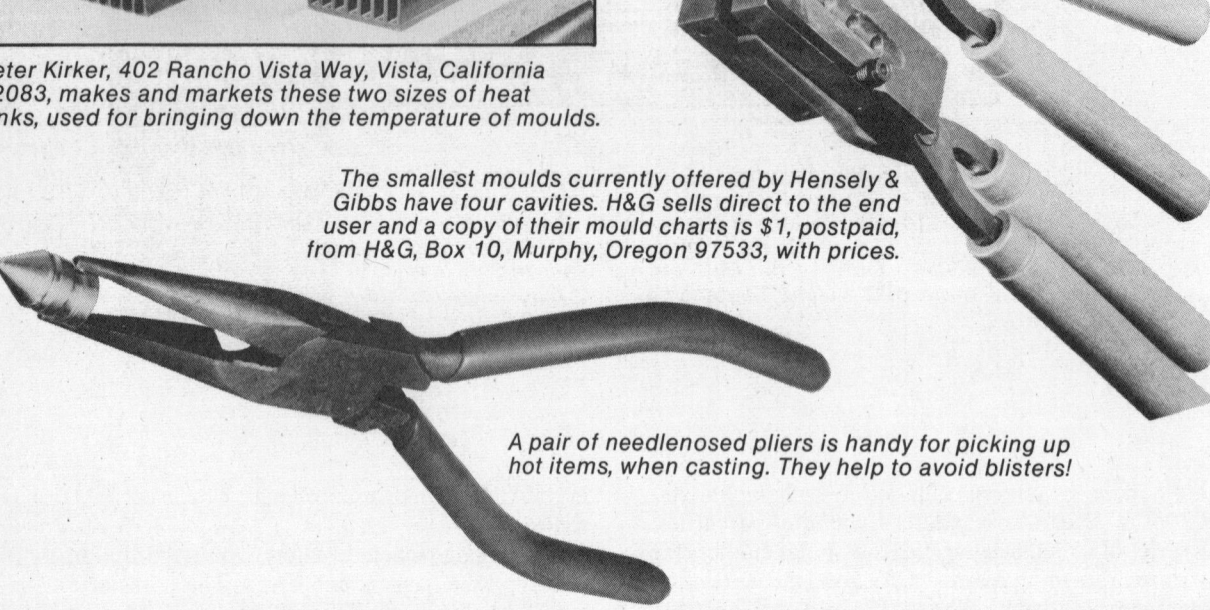

The smallest moulds currently offered by Hensely & Gibbs have four cavities. H&G sells direct to the end user and a copy of their mould charts is $1, postpaid, from H&G, Box 10, Murphy, Oregon 97533, with prices.

A pair of needlenosed pliers is handy for picking up hot items, when casting. They help to avoid blisters!

extruded through the bleed hole to be snipped off and set aside for recycling after completion of the bullet. The other approach utilizes a core-swaging die that extrudes surplus core metal through a bleed hole as a preliminary step in production.

The completed core and jacket can be weighed together on a reloader's scale to determine the weight after completion. In that manner, the weight of the core can be adjusted to result in a precise selected final weight.

Depending upon the equipment and dies in use, the next step may be to seat the core within the jacket. With other equipment, the seating of the core and final shaping of the bullet nose may be performed simultaneously in the nose-forming die as the final step.

With some dies and equipment, after completion of the nose-forming, the bullet is knocked out of the die by striking an ejector punch, integral to the die, with a non-marring mallet or similar device. Other presses and die sets eject the completed bullet by means of press leverage when the operating handle is reversed. Of the two approaches, I favor the latter.

Apart from the use of jackets, it is also possible to make up cast bullets, lube/size them in the usual manner, then use a press and swaging dies to form them into some other configuration. As the lube is incompressible and has no way of escaping, it remains in place in the lube grooves, serving to prevent bore fouling when the bullet is fired. This approach is one I've found to work quite well and it bypasses the not-inconsiderable cost of the jackets.

Be aware, however, that you must have either a copper-alloy jacket or a suitable bullet lube in the grooves commonly found in cast bullets, so as to prevent lead-fouling of the bore.

I have worked up a bullet design I call the *Spelunker*—chiefly because, when it hits, it goes, "Spe-lunk!" and excavates a cave of impressive dimensions. It can be thought of as sort of super-hollow point design. The nose cavity is dual-angled and typical specifications are given in the accompanying line drawing. The principle is that the obtuse-angled forward cavity initiates expansion at the instant of impact and, as it expands, the acute-angled rearward cavity opens. The effect can be accelerated by filling the entire cavity with a suitable material, such as beeswax, paraffin wax or the like.

I have made jacketed Spelunkers and found they worked quite well. Although the strong point of the design — quite obviously — is not its favorable ballistic coefficient, the design does not seem to be inherently inaccurate at the greater distances. As an example, using a Model 700 Remington Classic rifle in caliber .350 Remington magnum, I loaded some 121-grain jacketed .38 Spelunkers ahead of 52.0 grains of IMR-4198 powder and, firing off the bench, put five of them into a group that measured .953 inch, between centers at one hundred yards. In that gun, velocity of that load averaged 3199 feet per second for 2750 foot pounds of energy and I've often speculated on what a remarkable load it would be for Idaho rockchucks.

The Spelunker was developed primarily as a bullet for

Lyman's #452424 is a 255-grain semi-wadcutter that was designed by the late Elmer Keith. It works well in the .45 Long Colt, .45 Auto Rim and — marginally, at least — in the .45 ACP, provided it feeds in latter.

Anchoring the casting furnace securely to the work surface is an excellent idea, to prevent tipping and spilling of the blistering-hot bullet alloy.

use in handguns, however. The shorter barrels often need all the help they can get for purposes of assuring reliable and extensive expansion. Accuracy out of handguns at typical handgun velocities is reasonably decent, although I've never gotten groups below one minute of angle — MOA: about 1.05 inches at one hundred yards — when firing Spelunkers in handguns.

Saeco — formerly known as SAECO, the initials standing for Santa Anita Engineering COmpany — recently was purchased by Redding Reloading Equipment (114 Starr Rd., Cortland, New York 13045) and the customary spelling now is with only the S capitalized. One of the Saeco bullet moulds, their nominal 9mm No. 371, produces a flat-tip bullet with a single grease groove at a weight of about 100 grains, as listed in current Redding/Saeco catalogs. I have an eight-cavity mould for that particular bullet and have had a lot of success with it.

The Saeco No.371 turns out to be exceptionally well adapted for production of unjacketed Spelunkers in .358-inch diameter. After casting, the bullets are lube/sized, then swaged to the typical Spelunker form. As a usual thing, I do this in an old Swag-O-Matic bulletmaking press, once produced by C-H Tool & Die. At the time of its introduction, in the early Sixties, it was intended for use with half-jackets, about .25 inch in length, leaving some amount of full-diameter soft lead in contact with the rifling. Unfortunately, that never worked out as well as it had been hoped. Bore fouling was the big problem. By using a lube/sized bullet to start with, the fouling is all but non-existent at normal pistol velocities.

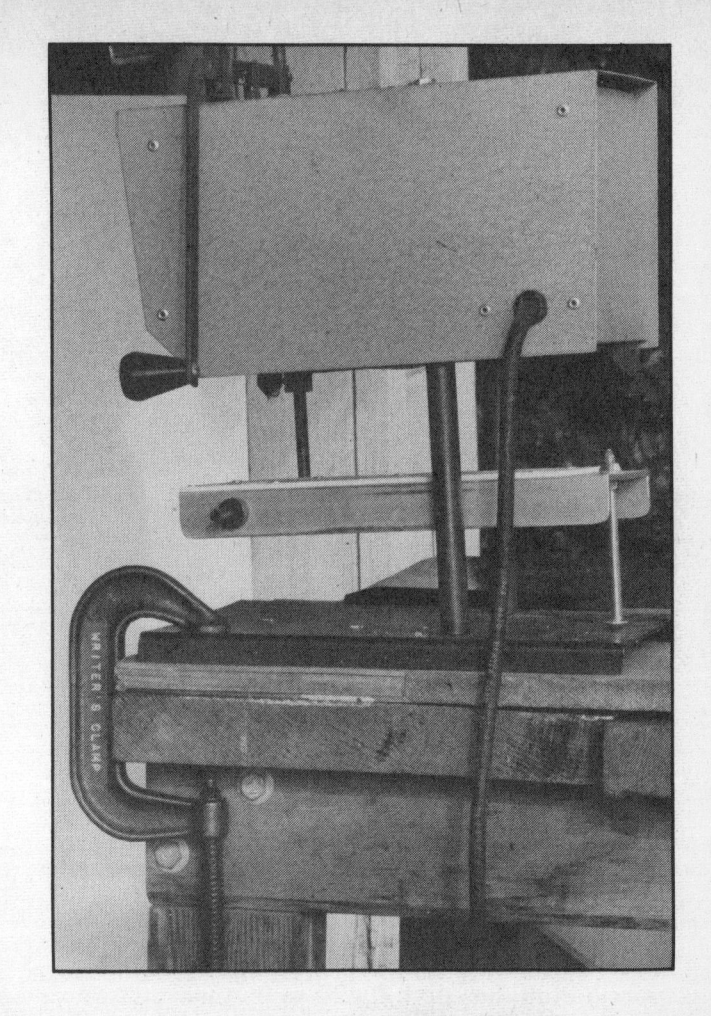

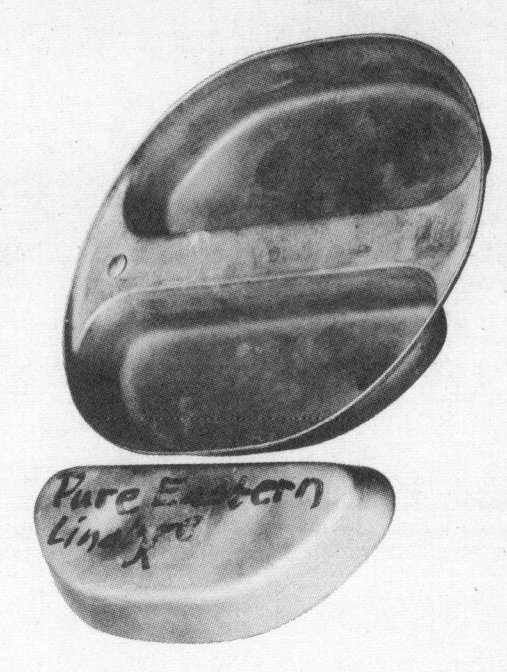

The top half of an old stainless steel GI mess kit is an excellent means of making up pigs of bullet alloy. The pigs weigh about six pounds and can be labeled.

GUN DIGEST BOOK OF HANDGUN RELOADING

C-H Tool & Die can furnish swaging die sets for the production of soft lead bullets with a zinc washer riveted onto the base. The washer minimizes fouling of the bore and the bullets expand enthusiastically, even at comparatively moderate muzzle velocities.

A propane-fired hand torch, such as this one by Bernz-O-Matic, readily available at most hardware stores, is useful for pre-warming mould cavities and for hastening the melting of a batch of bullet alloy.

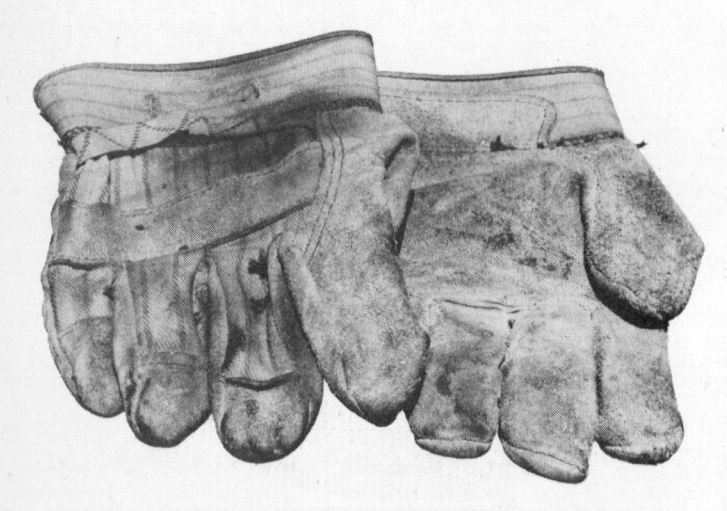

A pair of leather-reinforced welder's gloves makes your hands quite impervious to random spatter, but reduces your manual dexterity to some degree and accelerates hand fatigue on extended casting sessions.

ZINC-BASE BULLETS

After C-H had been purchased and moved to its present location in Owen, Wisconsin, their Tony Sailer did some further work with the Swag-O-Matic, using small zinc washers that were riveted to the base of the bullet by the extrusion of core metal through the central hole in the washer. Unlike the half-jackets, this did keep the bore fouling problem under some reasonable degree of control.

C-H continues to offer the zinc washers in caliber .38 size only, along with sets of Type Z dies that can be used in a sturdy, conventional reloading press for production of the zinc-based bullets. The zinc washers are quite moderate in cost and offer still a further interesting avenue to explore by way of slightly offtrail bulletmaking.

REMANUFACTURED BULLETS

There is at least one further intriguing possibility for the owner/operator of bullet swaging equipment and that is the remanufacturing of existing bullets, whether home-produced or the purchased output of one of the major makers.

As an example, a few years ago, when the 10mm Bren Ten pistol was one of the newest things around and the only ammo and bullets for it were the 200-grain FMJ/FP Norma production, I had a hard need for some JHP bullets to use in it, preferably somewhat lighter in weight. I had the set of .400-inch swaging dies for my Corbin CSP-1 press, the latterday counterpart of the earlier Mity-Mite press, with a most helpful upper brace to convert the original C-type frame into the vastly stronger O-frame. With the set came

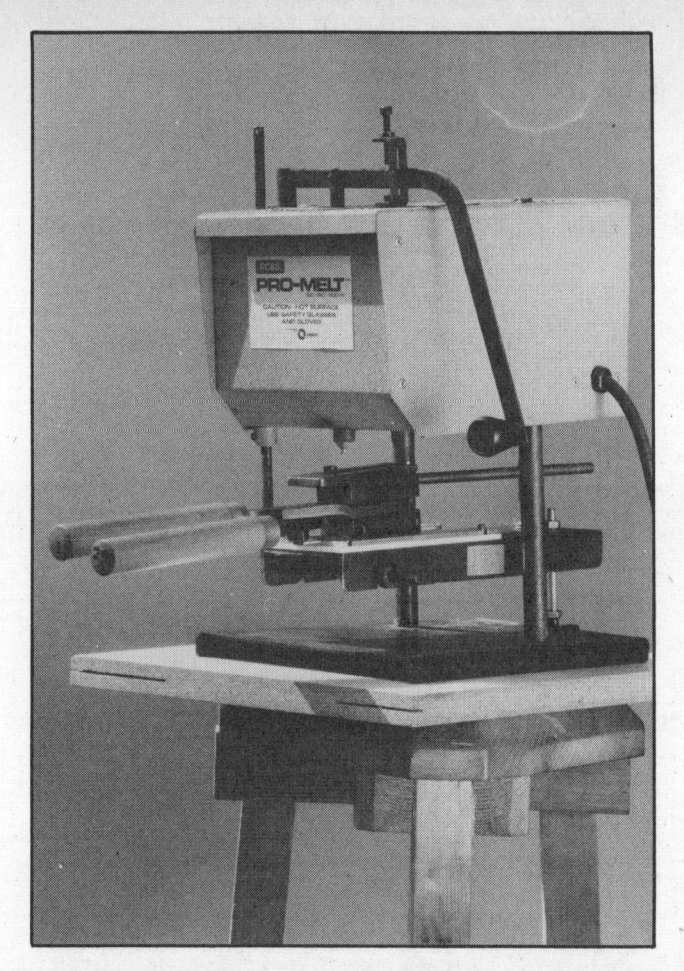

This RCBS Pro-Melt bullet casting furnace has been equipped with a mould guide now available from Hensley & Gibbs, including an adjustable stop bar.

The Saeco lube/sizer is pressurized by giving the top handle a few turns. After that, you can lube/size a good quantity of bullets before having to do it again.

a draw die to draw the readily available .44 jackets down to the proper diameter for producing 10mm bullets. At the time and down to the present, ready-made 10mm jackets were about as commonly encountered in the marketplace as custom aqualungs for snail darters.

As it turned out, it was quite a simple matter to take the 170-grain Sierra JHP bullet, in .410 inch for use in .41 magnums, and pass it through Corbin's draw die in the CSP-1 press, after rolling the bullets across the pad lightly treated with swaging lube. That left a bullet that one might assume could be used for loading the 10mm cartridge. It could, all right, but one really vital step was still needed. What you had to do was to give the slimmed-down bullets one final pass into the regular nose-forming die of the .400-inch set.

You see, when you put a jacketed bullet through the draw die, the jacket, being somewhat elastic, springs back after it passes through. The lead core, on the other hand, being not in the least, tiniest bit elastic, does not spring back at all you and you lose that intimate contact between jacket and core that is so extremely important in obtaining decent performance out of the modified bullets. Doing it with a final bump-up in the nose-forming die made all the difference in the world and the resulting bullets performed superbly, as did the ones made up from scratch with the Corbin dies. What it did was to give access to a generous supply of 170-grain JHP bullets for the 10mm, produced at the bare minimum of time and effort.

Speaking of Corbin, we have some discussion from the good gent, pertaining to swaged bullets and their production. I propose to crank it into the discussion at this point. David R. Corbin can be reached at Corbin Manufacturing & Supply, Inc., (Box 2659, White City, Oregon 97503), should you wish to address any question to him regarding his portion of the chapter at hand. Off the top of my head, I can't think of anyone else more apt to be able to supply the last possible details on bullet swaging. — *Dean A. Grennell*

BULLET SWAGING FOR HANDGUNNERS

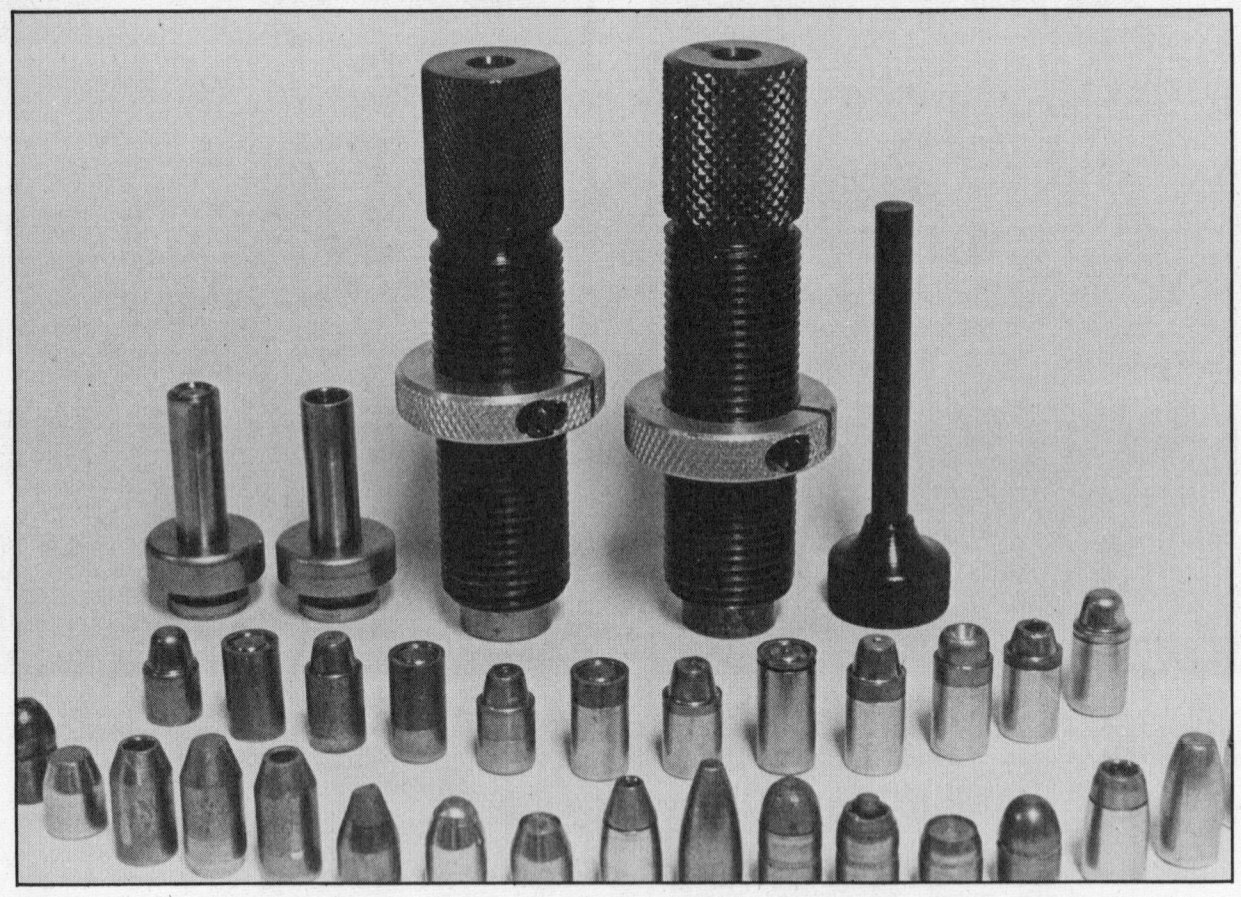

Here's a pair of swage dies for use in any standard reloading press of suitable leverage and heavy-duty construction, along with some of the possible bullet design variations that can be made by simply changing the punches or the swaging techniques. Swaging offers a broad assortment of possible variations.

IT'S BEEN nearly forty years since the first big flurry of interest in home bullet swaging. The idea is great: Instead of melting the bullet material and pouring it at high temperature into a metal mould, you use a strong pressure vessel with a cavity shaped exactly like the bullet you want, applying enough pressure so the material flows at room temperature.

It is only now, after nearly thirty-five more years, that swaging is experiencing the kind of explosive growth you'd think some hot new technology would receive. The average reloader, much less the average shooter, has only a vague and incomplete conception of bullet swaging. Few gun writers, with the notable exception of Dean Grennell, know enough about swaging to explain its secrets. Why the mystery?

Part of the reason you don't see bullet swaging as widely practiced as bullet casting is the false start swaging got during the period after World War II began. A number of reloading manufacturers dabbled with swaging equipment sales, but not enough of them made a complete commitment to writing the information, stocking the supplies, and developing all the tremendous design power that swaging gives to a handloader.

It would be similar to a situation where you could buy all the reloading dies you wanted in a couple of standard calibers only and no one told you how to use them or offered a press to use them in. People got tired of searching for all the pieces.

Another reason is that swaging *is* so powerful and most shooters were just barely getting their feet wet in handload-

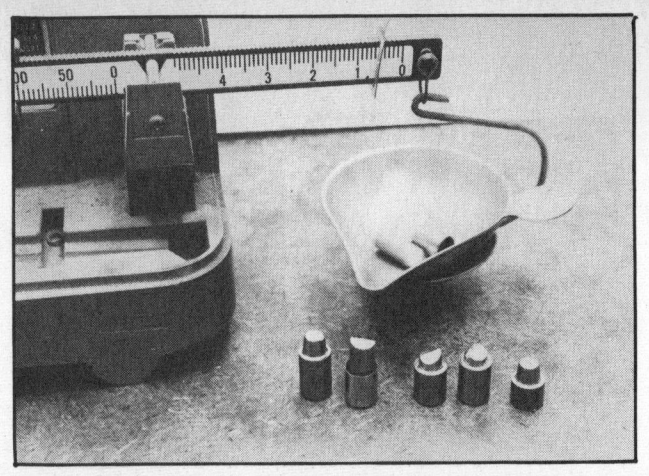

The core and jacket are weighed together to determine the final weight of the bullet. Length/weight of the core can be adjusted to arrive at a desired weight.

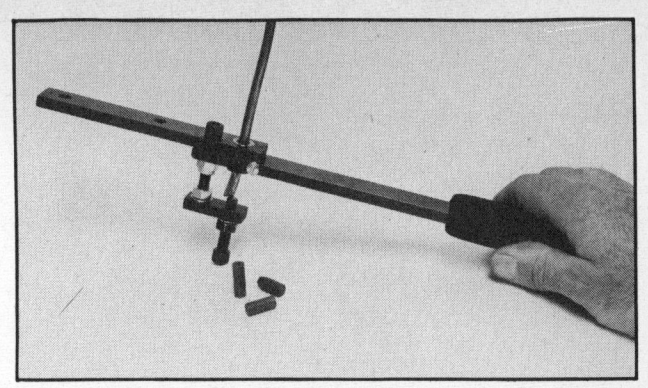

An adjustable core cutter snips predetermined lengths of lead wire, in a diameter suitable for the caliber.

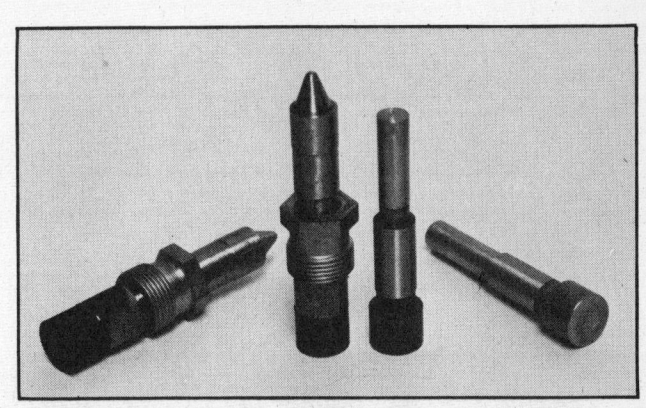

When making hollow point bullets, a punch with a cone-shaped tip — such as the two at left — is used to seat the core into the jacket, rather than the flat tip punches illustrated at the right.

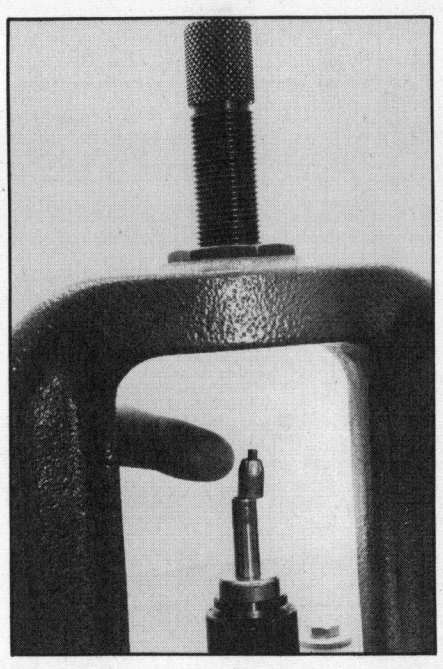

A probe-pointed hollow cavity pistol bullet, swaged in two steps in a regular loading press die set. You'd pay a premium price for similar bullets at the store.

ing with the most simple kinds of equipment. I think it's safe to say that today's handloader is far more knowledgable, in general, about the technical aspects of his guns and handloads than the shooter of the 1940-1960 era. More and better books, better equipment, more emphasis on high performance and the general knowledge that such performance is possible all combine to give today's handloader a big booster shot of interest in the more advanced fronts of handloading. He's more likely to want to try something that gives him more control, and frees him from arbitrary factory offerings.

Handgunners in particular have also seen the specter of ammunition bans in the past: various back-door methods of making it difficult to obtain handgun supplies have been attempted, and probably will be tried again by those who don't think the rest of the world is responsible enough to be trusted with a handgun. So, the ability to manufacture your own bullets to replace the usual factory supply is a strong incentive. One doesn't have to be a bona fide Survivalist with a capital S to see the logic behind backing up your supply lines!

Finally, because of characters like Dean Grennell who have diligently inspired thousands of shooters to involve themselves more deeply in their handloading and learn to push the limits of their ability to new dimensions, a huge market has developed for more exotic kinds of bullets: higher accuracy, greater explosive effect, higher speeds (due also to the Contender pistol and the rifle-like loads now being used by handgunners) and generally more precisely controlled terminal performance are all eagerly sought by the advanced handloaders and shooters. This has been responsible for a tidal wave of new, small custom bullet firms that turn to bullet swaging equipment, just as the major firms do.

Because there are so many custom bullet makers today, more people have been exposed to the ideas of exotic bullet performance. Thus, the cycle has fueled its own engine and produced an even larger market for even more custom bulletmakers. Since the kind of market is ideally suited to a

These .25 ACP bullets were swaged with jackets made from empty .22 LR cases and soft lead cores. Compare them to factory solids at right, expansionwise.

In a test medium that produced double-diameter expansion with home-brewed bullets, these factory .25 ACP FMJ bullets show little or no expansion.

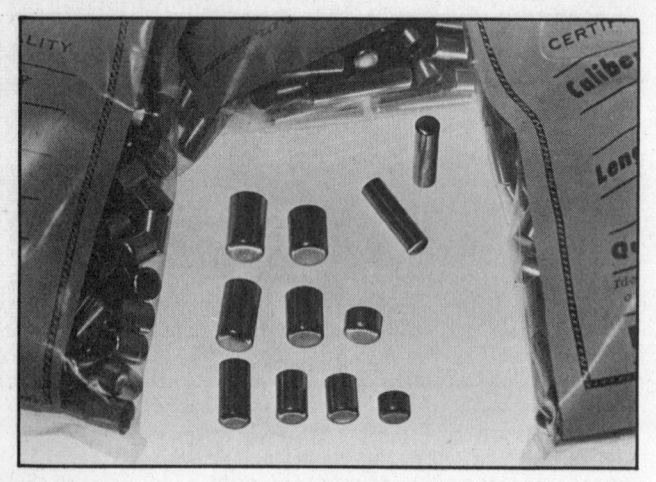

Bullet jackets usually come 250 or 500 to the bag, in .38, .44 and .45 sizes and assorted lengths.

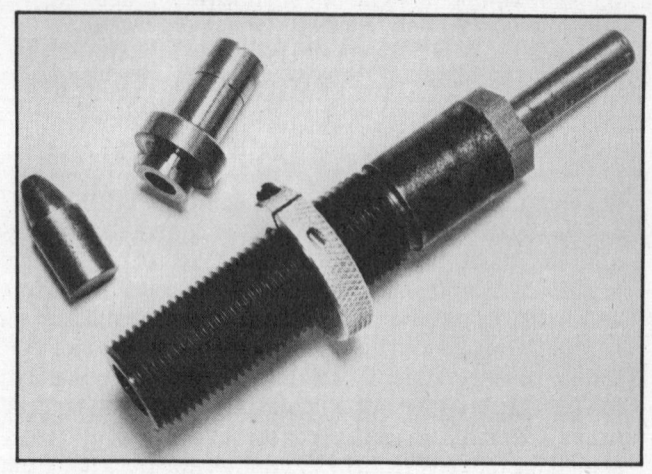

The Corbin Econo-Swage is made in standard sizes and permits low-cost bullet production in standard press.

Here's a better look at the unusual hollow cavity punch that produced probe-point bullet on preceding page.

person working at home with a fairly small investment in equipment, and since we live in a period when many people are either reconsidering whether they want a traditional job-mortgage-retirement lifestyle (and many others are finding that security doesn't live up to its advertising!) or the cost of living has made it necessary to bring in extra income, turning a hobby investment in bullet swaging equipment into a weekend business or even a full time career is becoming fairly common.

Custom bulletmakers abound, yet there is a seemingly endless unfilled market for more of their products. This market is unusual in that it doesn't attract much major competition. The reason is the volume of unusual bullets of any given type that can be sold each year. It is too low for any large concern like Sierra, Speer or Hornady to develop economically, yet it is just right for the kind of output that a person can develop on low-cost hand presses or moderate-cost power machinery. Anyone trying to make more, faster

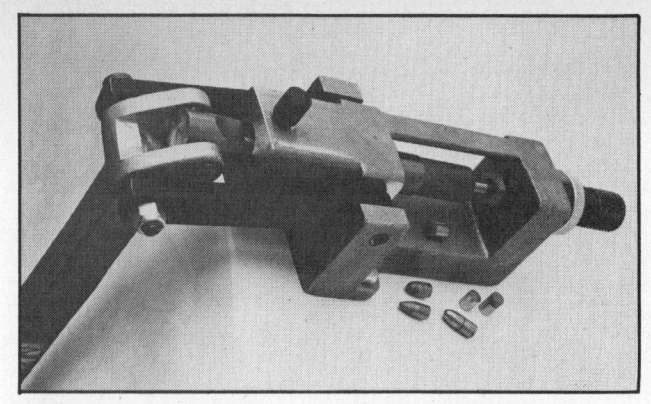

Bullet swaging presses, such as this Corbin Model CSP-1, have shorter strokes, more leverage and auto ejection of completed workpiece on the backstroke.

Seated cores are processed into finished bullets in one pass through the point-forming die. Shown are the flat-ended seated cores and completed handgun soft points.

The lead core is put into the jacket and the assembly swaged into a compact cylinder of correct diameter in the core-seating die, per the two samples at right.

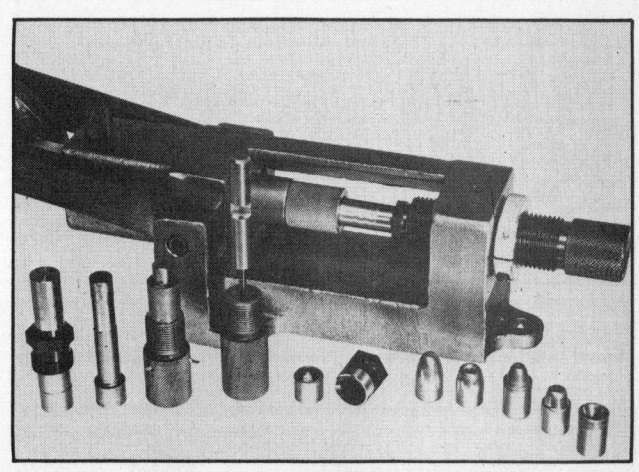

Typically, a set of three dies is used to make the more advanced handgun and rifle bullets. This is the Model CSP-1 press, with some typical dies to fit in.

and cheaper, would only flood the tiny niche and go broke trying to pay for his advertising.

Yet, a family enterprise expecting a reasonable lifestyle from the sales of perhaps three different offerings in custom bullets can — and does — find that the number of people willing to buy special-purpose, custom bullets is more than enough for them. Since there are so many possible calibers, weights, styles and ideas to use in making bullets that sell from 10,000 to a maximum of about 100,000 per year — spread out in individual one- or two-box sales all over the country — it seems that the saturation point — where there isn't room for one more custom bullet maker — is way over the hill and price competition will herald its approach.

That's the state of bullet swaging. Today. When my brother and I became interested in swaging many years ago, the two of us could make all the dies we could sell, using the equipment we could fit into half of a one-car gar-

Lead bullets can be swaged in the same dies that make jacketed types. Here, a lubricated lead core is being put into a press to make a bullet as shown.

GUN DIGEST BOOK OF HANDGUN RELOADING

An assortment of swaged bullets, from the simple, one-step semi-wadcutter to advanced handgun styles.

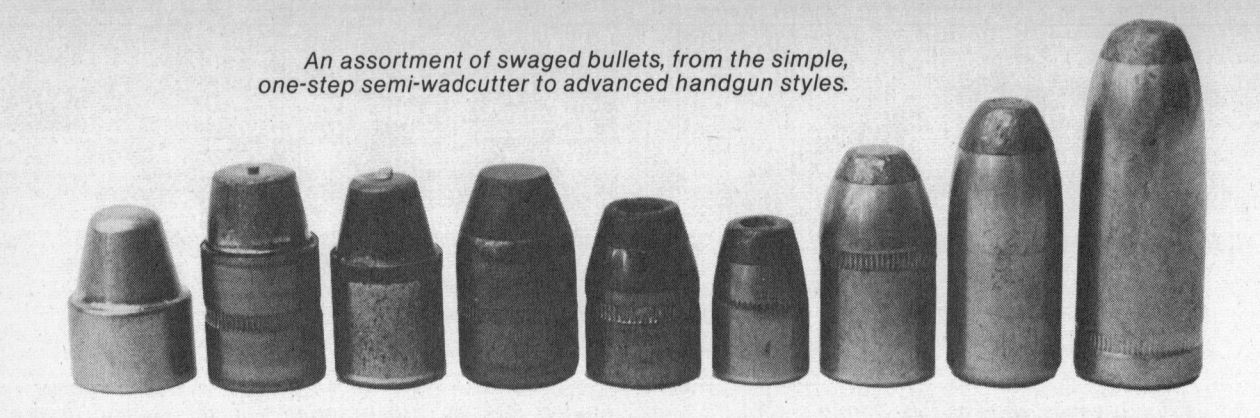

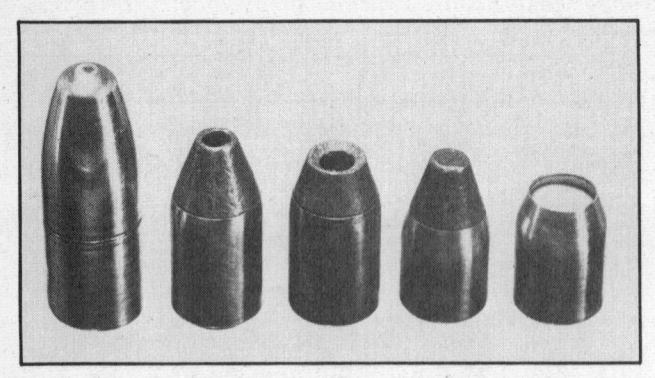

The ogive or nose curve can be formed in the jacket if the die has a smoothly curved interior to match. Bullets here were finished in the point-forming die.

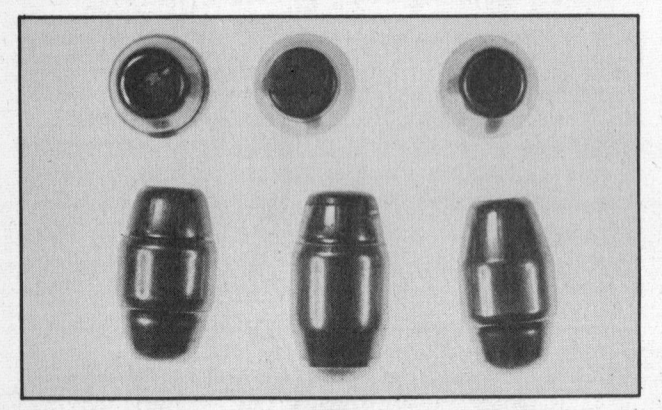

An interesting group of rebated boat tail .45 pistol bullets, all of which shot well, with the bullet at right performing best of all, upping ballistic coefficient.

Corbin's Core-Bond can be used to form a permanent joining of the core to the inner jacket wall, as here.

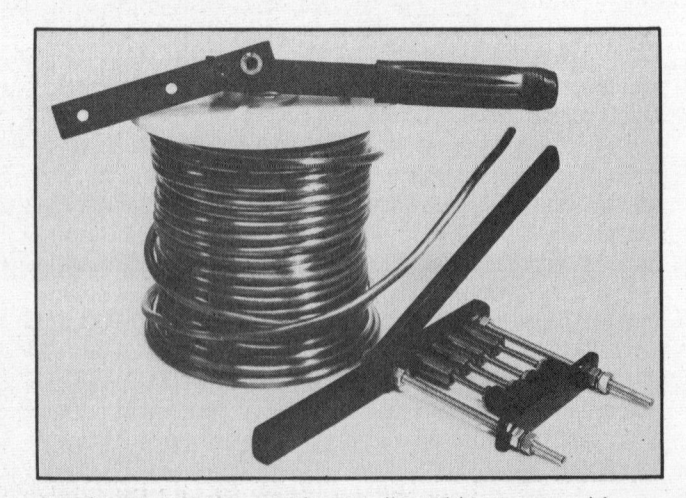

Lead cores can be cast in an adjustable core mould or cut from spools of lead wire by core cutters.

age at home. In a few years, we had backlogs that took two and three years to fill. Today, with a modern, computerized plant and offices, and a staff turning out tens of thousands of swaging tools every year, seven books in print about swaging and all the materials and supplies available for any caliber in the .14 to 20mm range, I'm amazed by the distance this ancient art has traveled.

Yet, at the same time, there is a tiny evangelical urge starving for converts: Why, with swaging so well developed and so many people making a living using the equipment, so many special bullets available to anyone who reads the advertising in gun magazines, do so many people still have no idea what swaging is all about or have totally obsolete ideas about its limitations or cost or difficulty?

Then I flip through a few firearms magazines and read the articles and I know. Test after test completely ignores the existence of bullets that could turn the results on their head. Story after story ends with a lament to the effect that it would be nice if there were some particular weight or style available, but, alas, you can count on one hand the bullets available from the mass production outfits. There's no mention of custom bullet makers; no idea at all of the possibility of a handloader whipping up his own design with a set of swage dies. Then I know exactly why there are

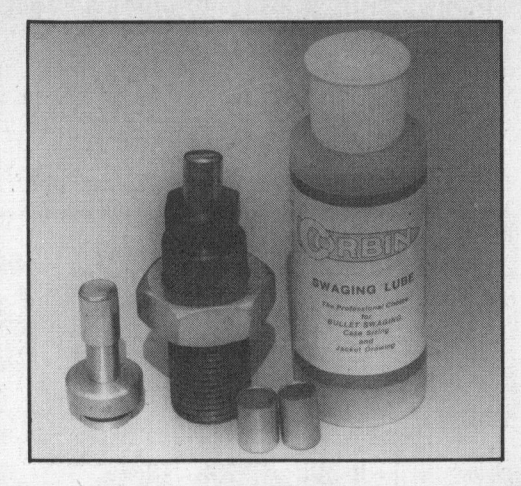

World Directory of Custom Bullet Makers is available from Corbin at $24.95 (Box 2659, White City, Oregon 97503). Information on bullet swaging is $1 from same source. Swaging lube eases drawing of jackets.

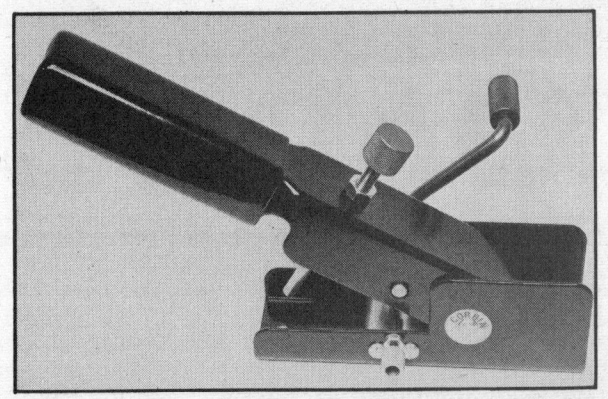

Corbin's canneluring tool applies cannelure grooves in bullets all the way from .22 up through .50 caliber.

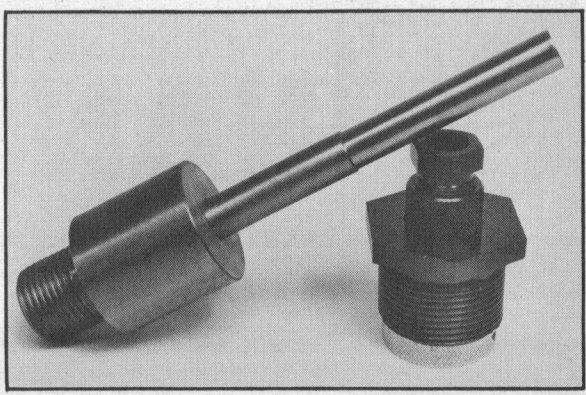

Jackets can be changed from one caliber to a smaller one by use of a draw die and suitable lubrication.

two parallel universes today: Those who are blowing away all the old ideas about firearms performance with their own custom bullets and those who haven't the foggiest notion of bullets that don't come from either a casting mould or their dealer's shelves.

You can count on one hand the writers who have any idea what bullet swaging can do. And, for better or worse, most of us get our information about what is possible in firearms today from those same writers. Until the rest of us point out that we can easily make any of the bullet styles or weights or calibers that the latest article complains about not being available, that big information void probably will continue to block the progress of thousands of handloaders who could benefit by making their own bullets.

Not, however, those of us who are happily curled up with this book! We've got a big head start. Whether for hobby use, to make a more dependable defense slug, or to bring in a few extra dollars in that growing custom bullet market — and, if nothing else, prove to a dubious spouse that shooting isn't a total drain on the family fortunes after all! — bullet swaging offers a unique opportunity during the Eighties, and probably far beyond, though that's another book! — *David R. Corbin*

Dave Corbin, president of Corbin Manufacturing Co., is an avid handgun experimenter and enjoys helping handloader develop new bullets to market.

CHAPTER 9

POWDERS FOR HANDGUN HANDLOADERS

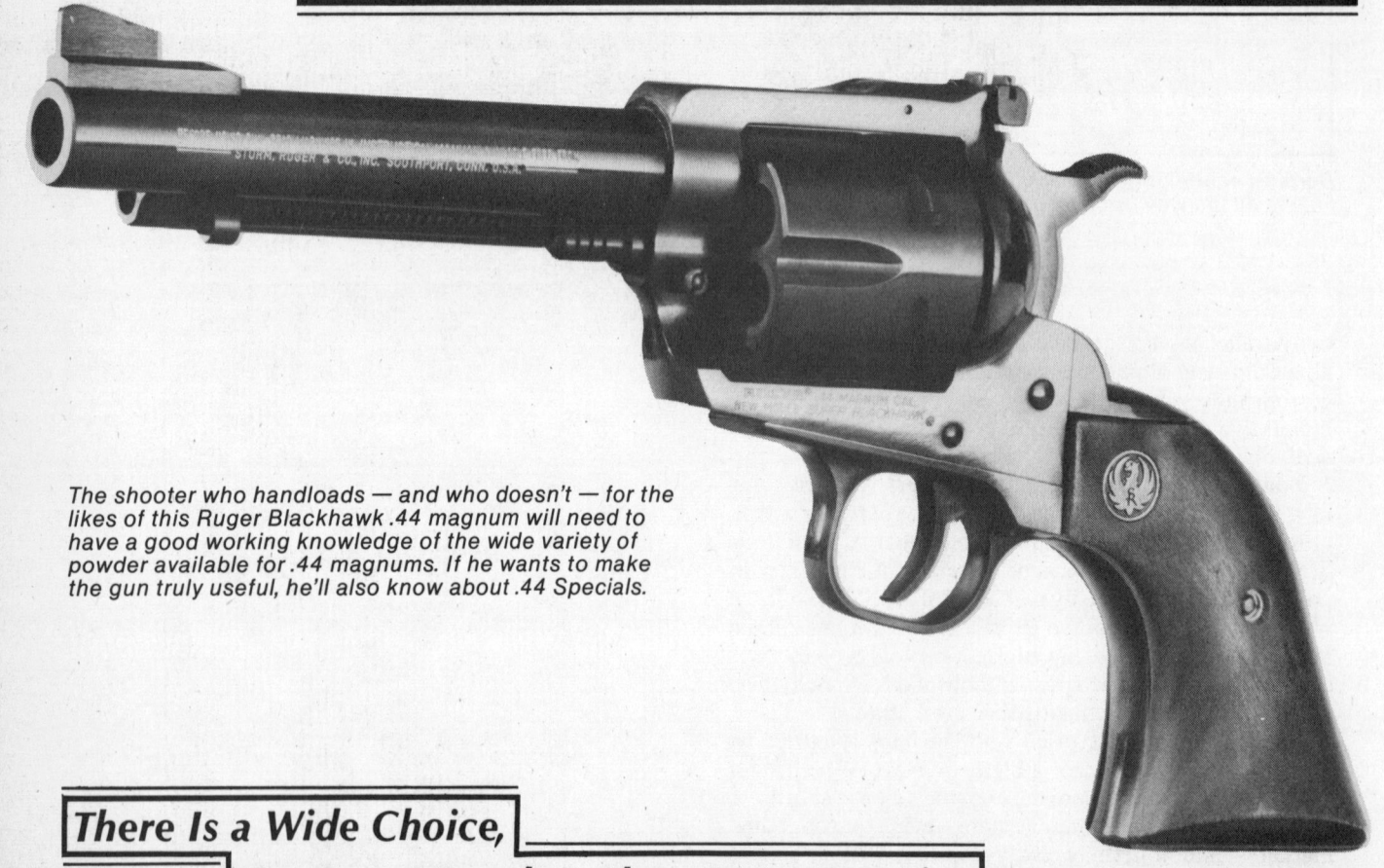

The shooter who handloads — and who doesn't — for the likes of this Ruger Blackhawk .44 magnum will need to have a good working knowledge of the wide variety of powder available for .44 magnums. If he wants to make the gun truly useful, he'll also know about .44 Specials.

There Is a Wide Choice, But Some Powders Show More Promise Than Others In Specific Handguns

AS WITH handgun bullets, the subject of suitable handgun propellents is clouded into gray areas by the current popularity of specialized and custom "handguns" chambered for what would otherwise be considered as rifle cartridges.

A good instance is one I have presently: a six-pound revolver chambered for the .45-70 cartridge! Powders suitable for reloading that venerable old military rifle caliber can range all the way up to the slow-burning Hodgdon's H4831, although this cannot be said to be a handgun reloader's powder in any traditional sense. Neither are any of the other slow-burning propellents usually meant for high-pressure bottleneck cartridges.

Therefore, this essay on powders for handgun handloading will be confined to those powders best suited to relatively straight-walled cartridges intended for revolver or semi-auto-type handguns. Even then, there is still considerable overlap in powder and labeling.

One of the most often used smokeless powders for magnum handgun reloading, 2400, is still labeled "Smokeless Rifle Powder." In fact, as a teen-age handloader, I was forbidden by the Gun Control Act of 1968 such powders as Bullseye or Unique, because their labels said they were handgun propellents. Yet, I could and did legally purchase many cans of 2400 for loading magnum revolver cartridges. Such are the workings of an overly bureaucratic government!

Currently there are five major powder suppliers with

Below: If you get into reloading for your handguns as deeply as many do, you'll end up procuring your powder in bulk. This is an economy-sized can of Hercules Herco — someone must be reloading a lot of 9mmP.

Above: An assortment of powders from the five major makers, all relatively slow-burning and useful in the big magnums. Left to right in top row: Hodgdon H4227, W-W 296 and Dupont IMR4227. On the bottom: Israeli-made Accurate Arms' #9, Hercules 2400 and Hodgdon's H110.

GUN DIGEST BOOK OF HANDGUN RELOADING

Left: Hercules Bullseye, long a mainstay of .38 Special and .45 ACP reloaders, is listed as the fastest-burning pistol powder made. It's truly stable stuff. Samples of the stuff made decades ago perform like yesterday's run.

Right: If forced to do so, a handloader could produce fine ammo for years with the Hercules lineup. The "Dots," Blue, Red and Green, were ostensibly shotgun stuff. All seven work well in various handguns.

nationwide distribution: Du Pont, Winchester, Hodgdon, Hercules and Accurate. Additionally, there are spotty supplies of some European propellents from time to time such as Norma, Alcan, Kemira, et al. Of those, Norma is the single brand most likely to be encountered on the American market.

With the above considerations in mind, a bit of research turns up approximately thirty smokeless propellents with burning rates making them suitable for traditional revolver and semi-auto cartridges. Of course, not all powders are meant for all cartridges or calibers.

Before getting into the actual powders, some mention should be made of the basic types. Smokeless powders can be divided into single- and double-base types. The full explanation of the differences would require a chemist's education, but the Speer Reloading Manual #10 puts it briefly as thus: "single-base powders are those of straight nitrocellulose and double-base have a significant percentage of nitroglycerine added."

Additionally, this helpful manual says most double-base powders are those made by Hercules, whereas only one by Du Pont is double-base: That is Hi-Skor 700X.

There are three types of smokeles powder forms: extruded

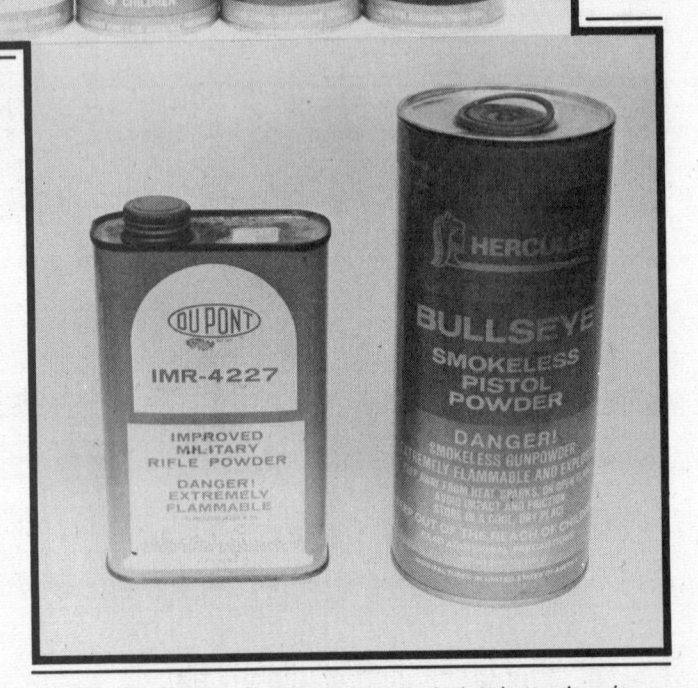

Above: One of the characteristics of pistol powders is demonstrated here. IMR4227 is one of the slowest of suitable powders. Bullseye is unquestionably the fastest of all. Burning rate, as well as volume, are important factors in selecting a powder for a particular purpose.

GUN DIGEST BOOK OF HANDGUN RELOADING

Below: As suggested by the variation in packaging, 2400 has been coming out of the Hercules plant for lots of years. The slow-burning 2400, originally intended for a variety of smaller rifle cases, was a favorite fuel of an intrepid handloader from Idaho — the late Elmer Keith.

Above: Lots of savvy loaders would argue long and hard if you tried to separate them from their Unique. There are few handgun cartridges that cannot be loaded well with an appropriate charge of this long-time favorite. Again, the chronology of packages suggests long service.

or tubular, spherical or ball, and flake. Interestingly, none of these helps to isolate a particular powder as good or bad. Our present handgun propellants encompass all types and forms and, in their specific applications, all do a creditable job.

Probably seven powders marketed by Hercules see the most use among handgun handloaders. In order of burning rate from fastest to slowest, they are Bullseye, Red Dot, Green Dot, Unique, Herco, Blue Dot and 2400. Two of those — Bullseye and Unique — are among the oldest of smokeless powders, having been concocted before the turn of the century. Bullseye sees use predominantly in light to medium loads and, as such, is amazingly economical. If 3.0 grains is used, as is common with the .38 Special, then a single can of powder should yield 2333 loads! Bullseye also is capable of extreme target accuracy in a wide variety of calibers. Of late, however, this powder has come under some criticism, following reports of damaged guns and unreasonable pressures with normal charges. Some say this happens due to double or even triple charging a case with Bullseye, because so little is required for maximum loads. Be that as it may, I have used Bullseye to reload many tens of thousands of rounds of revolver and semi-auto ammunition with nary a hitch.

Hercules' other old-timer, Unique, is aptly named. It is unique, because it can be used with good results in just about any handgun handloading application imaginable. It may not be the best choice for all those uses, but it certainly

GUN DIGEST BOOK OF HANDGUN RELOADING

121

Above: The Ruger Bisley revolver in .41 magnum is a magnum in every sense of the word — strong, sturdy and accurate in the extreme. Good handloads that achieve true magnum potential almost invariably use one of the slow-burners: IMR4227, H110, W296, AA9, 2400.

can give adequate results. Ask most any experienced handgun reloader what would be his choice of powders, if limited to one. Odds are he will say, "Unique."

There is one application for which Unique has reached legendary status, but this writer feels Hercules Red Dot is the better choice. That is in reloading the venerable .45 Colt with lead bullets at roughly the same ballistics of factory ammunition. Unique has to be the most-used propellent in .45 Colt, but in machine-rest testing done over the years, Red Dot has given better accuracy from a variety of revolvers. With bullets weighing from 225 to 260 grains, I rely invariably on 6.0 grains of Red Dot for the .45 Colt.

Hercules also has that favorite of many magnum handgunners, 2400. Having lost some of its popularity in recent years to cleaner-burning ball or spherical powders, 2400 is still a good choice for the big magnums. It does often burn dirty, leaving scorched powder particles in the cylinder, but it also offers accuracy and velocity. Some experienced reloaders consider it the best magnum handgun propellent, when cast bullets are used.

Above: While certain of the other Winchester powders have use in handgun loads, these two are the ones we commonly associate with beltgun loads. Winchester 296 is a slow magnum powder. 231 is a super-speedy target type fuel. Winchester offers no medium-speed powder.

122　　　　　　　　　　　　　　　　　　　　　　　　　　　　　　　　　GUN DIGEST BOOK OF HANDGUN RELOADING

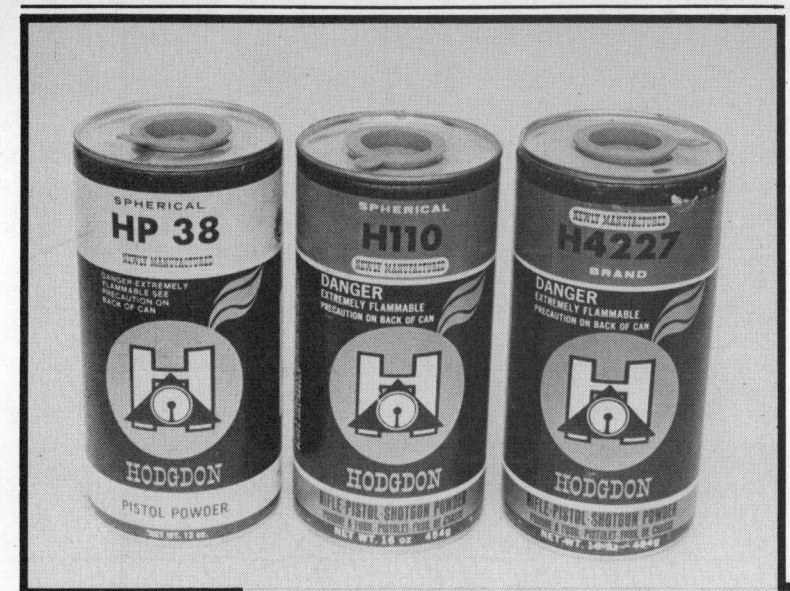

Left: The Hodgdon line of powders primarily for pistols. HP38 is a fast one which is much like Bullseye, a great target round powder. H110 and H4227 are both on the slow end of the scale. Author Dean Grennell prefers H110 with different bullets in his beloved .41 magnums.

Above: There are some distinct similarities in the burning rates of these two powders. Do not, however, even think of swapping data between the two. That is a dangerous practice — avoid it like a case of typhoid.

Above: In similar vein, HP38 and W31 are nearly the same in burning rate, but the data quoted shouldn't be changed. Use the manufacturer's quoted data always. It goes without saying that they're both good.

The other Hercules powders have more shotgun than handgun application, but still find special niches for themselves. An example is Blue Dot which is often touted as a "special" 9mm Luger powder.

Du Pont produces several faster burning smokeless propellants, most of which have their first intended purposes in shotguns or small-capacity rifle cartridges. However, their use in handguns is still widespread. They range through Hi-Skor 700X, 800X, SR-7625, SR-4756 and IMR-4227. That last number is the one most often found in modern reloading manual formulas for modern magnum cartridges. In fact, my own all-time accurate .44 magnum handload uses 27.0 grains of IMR-4227 powder under the 180-grain Sierra JHC bullet. From a machine rest at twenty-five yards, five-shot groups as small as 1.00 inch are possible. Du Pont's other faster-burning propellants are found here and there in most modern reloading manuals, but are not recommended universally in a wide range of applications.

Winchester is the company that is easiest to cover in regard to handgun propellants. At this time, they offer two whose purpose is entirely in the short-gun area.

However, what the big red "W" boys lack in selection they more than make up in quality. On the fast side is W231, which is just a tad slower than Bullseye. W231 is a ball powder that has been flattened. It powder measures well and is capable of some of the best handgun accuracy obtainable. It is giving Bullseye a run for its money with

GUN DIGEST BOOK OF HANDGUN RELOADING

Above: Handgun makers produce new models all the time. And as fast as they make them, custom gunsmiths are modifying them. This pistol is a Model 645 S&W, modified into a single-action 745 by ace pistolsmtih Wayne Novak. The fancy automatic will likely shoot well with handloads.

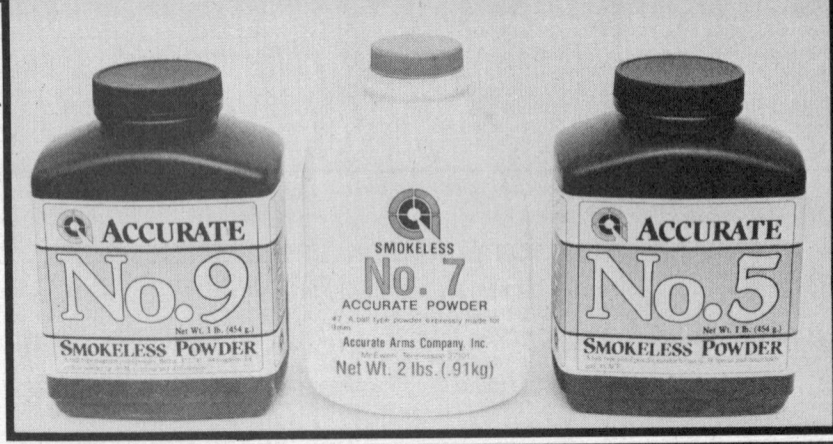

Above: Slow, medium and fast-burning powders from a new importer, Accurate Arms. #s 9, 7 and 5 cover the range of speeds necessary for most handgun reloading. The powders are produced in Israel and imported into this country by the Tennessee firm. This is good stuff!

those shooters who use milder loads. On the slow side is Winchester 296, which is becoming well known as a full-bore magnum handgun powder.

However, W296 is not for the experimentally inclined. Winchester cautions in factory literature not to try reduced loads with it, as varying from their published data can result in dangerous pressures. With that firmly in mind, W296 can offer fine accuracy in the .357, .41, .44 magnums, .357 Maximum and is one of the few powders suitable for the super .454 Casull.

From Hodgdon, the powder supplier located in Kansas, there are six faster numbers suitable for handgun reloading. Three of these are primarily shotgun powders, but their burning rates put them in the realm of handgunners. These are Trap 100, HS-6 and HS-7; all are "spherical" powders. Trap 100 lies between Bullseye and Unique in burning rate, while the other two are on the slow side of Unique. In practical effect, that makes Trap 100 a light target load number and the other two more suitable for full charge service and magnum velocity ammunition. Except in the Hodgdon reloading manual, data for these three is relatively uncommon.

Hodgdon's other three suitable handgun propellants are HP-38, H110 and H4227. The first two again are spherical in type, while the last is extruded. HP-38 is quite similar to Winchester 231 in burning rate and shares its other good characteristics in regard to accuracy. H-110 is similar to W296 for magnum handgun loads and as the similar designations would lead one to believe, H4227 and IMR-4227 are near clones. However, being similar does not mean identical. Consult recognized reloading manuals and sources for complete data and instructions.

The youngest powder supplier is Tennessee-based Accurate Arms Company. This firm's line of powders, all imported from Israel, includes three intended specifically for handgun reloading. Those are Accurate #9, #7 and #5. The first is their slowest burning and rivals Winchester's 296 for application. It is for full-charge magnums. In the middle is #7, which is somewhere between Unique and Blue Dot in burning rate. That makes it most suitable for full-charge loads in the non-magnum handgun calibers and for heavy bullets in some of the magnums.

Interestingly, Accurate Arms Company touts #7 as the best powder for the new up-and-coming 10mm Auto caliber. Accurate #5 is currently the company's fast propellant. It is roughly on a par with Unique and has the same general applications. According to Accurate officials, a #3 is slated for the future and will be designed to rival Bullseye and W231.

A fourth Accurate powder that has limited handgunning ability is their 5744. This is an extruded powder that is a bit on the slow side for most handgun applications. However, the company's manual lists its use with moderate to heavy lead bullets in large-capacity magnum cases.

If one cares to glance over the assortment of handgun propellants generally available today, the selection is impressive. Our American handloading needs are covered in spades. As one who stocks most of the propellants listed herein, I felt consternation upon visiting Germany a few years ago. There, while in conversation with avid German handloaders, I learned that they are restricted by law to only one kilo (about two pounds) of smokeless powder in their possession at one time.

I returned from that trip not only grateful for the vast array of powders we have, but for the right to have them. May we all strive to keep it that way. — *Mike Venturino*

Proficiency with handguns in today's competitive scene demands lots of reloads: Jack Mitchell hard at work in an IPSC practice session. The gun is a magnificently reworked Colt by gunsmith Art Leckie. Mitchell has a favorite reload recipe built around the H&G #68 bullet.

CHAPTER 10

The Indispensable Primer —
That Part Of A Handloaded Cartridge About Which We Can Say...

"THE BUCK STARTS HERE"

They all look so much alike, now don't they? Handgun primers are so similar in their appearance, that you'd be tempted to use them interchangeably. Except for the obvious difference in size between the large and small, primers are astonishingly alike. But one fact should be indelibly stamped in the awareness of all handloaders. You cannot identify a primer by appearance and the only logical means of identification is the original packaging.

TO HAVE a little long-distance fun with one of my shooting buddies, I asked in a letter whether he'd loaded any pewter loads lately. His wife read the letter aloud while he loaded ammo; and all along, muttered, "Pewter loads? What in the world does Ken mean by pewter loads? I never heard of anything called *pewter loads!*"

Fortunately for my scalp, I was on the other side of the continent when she got the end of my letter, which said no more about pewter or loads — except for the corny PS: "Oh, yes, you shoot 'em, they go *Pewt!*"

I'd learned a special fact worth knowing about primers before I pulled that prank on my buddy. I had carelessly loaded two "pewter rounds" in a batch of ammo for my 7½-inch Colt SAA, and that's how they'd sounded when the hammer dropped on them. More germane and enlightening than how they sounded was the fact that those primers threw the bullets forward hard enough to lodge them tight in the leade, but not far enough to clear the cylinder.

My Peacemaker was locked up tighter than a solitary cell in Lubyanka Prison. I was in the bush, a long way from tools, so the gun was out of action until I hammered those bullets loose by shaking the base pin back and forth inside the barrel with my thumb over the muzzle.

Primers are powerful little devils. Phil Sharpe's book shows a photo of a government .30/06 cartridge that had no flash hole to vent primer flame into the powder cavity. That cartridge had been fired in a rifle and, of course, only the primer had gone off. The bullet, neck and shoulder looked like new. The head of the case and the thicker, harder brass at the the rear of case also seemed unchanged. But the thinner, softer body wall behind the shoulder buckled into neat ripples when the force of that frustrated but stubborn primer drove the rear of the case forward while the front end was immovable in the front of the chamber.

Primers are safe enough when you handle them right. Bless them and give thanks for them, but never take them for granted. Also give fervent thanks for the neat little boxes that primers come in — and no matter how much nuisance these boxes might seem to be at times, *never* store primers apart from them. If you want to protect them from water or moisture, shroud your primers — still in their little boxes — in plastic wrap or inter them in sealable plastic bags or boxes. Saran or other plastic wrap inside a taped or glued heavy paper wrapper or a cardboard box should be good for years of dry storage.

Those carefully designed little boxes are vital to both safe handling and dependale identification. Different brands and sorts of primers are so similar, they're practically impossible to identify once you lose track of exactly which boxes they came in. If you know for sure that you have only two kinds of primers on the premises, you can tell Large Pistol from Small Rifle or Small Pistol, but that's about as far as it goes except in special instances.

Only one brand still uses two-legged anvils, as far as I know, so you can sometimes distinguish this brand of Small Pistol or Large Pistol from another brand of the same size — if by some other means you can be sure they aren't rifle primers. The thin paper disc between the anvils and pellets of some primers may be one color in one brand,

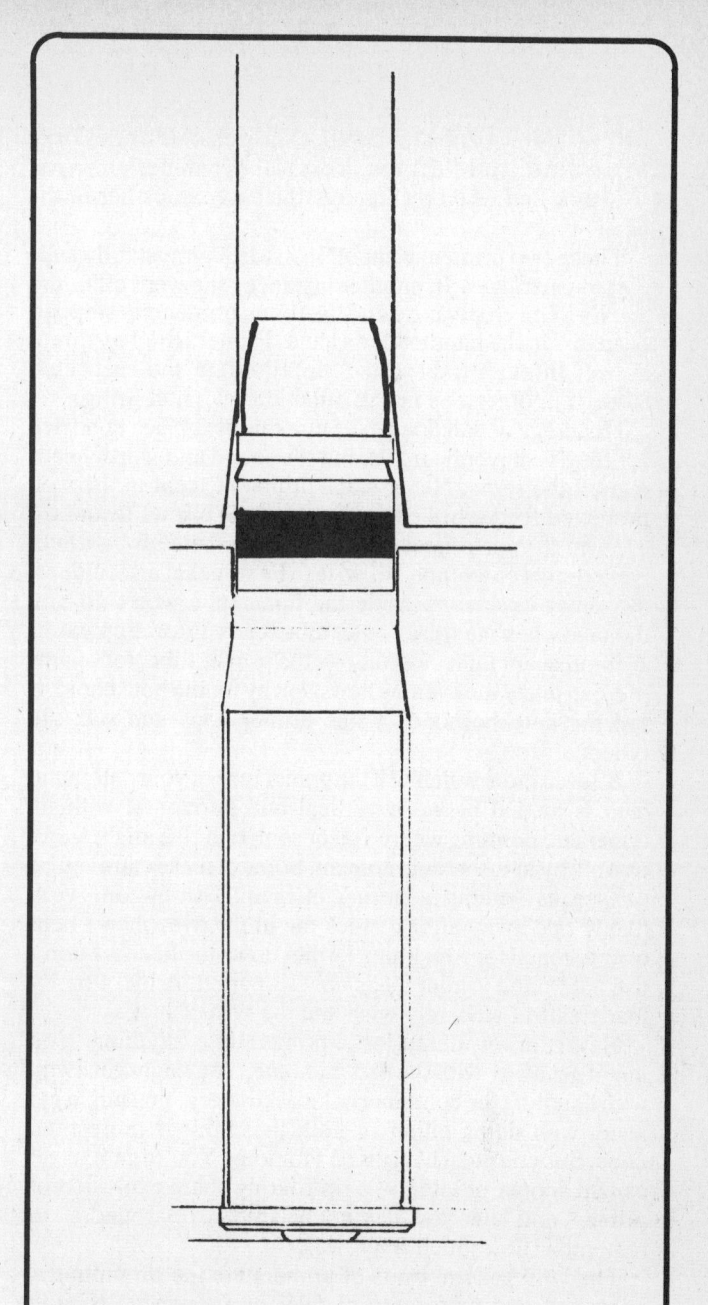

Here's a sectioned drawing of one of the author's pet "pewter" loads. The primed, but powderless round will drive the bullet out of the case and into the barrel and that ties up the gun. It is a testament to the power of the primer and carelessness of the handloader. See text.

another color in another brand; but don't count on it. *The only dependable identifier of any primer is the label on the box it came in* — as long as the primer is still in the box or so recently out of it that you know which box it came in. Besides, the box carries the lot number.

Never prime cases with a tool or setup that uses impact — however light — to seat primers. In several instances, seated primers have been set off by vibrations transmitted through the case, with nothing in direct contact with the rear of the case or the base of the primer cup. In one instance, a handloader had to have the jagged head of a case removed from deep in his thigh. In another, a shooter's wife — sitting behind the boltless breech of his rifle while he tried to drive a stuck round out of the chamber with a rod — was killed when the head of the case struck her in the head.

Those two primers went off in loaded, almost fully supported cartridges. In another instance, one went off in the pocket of an empty case and the flying primer cup stopped deep inside the handloader's hand. Primers fired by direct impact through transmitted vibration are the suspected culprits in otherwise unexplainable accidental firings.

Hundreds of handloaders ignore this next piece of advice (or they've never heard it), but it's sound and worth heeding, all the same. Never stack primers except in the well protected feed tubes of some automatic primer feeders.

Even if your primer tube is well protected by a thick-walled steel outer tube, it's wise to be sure that an accidental primer explosion inside the inner tube won't do any damage when the sturdy outer tube sends its force upward,. A fluorescent light close above the primer tube, for example, can make the evening newsworthy for the household, if not the neighborhood, if one primer pops and sets the others off too.

Be even more watchful if the primer tube of your automatic feed is angled between vertical and horizontal with its upper end pointing where you or someone else might stand or walk by at the wrong moment. Some of these can even be dangerous without a primer in them. I threw one very handy one away after it stuck me in the eye when I bent over to reach for something farther down the bench. I don't use automatic primer feeds now — period. A lot of handloaders do. I sincerely wish 'em the best of luck.

Never, never, dump loose primers into anything. You might get away with the first dumping. You might get away with handling the container of loose primers. You might get away with doing either or both these things more than once. But you most likely won't for long. You're sure to get caught sooner or later by a rip-roaring chain explosion of primers and later you may not be able — or around — to remember how much fun it was.

The vital general traits of primers are the promptness, adequacy and uniformity of their performance. "Going off" isn't good enough; they have to go off immediately and instantly, with no discernible or variable delay after the firing pin hits 'em. They must cut loose with enough heat and fury to do a peck of work in a piece of a second — and each must be more like its peers than any pea is like any other in any pod.

No one primer can be all these things to all ammo. The size of the cartridge case and its powder charge, the granulation of the powder, the force of the firing-pin impact, even the chamber pressure of the load — all bear significantly on the primer's chemical and physical characteristics.

To be any good at all to a shooter, primers have to be very sensitive to impact. Impact, after all, is what both firearm design and primer design ordain as the means of setting 'em off. At the same time, you have to be able to handle them — notably to seat them in the primer pockets of empty cases. So they're designed to be seated safely with a strong but steady *push* and to go off easily under a moderate but sharp *impact*.

Around this primer, there's the tell-tale indication of U.S. military origin, the crimp. Primers are crimped into place for reasons of shelf life and functioning in typical automatic weapons. The crimp means that the cartridge is harder to decap. It also means that the crimp must be removed before reloading. Removal can be overdone.

These three properties, then, are the keys to a general grasp of what primers are all about: (a) they're powerful enough to inflict pain or damage, directly or by flinging small hard objects at you; (b) they're delicately sensitive to sharp impacts a good bit lighter than a carpenter uses to drive tacks; (c) they're safe enough under strong but steady shoves.

Safe handling is but the prologue to the primer's short life. Here's basically what happens during a primer's moment of glory.

The firing pin crushes the fragile and sensitive primer pellet against the primer anvil. The legs of the anvil rest on the floor of the primer pocket, so the anvil resists the thrust of the firing pin. The pellet of priming compound, crushed under the blow, produces a chemical reaction.

The crushed pellet reacts violently, generating flame and fiery particles that spew through the flash hole to stir, heat and ignite powder kernels inside the case. The primer's ability to stir the powder is as important as its heat, since it has to ignite more than just the few powder kernels near the flash hole. Lighting off larger kernels, with a larger ratio of volume to surface, takes more flame and more time as well as more turbulence to stir them.

This is why we have magnum primers — to ignite heavy charges of large kernels, especially when the burning of the powder is controlled by a coating that delays or slows its burning. Smaller kernels heat up faster and ignite more easily, so regular or standard primers are theoretically milder than magnum primers (but some makers' regular primers are as hot as others' magnums). On a cold day, powder needs more heat to bring it up to its ignition point — another reason behind magnum primers.

Before you seat a primer, the pellet of priming compound is under little or no stress between the inside of the cup and the apex of the anvil. This pellet is both delicate and sensitive. To make the primer safe to handle before it's seated, the pellet should be under no stress — but consistent primer performance requires that it be under slight stress in the seated primer.

Proper seating therefore requires that the legs of the anvil be in solid contact with the bottom of the primer pocket and that you then seat the primer a tenth of a hair deeper, exerting half a smidgen more force on the pellet but not enough to crack it or crumble it.

Primer cups and primer pockets vary some tiny amount from their nominal diameters (large primers are 0.210-inch in diameter, small primers 0.175-inch). The force to seat them also varies slightly from primer to primer. Also, they vary slightly in depth from front to back and some anvils' legs protrude more from the mouth of the primer cup. This is why you want to feel the primer take its seat in the pocket. Automatic seaters and tools that seat primers at the highest mechanical advantage of their linkage give you little or none of this feel.

The otherwise very sensitive priming compound has to be virtually impervious to a strong, steady *push* to be seatable. Yet after it's seated, it must be even more sensitive to a small, sharp *impact* than it was when the primer was loose in its little box. It has to survive until the tip of the firing pin crushes it, then it must go off readily and with as much impatient fury as any of its classmates.

The modern handgun primer, this parvus paragon of pyrotechnic potency, didn't spring suddenly into being — nor did it evolve easily or randomly. The need for it was obvious centuries before anyone thought of this ultimate replacement for the burning stick or live ember applied by hand to a hole in the butt of the Fourteenth-Century hand cannon's barrel. The later punk string of the matchlock gave way to sparks generated mechanically.

Fairly late in the history of firearms, a relatively complex mechanism struck flint against steel in a basically primitive system before the first primers came along in the form of percussion caps stuck over the tips of nipples. Even after the primer moved indoors — inside the case — there was plenty of room left for improvement. Chemical and

Firing pins don't always hit primers dead center. Look close at the mark on the primer of the .380 case on the left and the badly aligned .45 case on the right. Primer anvils have to be designed and manufactured to allow for variances in the handguns in which they'll be fired.

In manufacture, the foiling disc (thin paper) lies flat on the pellet of priming compound until the anvil is inserted in the cup. Then the disc buckles somewhat around the apex of the anvil as it's pressed slightly into the pellet. The remarkable photograph was taken on a 1/100th grid.

physical modifications in primers have continued as further improvements are being planned on drawing boards and computers.

Shortly after the turn of the Nineteenth Century, a Scottish minister, the Reverend Alexander John Forsyth (1768-1843), developed the first percussion lock and priming compounds. He experimented with a number of metallic fulminates and other compounds before adopting fulminate of mercury and potassium chlorate for his patent of 1807.

Design attention appears to have been focused more on the containers and mechanisms for Forsyth's igniter for some time, with less attention to its basic chemistry. The first primers were percussion caps combined into the heads of early metallic cartridges like the Martin and the Morse. In some early centerfire cases, as in today's rimfires, the priming compound was inserted from the mouth and wasn't easily replaceable.

Even after the American Hiram Berdan and the British Colonel Edward Boxer invented the replaceable primers of today, the primary initiator in their priming compounds was fulminate of mercury. Surplus foreign military ammunition primed with mercuric or corrosive compounds is still being sold — even though nonmercuric, noncorrosive priming isn't exactly last week's news. As initiators go (ignoring side effects), Brother Forsyth's fulminate of mercury is probably still the best.

Incidentally, notice the irony: An American invented the primer favored today in Europe and a Briton invented the one favored in this country. Both men concentrated on mechanics — the means of containing, inserting and firing primers — not on primer chemistry. Others evolved nonmercuric, noncorrosive priming compounds.

The American handloader is the reason our ammo uses Boxer primers. Except for ease of reloading, the Berdan primer has some advantages over the Boxer, I'm told. Only one American maker use Berdan primers in ammunition for domestic use: Speer's unreloadable Blazer ammo uses nonstandard Berdans.

Mercuric priming, consisting mainly of fulminate of mercury and potassium chlorate, did nasty things to cases. Handloaders noticed that their brass got brittle overnight after only one firing. Today, we think "brittle" when a case neck splits too easily. When mercuric primers and brass cases tried to work together, this word had a much meaner meaning for handloaders.

Mercury especially weakened the case's harder brass in the head, around the primer pocket. Some cases fired with mercuric primers became so brittle and weak that the second firing blew out the entire web — the floor of the

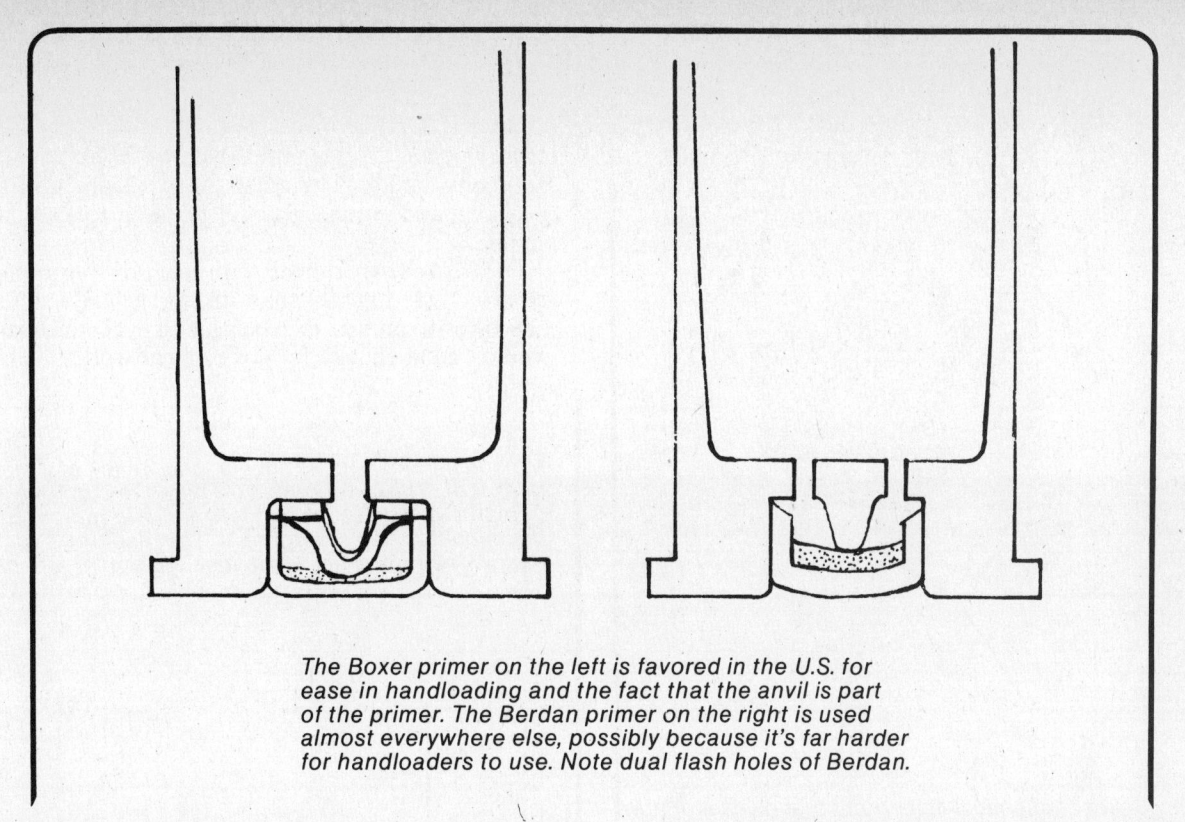

The Boxer primer on the left is favored in the U.S. for ease in handloading and the fact that the anvil is part of the primer. The Berdan primer on the right is used almost everywhere else, possibly because it's far harder for handloaders to use. Note dual flash holes of Berdan.

primer pocket — to leave only the thin, soft primer cup to withstand chamber pressure. That was a tad worse than losing a case when its neck split.

In a high-pressure cartridge, the mercury weakened the brass worse than in a low-pressure cartridge. The first firing drove mercury deeper among the crystals of the brass, weakening their intercrystalline bond. Then the second firing put a lot more pressure on this weakened brass. Deterioration began the day of the first firing and was noticeable the next day. As time went by, the brass got still weaker and more brittle.

American ammo makers used to print warnings on their boxes and in special brochures that their cases couldn't be reloaded because "all smokeless powders are injurious to brass shells."

An old Winchester brochure said, "Experiment shows that after the first firing with smokeless powder, the metal of the shell undergoes a slow but decided change, the exact nature of which the best experts have as yet failed to determine...If fired shells are allowed to set two or three days... the metal becomes brittle and rupture of the shells at the next discharge is probable..." This much was true, but the wrong suspect took the rap for it in those days.

The brochure went on to warn: "Experiments show that these problems are charcteristic of all smokeles powders and are in no way due to the material used in the shells, the process of manufacture or the kind of gun used...." Nobody seems to have suspected the primer used each time anyone fired a charge of smokeless powder, so everybody blamed the powder.

Army research ferreted out the cause in 1879 and the Army immediately turned to nonmercuric primers. One dependable mixture of that era was Frankford Arsenal's composition of potassium chlorate, antimony sulfide and sulfur. In time, both the potassium chlorate and the sulfur were found to cause serious barrel corrosion. "Nonmercuric" doesn't mean "noncorrosive."

In 1900, Germany began trying to develop a priming

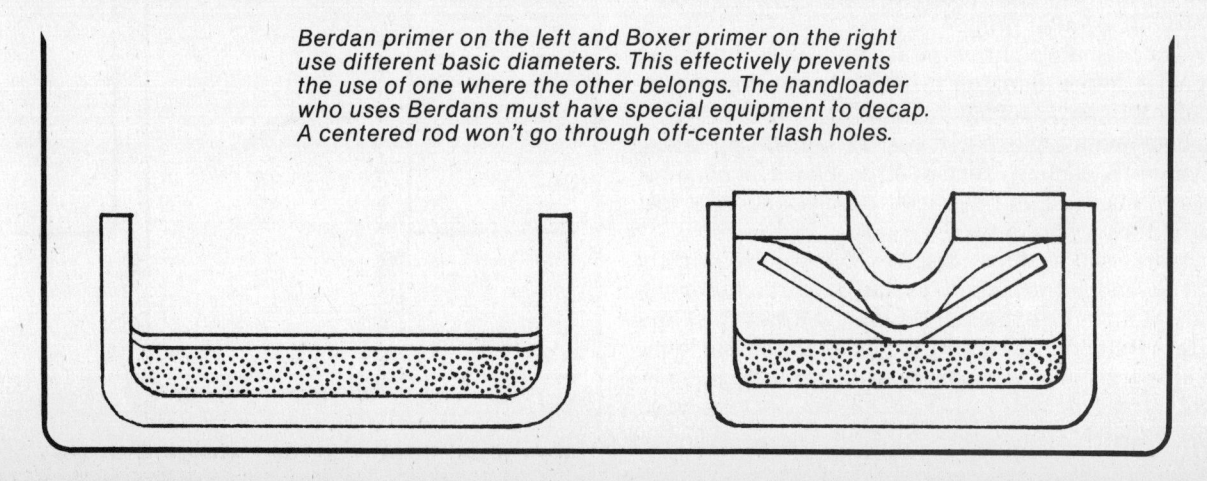

Berdan primer on the left and Boxer primer on the right use different basic diameters. This effectively prevents the use of one where the other belongs. The handloader who uses Berdans must have special equipment to decap. A centered rod won't go through off-center flash holes.

GUN DIGEST BOOK OF HANDGUN RELOADING 131

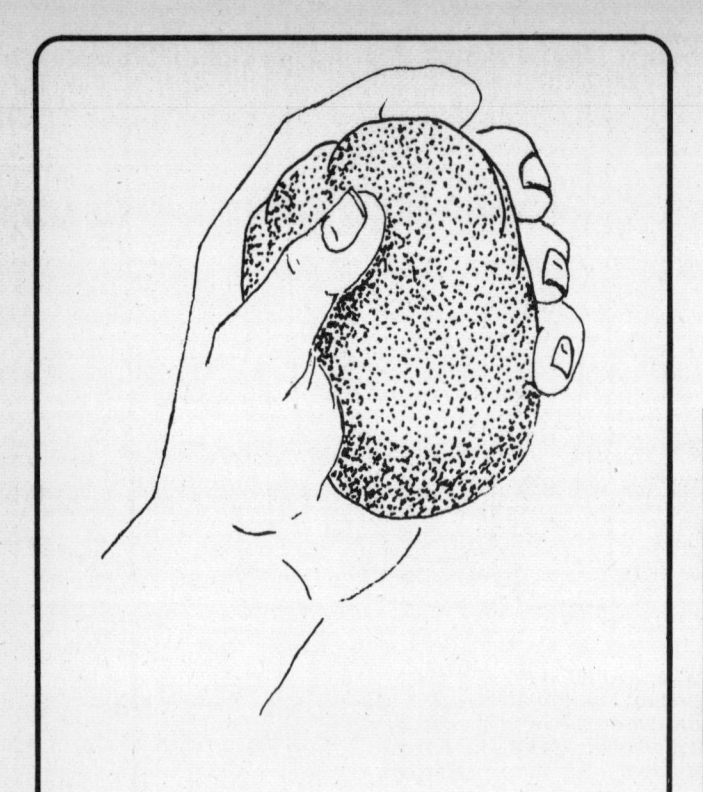

The tiny pellet of chemical compound that gives a primer its short life of fire and force starts as a moist clayey glob. The first step in forming the glob into pellets is to push it down into a finely-machined plate with holes of just the right diameter, thus forming uniform pellets.

compound that wouldn't cause rust in barrels. They made their first no-rust primers the next year, still using fulminate of mercury with a mixture of barium nitrate, antimony sulfide, picric acid and ground glass. In 1917, a rash of misfires and the press of war production led the Army to replace the old Frankford Arsenal priming mixture with Winchester's composition of potassium chlorate, antimony sulfide, lead sulfocyanide and TNT.

Handloaders knew nothing of these and later developments until 1922, when Dr. Wilbert J. Huff published his report of the Bureau of Mines research study that he had conducted — at the request of the War Department — to track down the chemical origin of ammunition-caused corrosion in firearms. He had found that the potassium chloride in priming compounds caused rust in barrels even though they were oiled after firing.

Dr. Huff's findings stirred the army and the ammunition companies into a flurry of research to develop a noncorrosive primer. Remington came up with the first — Kleanbore priming, introduced in 1927. Others came along right behind Kleanbore, but they all contained fulminate of mercury. They weren't sensitive enough and they lost strength after a year or two.

Hangfires and misfires caused by deteriorated primers stoked up another search — to find a substitute for the offender, fulminate of mercury. Edmund Ritter von Herz and Hans Rathburg developed, patented and sold to Remington two priming compounds with no mercuric fulminate.

Both compounds included guanyl nitroaminoguanyltetracene, lead trinitroresorcinate, barium nitrate and antimony sulfide. One of their priming compounds sometimes used calcium silicide instead of, or in addition to, the antimony sulfide. The other used calcium silicide, lead peroxide and ground glass in addition to the antimony sulfide.

The Army's experiments with noncorrosive primer mixtures led only to problems with storage and ignition. The noncorrosive primers then being used in commercial ammo weren't good enough for government work. The Army's

Making a primer cup is simple: one whack of a machine does it all. The outer punch (A) cuts the cup from sheet of material, then stops and lets an inner punch (B) drive the cup into a die that forms it. The resulting cup has thinner, springy walls which lock the primer into place.

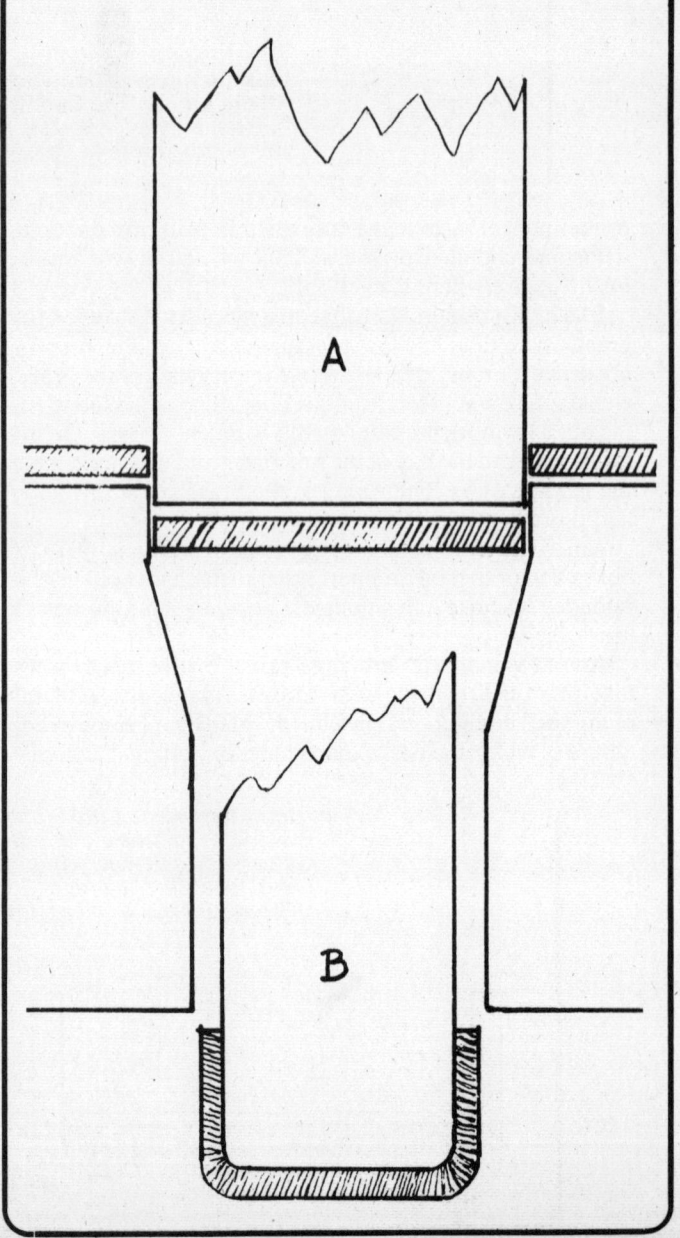

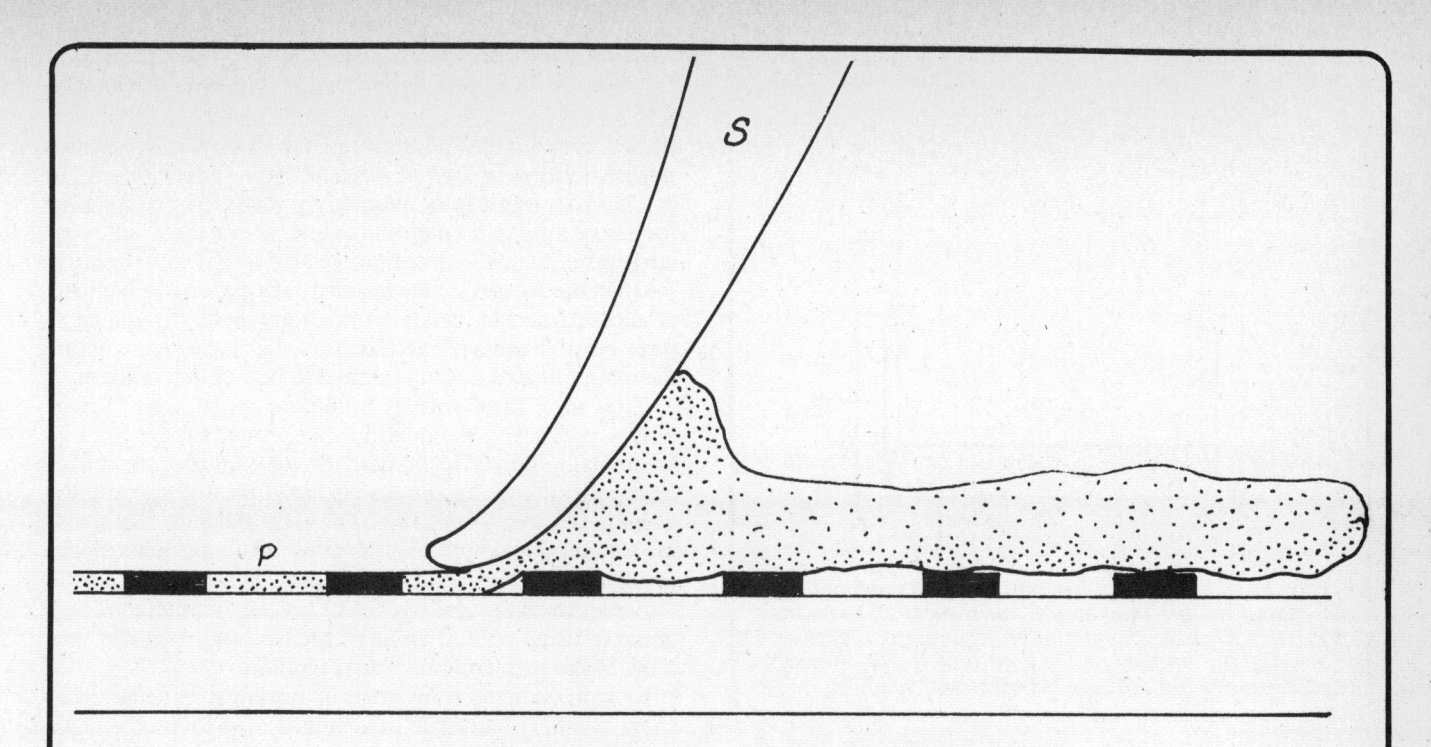

Precisely-dimensioned holes in a steel plate of controlled thickness determine how much compound goes into each primer. A squeegee is used to force the mixture into the holes — a dead-flat backing plate behind. It's a simple but precise operation that aligns uniform amounts of...

stiff specs for storage and reliability kept them out of military ammo. Not until the late Forties and early Fifties did noncorrosives finally replace corrosives in American military ammunition.

In modern centerfire boxer primers, still another compound, lead styphnate, is the basic component of the pellet's chemistry. After dropping its early research into noncorrosives, the Army later considered this new generation of igniters. Foreign patents had established much of the basic chemistry, but there were problems.

Making lead styphnate was relatively simple; a vital earlier step — making styphnic acid — was more difficult. Around 1950, Speer hired Dr. Victor Jasaitis, a European chemist and expert in the production of styphnic acid, as their top primer scientist. Though Speer's primers weren't the first to use lead styphnate, the work of Dr. Jasaitis and the addition of an entirely new manufacturer of primers were two dramatic waves in the sea of American primer research. Compounds based on lead styphnate replaced the old primer formulas.

This great boon didn't come without cost to handloaders, however. A classic powder that many of us still mourn, Hercules Hi-Vel No. 2, became a casualty of the new primer chemistry. A great powder with older primers, it is

...the compound into rows. A plate of holes filled with priming mixture is then aligned with another plate with matching holes filled with empty cups. When everything is properly aligned and backing plates are in place, tiny individual rams push the compound into the primer cups.

GUN DIGEST BOOK OF HANDGUN RELOADING

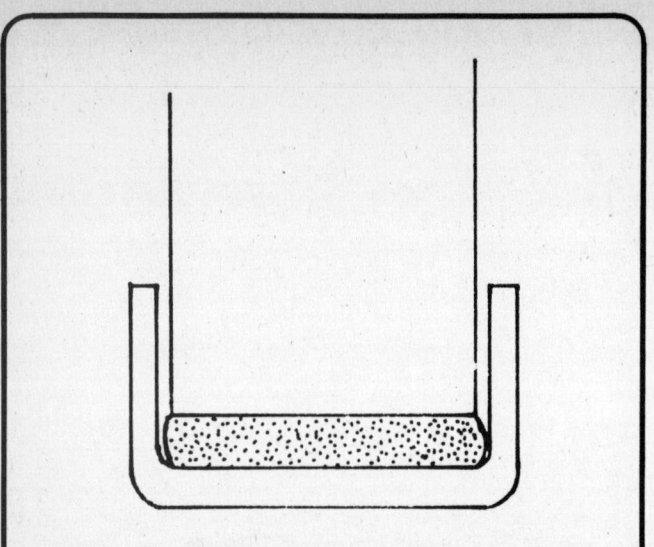

In the cup, the primer pellet is not quite a perfect fit. It's actually a little thicker and narrower than the base of the cup's inside. In yet another operation, the pellet is pressed out and formed to a perfect fit with the cup's inner surface. The process is performed by still...

chemically incompatible with the new ones. It's for other reasons, however, that primer chemists are working on compounds to replace lead styphnate.

Another great chemical advance of this era is the testing and acceptance control that primer makers use today. In olden days, they took whatever grades and purities of their basic chemicals their suppliers sent to them. Now, with better ways to produce those chemicals at the beginning and better ways for the buyers of these chemicals to test them when they arrive, better grades of chemicals are going into priming compounds. Sometimes, little differences mean a lot.

Nothing in or about a primer is simple or insignificant. Oh, sure: It's the smallest component of a load, an intermediary that translates the mechanical impact of the firing pin into the flame that lights off the real energy component — the powder. But as small and as secondary as it is, it must be as consistent, safe and dependable as the gun and every other component of the load.

A good primer — even a superb primer — merely does its job quickly and well and that's that. The burden of ammunition performance rests on the other load components and how well they're put together. But a bad primer turns the excellence of all the other load components into rubbish.

Making something so small, in such large numbers, to be both consistently excellent and reasonably affordable, is to my mind a major feat in the manufacture of ammunition. Making this "simple" device combines chemistry, mechanics, intricate assembly, meticulous inspection and careful packaging and handling — just to make *acceptable* primers.

The three main parts of the primer are the cup, the anvil inside it and the pellet of priming compound sandwiched between them. The only other part is a disc of shellacked paper between the pellet and the anvil. About fifteen to twenty steps make these parts and put them together. The three main phases are: (a) making the cups and anvils, (b) mixing the priming compound and forming it into tiny pellets, and (c) putting 'em all together. Each of these phases comprises several steps.

The strips of brass for making cups have to be soft, those for anvils hard. Cup material is annealed beforehand and anvil material cold-worked beforehand to give them nearly the right hardnesses for the finished cups and anvils. Neither is annealed later and each is worked only slightly in manufacture. The thickness of each strip also has to be just right for the dimensions of the cup or anvil to be within tolerances.

Cups have to be soft for minimum resistance to firing pins — yet not soft enough for hot propellant gas to perforate them. If they're too hard, they retard ignition, make load performance erratic, cause hangfires and sometimes even make cartridges misfire. The cups of handgun primers are thinner and softer than the cups of rifle primers, since handgun firing pins and chamber pressures have less clout.

Anvils must be hard but not brittle, for maximum resistance to firing pins. If an anvil breaks or gives under the blow of the firing pin, it retards ignition.

In contrast to the many steps in forming a cartridge case (next chapter), only one mechanical step forms the cup, another the anvil. Presses punch cup discs from strips of brass or gilding metal in a process called blanking and cupping. Unlike cases, which undergo several drawings and a few trimmings, primer brass becomes finished cups in one combined punching and drawing and no trimming afterward. This operation doesn't work the base of the cup but does draw its skirt a shade thinner. After a bath to clean off

...another punch. It forces a disc of foiling paper, which it has just cut from a sheet, down onto the loose pellet of priming compound. The paper keeps bits of priming compound from sticking to the face of the punch when it's withdrawn. Paper stays in place for primer's life.

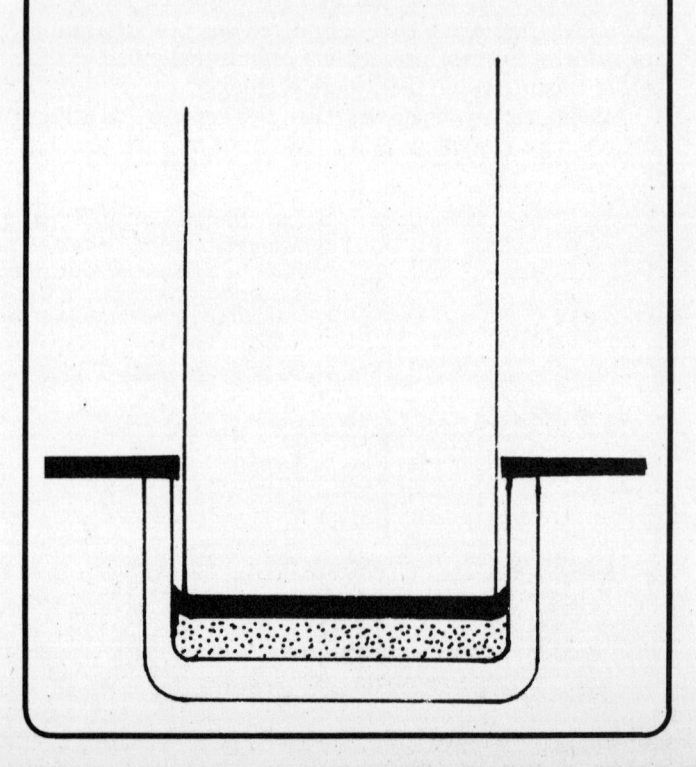

machine lubricant, the cups are ready for their priming compound and anvils.

A separate but similar operation forms the anvils. After they're punched from a strip of hard brass and stamped to shape, they also get a cleansing bath. The stamping of the anvil from a flat strip into a clover or butterfly also forms the apex to just the right shape — not sharp, so it won't poke into the priming pellet, but broad enough to tolerate off-center impact from the firing pin. This added working of the metal also further hardens it, with the apex harder than the legs — just short of brittle.

Mixing the ingredients that give the primer its chemical fire and fury is a different kind of operation. If the color of the dough didn't give it away, you'd think the worker was kneading bread or biscuits. The primer mixture that I watched being made was green, like colored modeling clay.

The workers making primer pellets put wads of their dough onto thin, smooth steel charging plates perforated with row upon row of precisely ranked and bored holes. Each plate forms several hundred primer pellets. The thickness of these charging plates makes the pellets the right thickness and their holes make them the right diameter. The operators spread out the dough, bearing down to pack mixture into every hole, then squeegee the excess off — leaving neat little discs of mixture in the holes of the charging plate.

The charging plate fits precisely over an almost identical steel foiling plate — a thicker plate filled with primer cups waiting mouth-upward in holes that match those in the

The stamping of the anvil is as simple as the blanking and cupping of the primer cup. A shaped punch cuts a flat sheet of metal. Then another special punch forces this tiny piece of metal into a die that forms the point of the anvil as well as bringing the flared legs down.

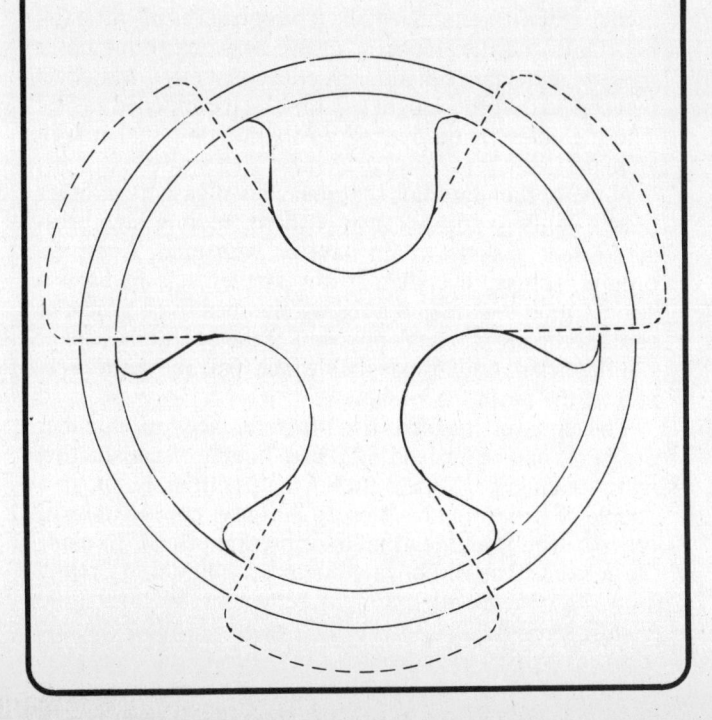

The very last step is a process by which the anvil gets pushed into the cup. The rounded "point" of the anvil presses gently into the pellet and the flat disc of paper buckles upward, away from the pellet, around the anvil. This leaves the anvil's legs ready to be seated into case.

charging plate. A third plate, also made to match these two, but with pins to match their holes, swings over to press the priming mixture out of the charging plate into the cups waiting in the foiling plate.

Another operation tamps the mixture in the cups. A machine punches discs from a strip of waterproof paper, pushes each disc down onto the mixture and packs the mixture into a hard cake in the bottom of the cup — with the disc of paper sealed over it. Next, these charged primers go into another perforated plate — into holes that align them with anvils. A press pushes the anvils into the cups and the primers are nearly ready for use.

I've skipped a few operations — the plating of the cups and the drying of the charged cups, for example — to simplify this description of how primers come to be primers. Also some of these processes are slightly different from plant to plant. At several points, of course, each part of the primer passes under the eyes of incredibly alert inspectors. I'd hate to be the mischievous son of one of these hawk-eyed ladies! They can spot flaws, at arm's length, that I can't see close up, even when they point them out to me.

There's a lot of powder in, and a lot of work behind, those wee boxes of shiny little goodies that we all take for granted, but can't get along without. — *Ken Howell*

CHAPTER 11

Brass: What It Is And Does — How It's Made And How We Handle...

...HANDLOADER'S GOLD

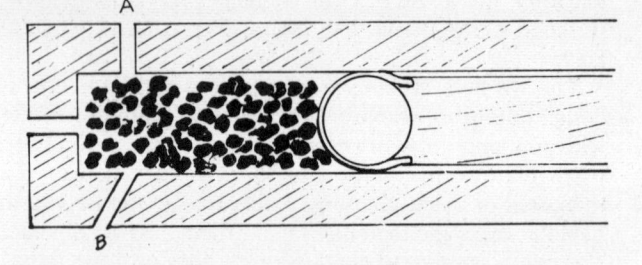

Early firearms needed a hole to introduce the powder-igniting flame into the breech from side, rear or at an angle. One of the advantages of modern brass is that it serves as a simple jig, housing all parts of the charge.

This sectioned round of .45 ACP brass was drawn longer by an experimenting handloader, but still serves to make a point — notice how case walls thicken near the base.

BACK WHEN the load had to go into the gun from the front, the spark that lit it off had to get in from the rear or the side. Except for that hole, the rear of the barrel or cylinder was solid. The breech, the gun's *locus sigilli* — its "place of the seal" — has long been also the locus of the gun's greatest weakness, just because that hole has to be back there to let in the spark that fires the gun.

Sealing the breech tight enough to contain the extremely high pressures of the propelling gas has always been a key problem in basic firearm design — even with the much lower breech pressures of muzzle-loading pistols and cap-and-ball revolvers. There was for these old guns no perfect way to let the spark come in from outside and not let gas get out from inside.

An easily opened breech — vital to fast reloading — wasn't possible without a gasket that could both seal the breech when it was shut and let it be opened easily after firing. Until the development of the brass cartridge case, these were demands that no material could meet. Materials that sealed the breech tight wouldn't let it be opened easily. Those that let it be opened easily couldn't seal it well enough.

None of the materials that make good gaskets in other places would work, because of the incredibly high pressures such gaskets would have to withstand. Even the ordinary pressure cooker in the kitchen has to have a strong, solid lockup and a tightly clamped thick gasket to contain the normal peak pressure of the operating cooker — fifteen or so pounds per square inch (psi) above the pressure of the air in the kitchen.

The units of measure used in references to chamber pressures can be misleading. There's some dissatisfaction with "pounds per square inch," the traditional unit, so a couple of new units have come into use: copper units of pressure (cup) and lead units of pressure (lup). In using these terms, the big boys confess that they don't really know how their pressure measurements relate to the old traditional units like pounds and square inches. Note the units of measure used in references to chamber and barrel pressures, but don't worry about how they relate to the weights of elephants or the sizes of postage stamps. Note them just so you won't find yourself trying to compare 30,000 cup with 30,000 psi.

By long and sometimes bitter experience, the general body of firearms experimenters have fairly well located the hedge that separates risk from safety. They could all just as well agree to list these pressures as so many megapuppies per square gnat hair — their comparisons and meanings would signify exactly the same two basic facts: Too much is too much, and enough less is as safe as a baby's crib.

The breech systems of today's handguns have to withstand internal pressures that can be as high as those of modern high-intensity rifle cartridges — some recent handgun types are chambered for rifle cartridges. Traditional handgun breeches like the automatic and the revolver must withstand somewhat lower peak pressures, but even loads in old handgun cartridges like the 9mm Parabellum reach peaks between 30,000 and 40,000 psi. Some loads for the .357, .41, and .44 magnums crank up 40,000 psi. Very-high-quality brass is the gasket that seals their breeches to hold these terrific pressures.

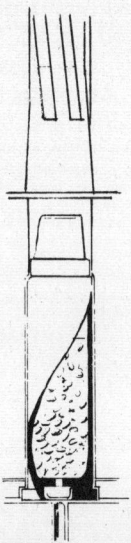

Perhaps least appreciated role of a cartridge case is the way in which it holds the primer in alignment with the firing pin and the bullet in line with the bore. Here, a sectioned revolver case, chamber and bore make it clear.

In effect and function, then, the brass cartridge case is a quickly and easily replaceable part of the breech — a part of the gun. Since the gun can't operate without it, the case is an even more vital part of an auto than the magazine. Loading and shooting an auto without a magazine is awkward, but generally possible (yes, even with most autos that have magazine safeties). No case means no shooting, however many magazines may sag your pockets.

Secondarily, the case happens to be a handy little capsule that holds all the components of the load, which it protects while it's apart from the gun and aligns properly with firing pin and bore when it's in the breech before and during firing. It is not just one thing but three: a gasket, a jig and a capsule — in that order of importance.

As a capsule, it's just a part of the load. If this were all it had to be, it wouldn't have to be precisely like any of the other rounds in the same batch of ammo. Its shape is unimportant to its role as a protective container for the primer and powder. In this role, its partner is the bullet — a stopper. A plain rubber plug or tight plastic cap might be a better stopper. Neither the case nor the bullet needs any special care for them to function well as a capsule. But of course the case isn't just a capsule.

Far more crucial are its functions as a breech gasket or seal and an internal breech jig. In both these roles, the brass case is as much a part of the gun as its most precisely fitted and vital steel parts. Each section and dimension of the case is as carefully designed and made for its function in the operation of the gun as if it were a single internal steel part in the works of the gun itself.

The exact material from which it's made, its dimensions, even its several hardnesses are as important — and as carefully ensured in manufacture — as the corresponding characteristics of the barrel, frame, firing pin, trigger or sear.

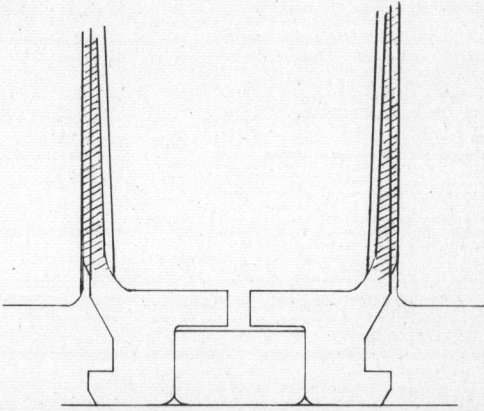

The faintly shaded areas toward the rear of a sectioned case illustrate the amount that the case will temporarily expand upon firing, thereby sealing the breech tightly.

To be worth a shuck as a gasket, the case must be in exactly the right place when the hammer comes down. And it has to drop into that exactly right position immediately when you shove it into the chamber. It can't require the time, attention, special tools or dexterity that you'd use to seat a faucet washer in the bathroom or a neoprene seal in an automatic transmission. The shapes of most bullets help the loaded cartridge slip into the chamber easier. But most crucial are the outside diameters of the case and the dimensions of its headspace system. Its outside diameters must be small enough to let it slip easily into firing position, yet large enough to leave almost no space between the case and the chamber walls. It must be able to slip in easily, yet come very close to sealing the breech even before it's fired.

The head of the case and the face of its primer must be in position against the face of the breech. The mating diameters of case and breech align the primer on the axis of the firing pin. The critical dimensions of the case's headspace system place and hold the primer against the face of the breech — and make sure that the case firmly resists the blow of the firing pin.

On the .38, .357, .41, .44s and the .45 Colt, for example, a flange or rim of about the right thickness rests against the rear of the chamber wall. On the .38 Auto and .38 Super, a skimpy flange called a *semirim* is supposed to rest against a ledge at the rear of a spur jutting from the top wall of the chamber — but it doesn't, so these fine .38s take a bum rap as inaccurate cartridges. When they're headspaced firmly and consistently, they're great.

On the 9mm and .45 ACP, the mouth of the case rests against a tiny annular ledge at the forward end of the chamber. On bottlenecked handgun cartridges inherited from rifles or inspired by rifle cartridges, the male cone slope of the shoulder rests on a matching female cone slope in the forward area of the chamber.

The dimensions and tolerances of the case determine how reliably it functions as gasket and jig. If it fits the chamber fairly snugly (yet is still moderately loose to allow easy insertion into the chamber), it's a good gasket and a fine jig. There's room for a few very small variations in each of its dimensions. The function of the case tolerates these variations, which is why we call 'em *tolerances*. But they have to be small — just a few thousandths or ten-thousandths of an inch.

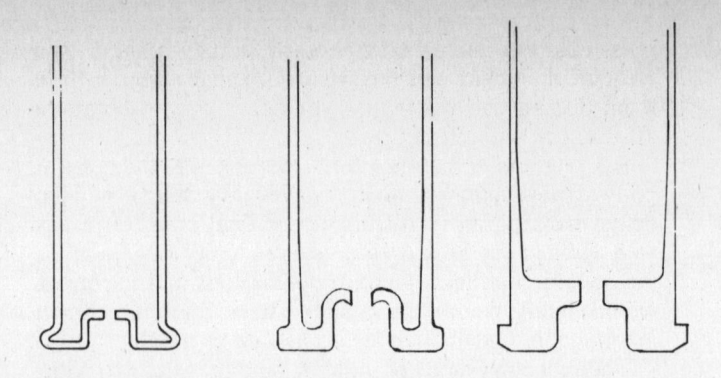

From the left: Folded, balloon and solid head cases, all sectioned for easy view. Never load the first and avoid using the second. Use modern data for the modern solids.

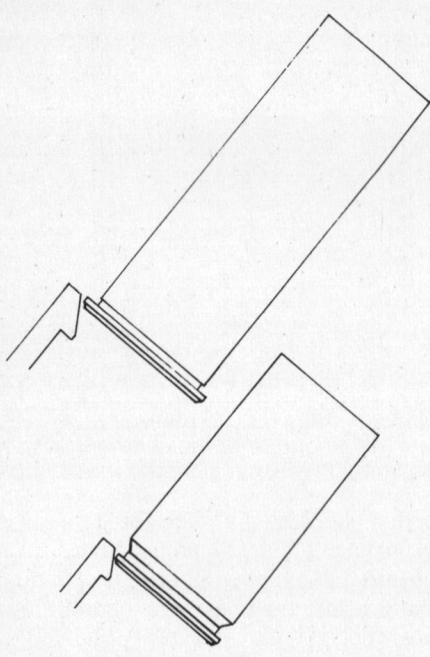

designers recognize how weak it was. A long time ago, they replaced it with the much stronger balloon-head case. But even that design and its still-too-low pressure ceiling handicapped loads for the highest-intensity cartridges, so designers made its thin head sections thicker. The result is today's robust solid-head case.

I doubt that you're likely to happen onto any folded-head cases that aren't too old and brittle to load. I also doubt that you'll find any new cases with anything except solid heads. All the new American cases that I've seen in the last twenty years or so have been solid-heads. But you may run across some old balloon-head cases and you're even more likely to find some old handload data for loads that were developed in balloon-head cases. So here come a few words of warning.

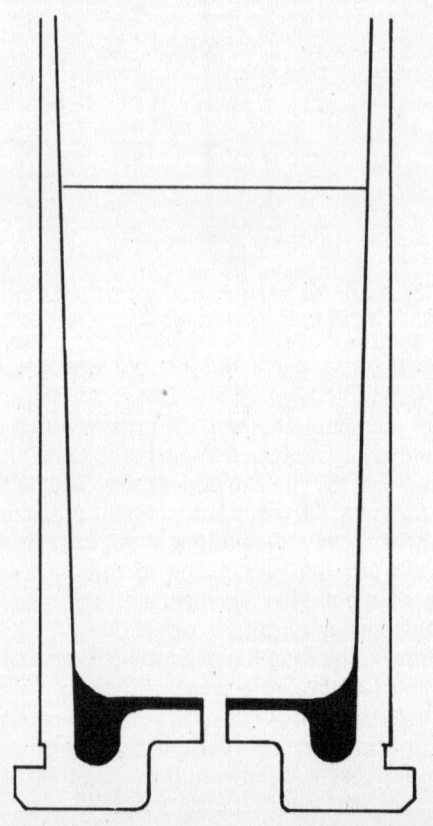

Brass has to come out of the chamber after the extreme pressures of firing. Two different kinds of rims are seen here — .44 magnum and .45 ACP — and their extractors.

Making a few cases with narrow tolerances to fit all six chambers of a single revolver is not nearly so simple as it might seem. Making billions of cases to match the chambers of millions of guns made by a hundred factories in a dozen countries is sure to be enough to give ammo executives galloping nightmares in stereo and Technicolor.

The case didn't spring into existence already perfected in its modern form. Its recent evolution shows growing awareness of its several functions and the demands that it must satisfy. The several headspacing systems are tracks left by creative explorers in the evolution of design. Sometimes more critical to a handloader are the three steps in internal head design — the folded, balloon and solid case heads.

Before smokeless powder and its dramatic increases in peak chamber pressures, the old folded-head case was good enough as a breech seal. But smokeless made case

Construction details of modern solid cases and the older balloon heads. The shaded area of the case is taken up with brass in the modern case, thereby reducing capacity.

Cases with folded heads are capacious, but they're too weak for smokeless loads. Any that are still around are suitable only for cartridge collections. However good they may look, they're too old to be any good for handloading anyway. If they've been fired, old primers have undoubtedly made them as brittle and weak as cornflakes. Don't even think of loading them.

Cases with balloon heads are roomy and strong enough for good moderate loads if they're not too old and haven't been fired with mercuric primers (but how will you know what kind of primers set 'em off forty years ago?). A good many should still be around, but the age of even the newest ones makes them only marginally acceptable for handloads — if they're any good at all.

The real problem with balloon-head cases is that the Elmer Keith .44 Special loads that eventually led to the .44 magnum were loads for balloon-head cases — and those loads are still in print. Elmer's classic .44 Special load — 18.5 grains of 2400 with a 250-grain cast bullet — is much too hot for the later solid-head .44 Special cases.

Although the heads of the newer cases are stronger, the powder capacity of the solid-head cases is too much smaller. So a maximum charge for the roomier balloon-head case fires up just too much pressure in solid-head cases — somewhat like .357 or .44 magnum charges in .38 or .44 Special brass. Remember, Elmer's 18.5 grains of 2400 was a *maximum* load in the roomier case. (UPS has just interrupted the writing of this opus to deliver a copy of the *RCBS Cast Bullet Manual number 1,* in which the 2400 loads for the Keith bullet in the .44 Special are 10.9 and 11.9 grains. That sounds about right. Either load is far shy of 18.5 grains; the differences lie in the design of the case head and the related reduction in net powder capacity.)

Don't think of the brass case as nothing more than a semiexpendable container for powder. Think of it as the removable, reusable, replaceable part of the gun that contains and aligns the load. Then you'll better discern its function and you'll see why so much care and handling are necessary in making and loading it.

Come to think of it, calling it *brass* borders on slander. Some copper-zinc alloys that do well enough as brass in mundane practice aren't good enough to be cartridge cases. The alloy used in cartridge brass (70% copper, 30% zinc) belongs to one of the higher castes in the world of brasses. A similar alloy used in making jewelry is called *casters' gold* or *jewelers' gold* (though gold it ain't). Considering what goes into cases — material, manufacture and prices — "handloaders' gold" might be a better name for 'em.

Ay, there's the rub — a problem worth Hamlet's worry. Cases have to be both plentiful and precisely made, both economical and consistently dimensioned. For no other part of the gun must be so readily replaceable, even expendable. Most of us try to keep a passel of them on hand all the time. I like to have a minimum of 100 matched cases for any rifle cartridge that I load and at least 500 for any handgun cartridge.

Nobody keeps anywhere near that many magazines on hand, for example, though he may own several autos. I have boxes of spare parts for my Colt GMs — but no more than four or five of any one gun part (except grips and magazines). Nobody needs so many and nobody I know could afford so many hammers, sears, triggers, even pins or springs. But we have to have gobs of brass cases — to support long shooting sessions between loadings and to allow for normal losses. Cases have to be precise, uniform and yet affordable if not economical in large quantities.

Now you see the problem. How can anyone use a material as costly as 70-30 brass, machine it to be as precise as cartridge cases have to be, consistently by the millions or billions, then advertise, distribute and sell them at prices you and I can afford to pay for the quantities we need? It's easy to see why the case is the most expensive component of the cartridge. It's interesting to see how the ammo companies turn strips or rods of 70-30 brass into "hulls" for you and me to load and reload.

Ammunition makers' methods vary a little but are generally alike. The traditional and usual method is to start with a wide, long strip of thick, annealed brass — though one or two makers use round bars instead. In the first step with the strip, a machine punches several discs out of the strip at a time. This punching operation at the same time stuffs the discs into dies to form wide, shallow, thick-wall, round-bottom brass cups. This operation is called blanking and cupping.

The strip was already annealed — softened slightly — to make it workable. Blanking and cupping harden the cups, so they go into the annealing furnace to be softened again for further working. If they come out too hard for the next operation, they can break punches in the next machine. Or the punches may ruin them. If they come out of the first annealing too soft, the next step doesn't harden them enough to come out of the several later annealings with the proper hardness at each step. A pickling bath — a solution of sulfuric acid — removes dirt and annealing scale, then a water bath cleans off the acid.

Other machines draw these cups — in a specialized punch-and-die operation — to make them deeper or longer. (Makers who start with a rod instead of a strip cut off solid cylinders, then draw them into cups.) The bottom of the cup stays thick, but the walls get thinner as each draw makes them longer. The first draw reduces the diameters of the cups and hardens them, so they get a second annealing, another acid bath and another water rinse, then another draw.

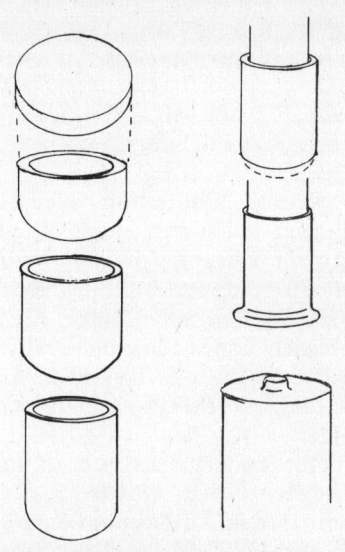

These are steps in the process that results in a modern cartridge case. From top left, punching a disc from brass sheet, then drawing it into a cup and swaging on a rim.

GUN DIGEST BOOK OF HANDGUN RELOADING

In the second draw, the die and punches are smaller in diameter, but the operation is essentially like the first draw: The cups get smaller in diameter and longer, with still thinner walls. To keep the thick head ends from thinning, another operation flattens and thickens the round bottom of the cup. An anvil punch inside the cup supports it while a hammer comes down hard on the end. This makes the outside of the head flat, the inside flat in the center with a surrounding fillet. This operation, called bumping, preforms the head area to give it enough material for final machining later. Bumping must occur after one of the earlier draws, before later draws can make the head end too thin for final machining.

After the second draw, the third annealing, pickling and rinse prepare the case cups for their third draw, which takes their diameters and wall thicknesses down a bit more. Then they're washed and dried to be ready for their first trimming.

After three draws and a head bumping, the cup has become a blind tube with walls thick at the base and tapering to very thin at the ragged uneven mouth. It's also visibly longer than the finished length of the case. Wall thickness at the mouth is a vital dimension: It has to be within dimension tolerances and very nearly the same all around.

Drawing keeps the wall thicknesses pretty even along most of the length of the now-long cup. Closer to the mouth, the wall gets uneven and ragged, with irregular tags called pig ears. The cup has to be fairly clean-lined to go obediently through the production machinery, so the pigs ears have to go. The first trimming cuts some of the excess length off the open end of the cup to make sure that later operations go smoothly.

After a fourth draw, the cups are again annealed, pickled, rinsed, and washed before getting trimmed again to dispose of the irregularities produced by the fourth draw. The cups now begin to look almost like cases. They're about the same hardness — which is soft at this point — from stem to stern. The heads have to be significantly harder, so they need a little special cold-working of their own.

A rigidly fixed internal anvil punch supports the cup against the impact of a pocketing ram. A small punch, integral to the face of the pocketing ram, makes a large, deep dimple in the center of the cup base. Pocketing forms only an approximation of what the finished primer pocket will be; later operations will form the pocket to its primer-seating dimensions.

After the pocketing, an operation called heading does the same thing again but adds a couple of touches to the further shaping of the pocket and the work-hardening of the head. The pocket-forming punch, a bit larger than the one used in the pocketing operation, makes the raw primer pocket a little bigger. The heading ram really smacks the end of the cup, making the brass flow under the impact — with two results that make the cup look almost ready to slip into a shell holder.

The impact of the header makes the brass flare out in the same way that a hatchet head flares the top of a tent peg — but with a smoother edge. The flare or flange becomes the raw material for the rim of a rimmed or semirimmed case; it'll all be trimmed away for a rimless case. On the face of the case head, the impact imprints the head stamp — designations for the maker and the cartridge on commercial cases, maker letters and year numbers for military hulls.

Forward, the case must be softer — much softer at the mouth — but the head must be hard. The face of the head must be the hardest part of the case. The rim, whether it's for headspacing or only for extraction, has to be hard and tough enough to withstand the pull of a strong extractor, even though high pressures and rough surfaces may lock the case tight in the chamber. Heading impact pounds the head brass hard enough to make it a little denser and harder.

Next, a venting punch pokes a flash hole through the web at the bottom of the primer pocket. The venting punch is the smallest and therefore most delicate punch used so far, so the hardness and thickness of the web brass are crucial. This tiny punch wears, bends and eventually breaks under the harsh and variable conditions of its job. From its beginning to its end, it makes correspondingly variable flash holes.

Diameters vary a little. More noticeable by far are the punch's effects on the web. I've never seen a crumbly web, but I imagine that the inspectors find some where hard brass has broken away under the venting punch, leaving a ragged flash hole. But I have seen a wide variety of burrs on the far side of the web, and some bulging of the web inward. What the inspectors let go through is pretty uniform.

Next, a pocket sizer swages the primer pocket out to its final dimensions. This operation works the head only a little, with no discernible change in head dimensions and only a bit more work hardening of the brass right around the primer pocket. Head turning, the final shaping of this end of the case, comes next.

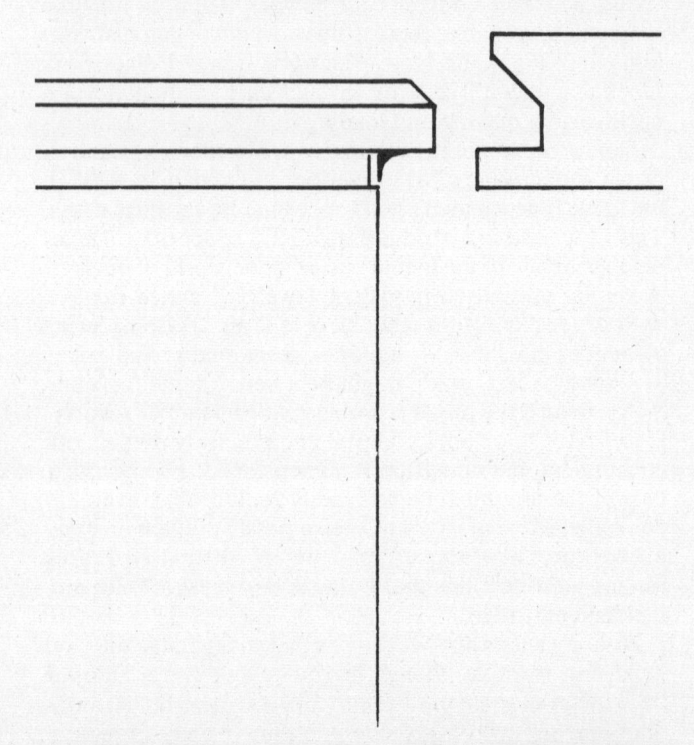

This drawing shows the manner in which a cutter is used in the final step of finishing the case. The small black area has to go to insure a right angle for headspacing.

Head dimensions and tolerances are critical, so head turning has to be precise. It must also leave smooth surfaces that need no further finishing. In only one pass, a specially shaped form cutter shaves the rim to the proper diameters and thickness, cuts the annular extractor groove of rimless cases to the right width and depth, and neatly bevels both rim and groove.

Machining leaves tiny brass chips in and on the case, so another bath cleans them off. Then another annealing softens especially the forward portion of the case to get it ready to be trimmed to its final length. A sizer ball or plug makes the mouth end slightly smaller than bullet diameter, then the slightly long case gets still another bath. Finally, the spinning case encounters the cutter that trims it to final length. The trimming cutter leaves fine burrs or "feathers" at the mouth, which handloaders remove later.

An acid pickling bath, a water rinse and a polishing bath clean the case again and give it the basic look of a finished case. But all the machine working of the case has left the mouth too hard for the brass to be properly resilient at the open end, so it's back into the flames for the bullet-gripping portion of the case. Annealing discolors the brass. American makers remove this discoloration; others leave it.

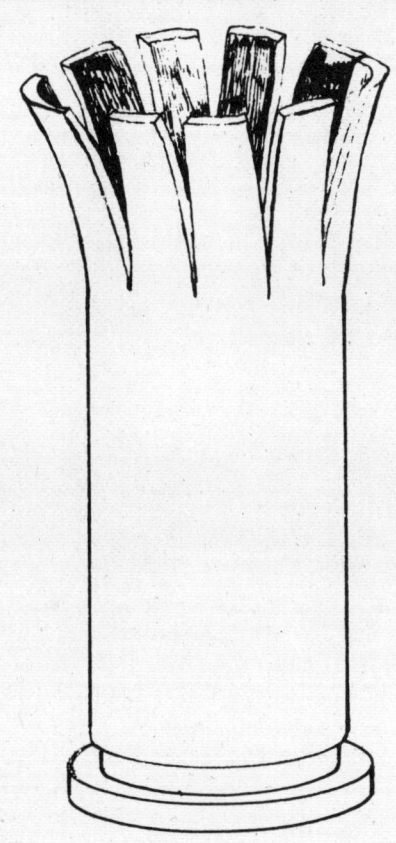

Made from a longer rifle case, this case wasn't annealed in the wildcatting process. Brass is much harder in the base of a rifle case and must be tempered by annealing.

Mouth annealing leaves the gripping portion of the case soft enough to be cold-worked several times — by firing, then by resizing, then both again and again — but another annealing may be necessary later. Mouth annealing is also advisable, if not necessary, whenever you form a case mouth to another caliber, as handloaders of most wildcat cartridges know. But we often form cases for one factory cartridge into the shape of another factory cartridge to make a wildcat case. This forming cold-works the brass, making it harder and more brittle.

Annealing softens the mouth end, to offset the cold-working and to retain the resilience of softer brass. If the mouth is to stay the same, but the body is to be blown out, annealing usually isn't necessary. Knowing when it's necessary to anneal and how to do it — especially how not to overdo it — marks the expert former of wildcat cases.

New brass that has never been sized or loaded can be formed into another shape with only the factory's annealing if forming it to the new shape doesn't work the mouth drastically. Brass that has been loaded and fired, has sat around for years, or has been passed along from one handloader to another probably ought to be annealed before it's formed for a different cartridge.

Proper annealing is so ticklish that some handloading experts flatly advise against it — but they're usually not wildcatters, so they get along with factory brass and see annealing only as a means of restoring resiliency to the mouths of tired cases. Buying new ready-formed brass isn't always an operative choice, even for some factory cartridges. It's never an option for the fellow who must load wildcat cartridges.

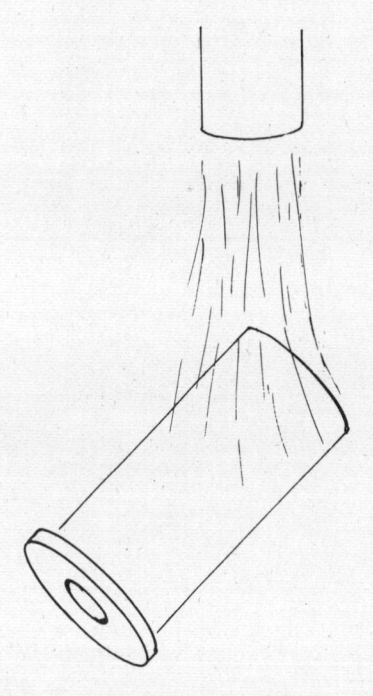

Annealing a rifle case is easier, but handgun cases can be annealed with a torch. The trick is to get just the right amount of heat on the mouth portion of the case.

So, despite some otherwise worthy advice, annealing can be inevitable if the brass is to be fit for handloading. And for storage — inadequately annealed brass can split in the box, if it sits around for a while before you call it up to active duty. But over-annealed brass is just as ruined — this probability is what lies behind the experts' concern about amateurs' annealing.

The wildcat case on the left may not require annealing if it is made from a moderately reshaped .45 ACP. If the parent case was a drastically shortened rifle case, then proper annealing is mandatory.

Over-annealing is a double danger. Only the mouth can be annealed at all. The rest of the case must retain its factory-pure hardness. The head (especially) can't be safely softened, so the annealing absolutely must be confined to the other end — a process that's ticklish enough with a case as long as the .30/06 and immeasurably trickier with handgun cases. But the mouth is easily over-annealed, too — get too hot, make it too soft and it's too weak to grip the bullet as it should.

Practice on ruined and junk cases before you try to anneal any that you don't want to ruin. Use those with mouth cracks, enlarged primer pockets, incipient or partial head separations, or Berdan primers, or those oddball cases you've picked up here and there and never planned to load anyway. Clean and polish them so you can see when the brass changes color with the steep heat you're going to turn on them.

When the brass around the mouth reaches a temperature of about 660-665°F (about 350°C), its surface becomes light blue — and this is as hot as you want to let it

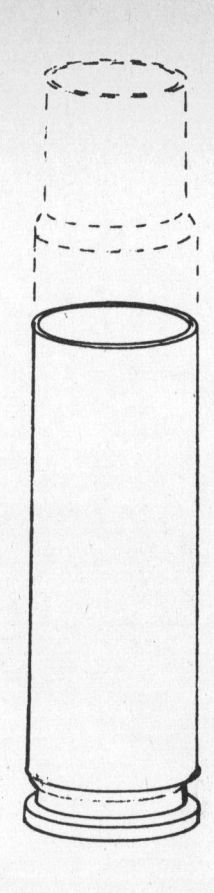

The case on the left was once like its mate, but was re-formed to the wildcat contour shown. The important thing to note is greater thickness in the neck and shoulder area. Also, a camera can't show the harder brass. Top: Bottleneck into a pistol case!

get. If you let the color run too far toward the other end of the case, you can ruin the head by making it too soft. If you let the color in the mouth area go beyond light blue and the shine disappears, you're on the thin edge of ruining the case. If you let the case get red, it's a goner. Squeeze the mouth with pliers, and you'll see how Charmin-soft it is. Remember two things: shine and light blue. Anything further is too much — and even these, too far from the mouth, mean too much heat.

Whether you use a small torch or a bullet metal pot to heat case mouths, a high temperature is better than a low temperature. (The torch offers one critical advantage over the pot: Lead puts out poisonous fumes at 900°F and hotter.) High heat brings the color into the mouth area quickly, while the head end is still safely cooler; low heat lets the head end get too hot while the mouth end is getting just hot enough. Therefore, safe mouth annealing takes high heat and a surprisingly short time — and he who dawdles over it ruins cases.

High heat is necessary to protect the head area, but its rapid heating of the mouth area risks over-annealing the mouth. When you feel uncomfortable heat transmitted to your thumb and forefinger through the brass, you may have already ruined that case (some people have a high heat tolerance in their fingertips, others none, so your personal jerk-and-cuss level is a poor way to judge when the case is safely annealed).

Annealing, of course, comprises not just one but two steps — heating and quenching. As soon as the color is right, drop the case immediately into cold water. Use a good-sized container for your coolant water. A bunch of hot cases, even little ones, can heat up a bucketful of tap water faster than you might think. If you plan to load them soon, dry them thoroughly with a jet of compressed air or a short soak in alcohol.(Alcohol displaces water, then evaporates quickly in open air.) Make sure that no drop of moisture remains in the flash hole.

Brass is valuable stuff to handloaders. Any dedicated pumper of the press handle sobs at the thought of good brass being ruined or thrown away unnecessarily (unless it's thrown where he can pick it up). It's "handloaders' gold," for sure. — *Ken Howell*

DOUBLE DATA

Chapter 12

Loading Data For The Typical Handgun Calibers, But From At Least TWO Guns

THIS CHAPTER is an effort to provide the reloader with a type of information I would have loved to have had when I started reloading. Most loading manuals run through a series of logical, sensible loads for the handgun calibers. With rare exceptions, the data they report is derived from firing tests in a single firearm. All too often, the test gun is not the brand and type of gun that the shooter is using. This means he can't get much of an idea what ammunition produced from loads from the particular manual will do in his gun.

You can get a little better idea from what you are about to read. In the case of each of the ten handgun calibers on which we will quote load data, the data will be provided for at least two guns. For .380 Auto, 9mmP Auto, .38 Super Auto, .44 Special and .45 Colt, the tables you will encounter shortly will show what happens when the handloaded ammunition is fired in two guns. In the case of the .38 Special, .357 magnum, .41 magnum, .44 magnum and .45 ACP, we deferred to the popularity of those cartridges and used *four* different guns. In effect, this doubled the double data.

With several notable exceptions, the guns were all currently available models that can be purchased in the marketplace by anyone. None were specially selected or modified guns provided by the manufacturers for the purpose of making their products look good. They were "stock." The only exception was the case of the two six-inch .38 Specials and the two autoloaders with Bar-Sto barrels.

You can't buy a new K-38 Model 14 S&W or Colt Officer's Model Match these days, but they are still around in great numbers. Since these two fine old guns were popular for so long on the Bullseye Circuit, we felt they deserved to be included. We also included a pair of autoloaders with barrels from Bar-Sto Precision Machine. One was a .38 Super, the other a .45 Commander. Among many — perhaps most — serious users of automatic pistols, the aftermarket barrel from Irv Stone is seen as a near necessity. It isn't surprising that he has produced and sold many thousands of them. Including Bar-Sto'ed guns in our test battery is fully justified.

Out of necessity, the popular loading manuals don't say a great deal about the specific accuracy of the loads they recommend. They don't, because they know full well that, if they say that a particular load combination produced a two-inch group in their test gun, readers would expect the same from their own guns. There are a lot of factors that contribute to making that impossible. One thing is the age and condition of the gun, but there is also the possiblility that the ammo loaded in a ballistics laboratory might be just a bit different than the stuff that Harry Handloader puts up in his garage. Therefore, the accuracy level is suggested, rather than stated.

In this chapter, we'll go out on the limb a little. As long as the reader understands and accepts the fact that the loads quoted produced the described results in the test guns *only* and that there is no intent to say that they will produce similar results in other guns, we'll tell you what happened in our guns. The same is true with respect to other quoted information — average velocity, extreme spread and standard deviation. We are describing the effect of ammo fired in our guns only.

A rack to hold the chronograph screens was built from thinwall conduit and square tubing. Positioned by means of a camera tripod, the rack insures that the screens are exactly the same distance apart in both directions.

Below: The familiar black box that is the Oehler Model 33's electronic heart. The window on the left is where the shooter reads direct and immediate velocities, plus other statistical data. See text for complete details.

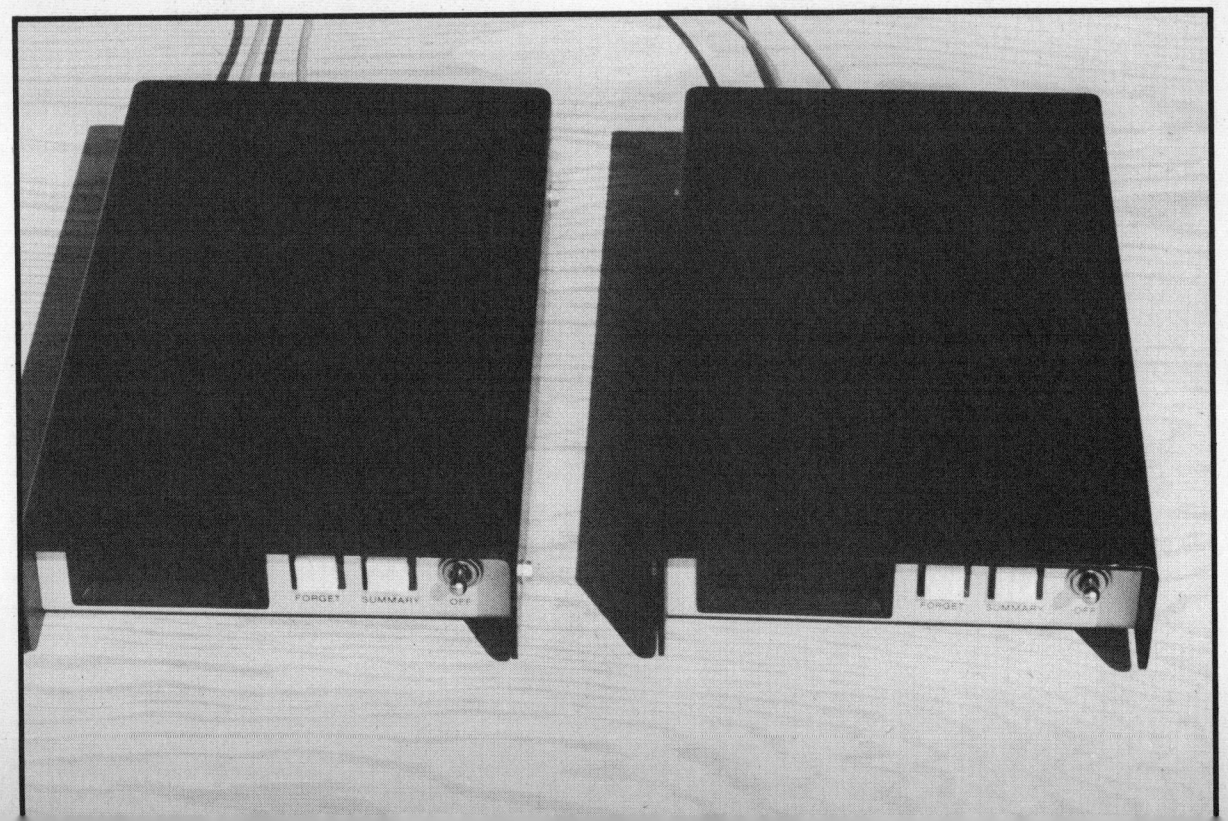

145

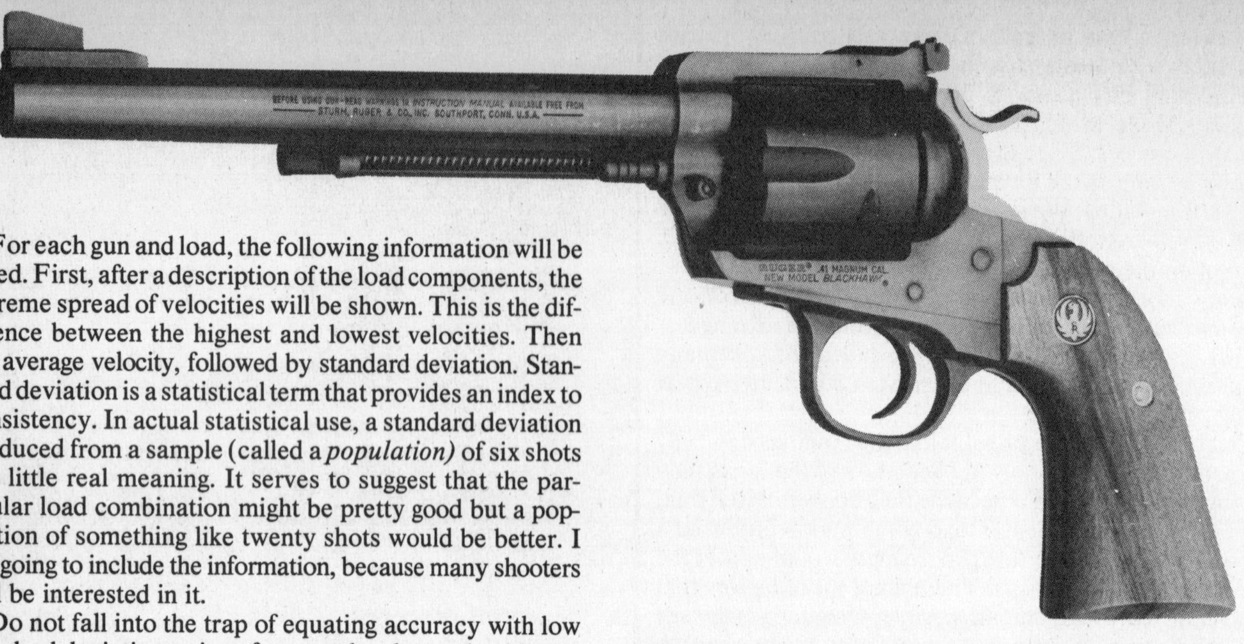

For each gun and load, the following information will be listed. First, after a description of the load components, the extreme spread of velocities will be shown. This is the difference between the highest and lowest velocities. Then the average velocity, followed by standard deviation. Standard deviation is a statistical term that provides an index to consistency. In actual statistical use, a standard deviation produced from a sample (called a *population*) of six shots has little real meaning. It serves to suggest that the particular load combination might be pretty good but a population of something like twenty shots would be better. I am going to include the information, because many shooters will be interested in it.

Do not fall into the trap of equating accuracy with low standard deviations, since few people who work with ammunition and chronographs ever have claimed that the tighter the group, the lower the standard deviation will be. 'Tain't so.

You won't find energy listed in foot-pounds or horsepower or calories or whatever. Muzzle energy is fairly easy to calculate, but you end up with a calculation from which too many shooters have made invalid conclusions for too long. The effect of a bullet striking a target most certainly is partially due to energy, but there is so much more that goes into it, that it doesn't make sense to pin powder and bullet choice on a theory. The effect of such theorizing almost always pushes things to the side of more velocity and that may not be the best way to go. Ergo, no energy figures.

This Ruger Bisley revolver in .41 magnum was used by Ransom to make the first set of inserts for this model. The revolver performs beautifully in a rest or handheld.

All of the test shooting was done with the two guns mounted in Ransom Rests. Made by Chuck Ransom over in Prescott, Arizona, the Ransom Rest is a device that holds the gun in such a way that careful manipulation of the Rest will return the gun to the same aiming point time after time. It holds the gun in rubber-faced inserts to avoid marring the finish. The inserts are firmly attached to a recoiling rocker arm that pivots upwards against the tension of a hefty spring and a friction fit to the base of the Rest.

When you handle the rest correctly, a shooter can determine the accuracy of gun/ammo combinations with a degree of reliability. It's also damned handy when you fire four hundred rounds of .44 magnum ammo in a day. Better that Chuck Ransom's "arm" take the beating than mine.

Below: The correct right side insert for S&W frames has been mounted on the rocker arm and a K-38 is being fitted to the shaped contour, after grips are removed.

Above: This view shows the left insert in author's hand. The contour of an S&W K-frame butt is clearly visible. Inserts are faced with a material that won't mar finish.

Mounting the base of the rest with proper rigidity is necessary to get good results. We bolted our two rests to two-inch planking, which was then mounted to a baseboard of plywood almost 2½ inches thick and topped with masonite. The entire board was C-clamped to the concrete shooting bench. It didn't move.

The centerlines of the two rests ended up exactly sixteen inches apart and parallel to each other. In order to get trouble-free results from the chronographs, the downrange sensing screens had to be positioned precisely. It was fairly easy to construct a rack from thin-wall conduit and square tubing which positioned a pair of start screens three feet from a pair of stop screens. The screens were sixteen inches apart, just like the mounted rests. The screens were Ken Oehler's reliable Sky-Screen IIIs and were linked to a pair of Oehler 33 chronographs. Many thousands of rounds of handloaded ammunition went over the chronographs and there wasn't a single problem.

Below: The left insert has been mounted to the three threaded rods on the rocker arm, leaving the revolver positioned in a sandwich between the two insert plates.

Above: An outer plate of rigid cast aluminum — marked A, B and C — goes over the three rods. This essentially finishes the sandwich. The last touch is tightening...

Our usual test procedure was to head for the range with a hefty supply of load-ready cases; usually about four hundred, all sized and primed. The necessary dies to complete the load were installed in a Ponsness-Warren Metalmatic P200 press. The first die was the new Lyman Multi-Expand Powder Charge Die mentioned in the chapter on loading organization. The press operator ran a primed case into the die, expanding it, then dumped a charge of powder from an old Hollywood powder measure screwed onto the die top. Then he rotated the case under appropriate dies for bullet seating and crimping.

In this way, we turned out ammunition as quickly as possible, but with no compromises in quality. In time, the procedure evolved to the point where one shooter produced the ammunition just about as fast as the other one could load and fire it, record the results and retrieve and replace the target. All shooting was done at twenty-five yards and each was a six-shot group.

...three "Star" nuts on the three rods. They have to be tightened evenly to be certain that the revolver will be held with consistent pressure. Gun is fired by the lever.

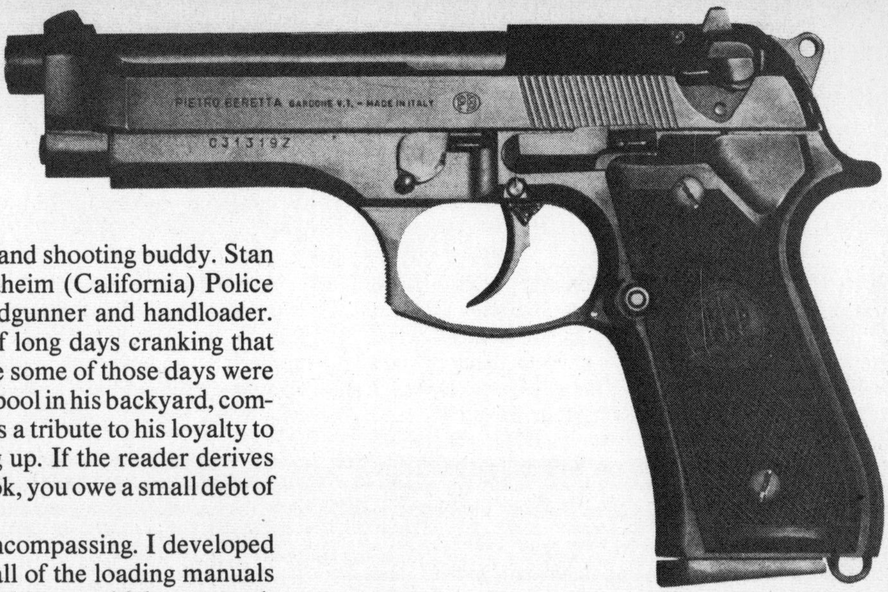

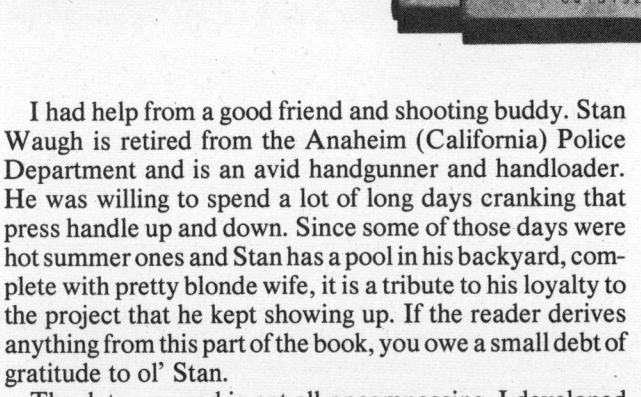

I had help from a good friend and shooting buddy. Stan Waugh is retired from the Anaheim (California) Police Department and is an avid handgunner and handloader. He was willing to spend a lot of long days cranking that press handle up and down. Since some of those days were hot summer ones and Stan has a pool in his backyard, complete with pretty blonde wife, it is a tribute to his loyalty to the project that he kept showing up. If the reader derives anything from this part of the book, you owe a small debt of gratitude to ol' Stan.

The data we used is not all-encompassing. I developed our loads by consulting nearly all of the loading manuals and picking powders and bullets that are widely accepted. There are other powders that may produce good — possibly better — results than you see here. Most of the time, the powder and charge chosen are what more than one handloading authority has suggested. While we looked at a few cast bullet loads, we spent more time and attention on ammo made with the incredible variety of fine jacketed pistol bullets made in the USA.

The following is a listing of ten pistol calibers, fifteen

The Beretta is one of the automatics used to develop load data. With the right combination of powder and a good bullet, this 92 is capable of target-grade accuracy.

pairs of guns, and one hell of a lot of powder, primers and bullets expended. We loaded and fired nearly 6000 rounds of handgun ammunition to bring you this data and we hope that you get something out of it.

But we also hope that you will understand if we shoot factory ammo for a few weeks. — *Wiley Clapp, with Stan Waugh.*

WARNING: This book contains suggested load data for use in the reloading of centerfire metallic cartridges. Developed under controlled conditions, no obvious signs of excessive pressure were noted when fired in the test guns, unless otherwise noted. Loads were developed in accordance with established procedures of reloading safety.

HOWEVER — as neither the publisher, authors nor production personnel have any control over equipment, components and techniques used in reloading by others, they do not and cannot assume any liability, either expressed or implied, for any injuries, damages or other undesirable consequences arising from or alleged to have arisen from use of this load data by others. Any and all such use is clearly and specifically at the risk and discretion of the reloader and/or shooter who fires such reloads.

Stan Waugh, who contributed hours of time and effort to the successful completion of this project, seems to be pleased to hear that it's over. Thanks a lot, pardner!

There is a glaring omission in the lineup of guns used for these tests — no Dan Wesson revolvers. Until the early part of 1987, Chuck Ransom didn't produce grip inserts for the Dan Wesson guns. I was able to talk him into doing so for the purposes of producing this book. Sadly enough, factors beyond the control of Ransom or myself intervened and the inserts haven't yet materialized. Since the rest is used for all other guns, I had to grudgingly forego their use in compiling the data.

For this unhappy circumstance, I must apologize to the legion of Dan Wesson fans and to the company. As serious revolvermen all know, Dan Wesson guns are fine firearms and shoot as well as anything that you can find. Sometimes, ...well, they're better.

.380 AUTO

THERE ARE thousands of .380 automatic pistols in use in the United States and, with rare exceptions, they are used primarily for defense. The cartridge itself is relatively easy to reload and all major loading manuals provide plenty of good data. While .380s don't have a reputation for gilt-edged accuracy, I know of at least one target model Browning that is regularly used in formal target shooting. The .380s can be loaded to acceptable levels of accuracy and they can serve as a beginning centerfire gun for novice shooters.

Gun #1 — Walther PPK

This particular PPK is one of the new ones made in the Interarms facility in the United States. It's a stainless steel version and has all that James Bond aura about it. Throughout the firing tests, the pistol performed beautifully.

Gun #2 — SIG Sauer 230

The 230 is a little larger than the PPK and a trifle less concealable. It has the unique SIG Sauer action, featuring a hammer drop lever and no manual safety. Like its mate, the 230 is a stainless steel gun and performed without malfunctions. That is rapidly becoming the norm with modern autos.

We tried several powders, but obtained clearly superior results with IMR's SR7625, followed closely by Accurate Arms' #5. Federal and Winchester cases were used, along with CCI #500 primers. I would love to report that the two little .380s were accurate in the extreme, but we only produced one group under two inches and the norm was probably well over three.

Part of the problem in obtaining accuracy in these little handguns is the fact that most of them, including our two test guns, are blowback guns. This means that they operate at relatively low pressure and use the weight of the slide and the power of the recoil spring to hold the action closed until pressure drops to an acceptable level. The recoil impulse will vary a great deal from one sort of ammo to the next and the breech will open at a different point.

The handloader who is primarily interested in accuracy is faced with the problem of building a load that will operate the action reliably, but has just the right recoil impulse to deliver consistent velocity figures. When there is a large extreme spread in velocity, it suggests that the load isn't right. The solution is a great deal of experimenting — or use of a locked breech .380 auto, like the new Colt .380.

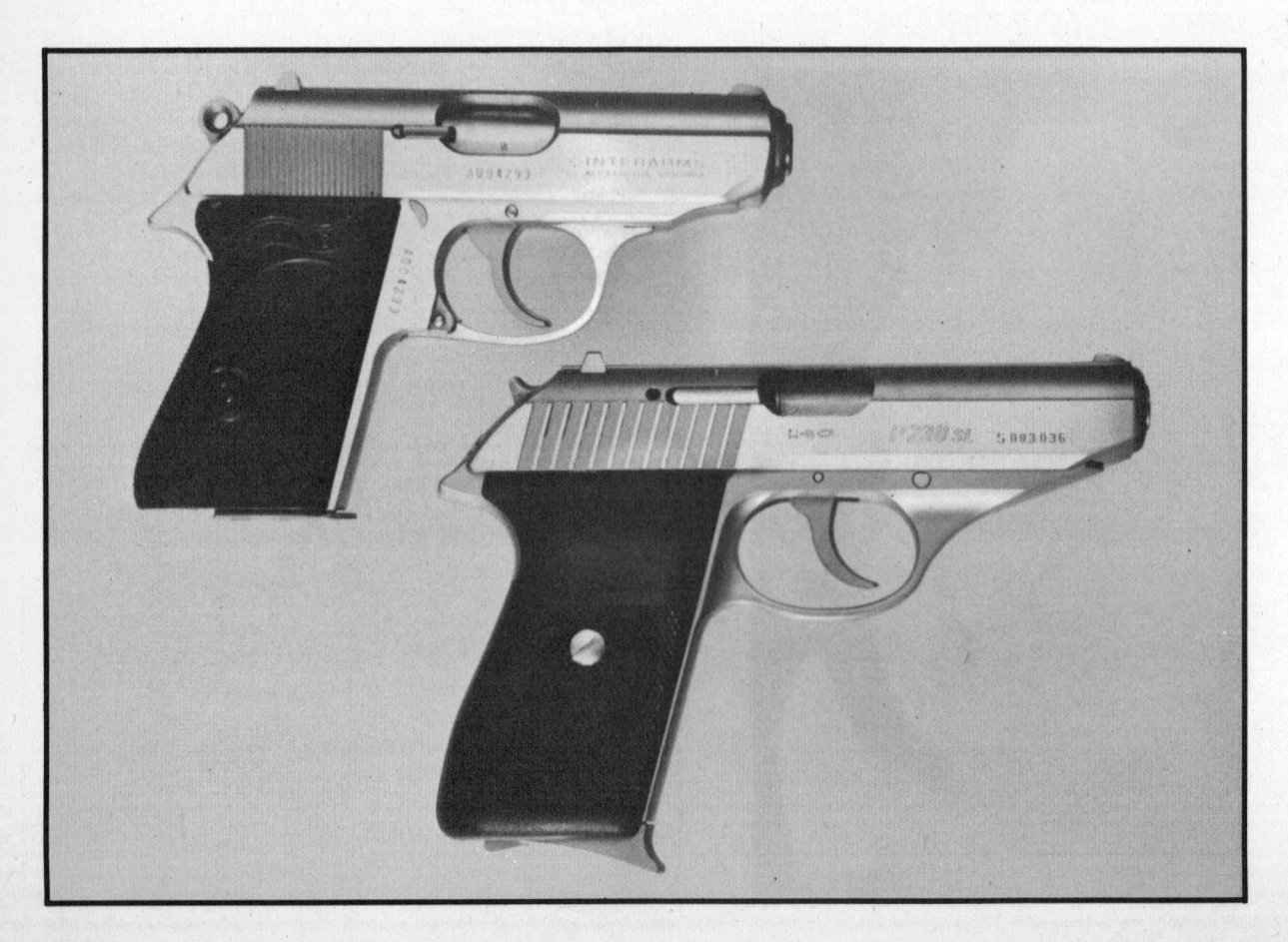

.380 AUTO LOAD DATA

BRASS	POWDER CHARGE (grains)	BULLET MAKE/WEIGHT/TYPE (grains)	ES	GUN #1 AVG	SD	GROUP (inches)	ES	GUN #2 AVG	SD	GROUP (inches)	COMMENTS
FEDERAL	3.0 SR7625	HORNADY 90 JHP	155	715	59	3+	143	707	55	3+	Large extreme spread
FEDERAL	3.2 SR7625	HORNADY 90 JHP	57	840	28	2½	87	789	29	3+	
FEDERAL	3.4 SR7625	HORNADY 90 JHP	85	921	30	3	126	873	48	3	Vertical stringing in #2
FEDERAL	3.0 SR7625	SIERRA 95 FMJ	63	759	22	3+	91	743	30	3+	
FEDERAL	3.2 SR7625	SIERRA 95 FMJ	69	812	28	2½	60	822	21	3	Gun #2 had five in 1½"
FEDERAL	3.4 SR7625	SIERRA 95 FMJ	90	880	35	3+	63	851	25	2½	
FEDERAL	3.0 SR7625	SPEER 100 JHP	31	839	18	3+	76	803	29	2¾	
FEDERAL	3.2 SR7625	SPEER 100 JHP	101	905	39	2	96	840	32	1¾	Most accurate load — Gun #2 measured 1.811"
FEDERAL	3.4 SR7625	SPEER 100 JHP	67	946	28	2+	52	931	19	2½	Second most accurate load & second highest velocity
FEDERAL	4.2 AA-5	SPEER 88 JHP	66	828	28	2¾	82	791	31	2+	
FEDERAL	4.4 AA-5	SPEER 88 JHP	48	862	19	2½	76	849	28	3	Best overall load with this powder
FEDERAL	4.6 AA-5	SPEER 88 JHP	57	896	26	4	68	905	29	3+	
FEDERAL	4.2 AA-5	HORNADY 90 JHP	99	836	36	2½	64	797	23	4+	
FEDERAL	4.4 AA-5	HORNADY 90 JHP	97	886	34	4+	42	855	15	3	
FEDERAL	4.6 AA-5	HORNADY 90 JHP	110	938	63	4	73	895	26	4+	
FEDERAL	4.2 AA-5	SIERRA 95 FMJ	118	763	82	4+	135	775	44	4+	High extreme spread in both guns
FEDERAL	4.4 AA-5	SIERRA 95 FMJ	89	829	30	6+	67	852	24	4+	
FEDERAL	4.6 AA-5	SIERRA 95 FMJ	153	885	62	3½	80	873	37	3½	
W-W	3.2 BULLSEYE	SPEER 88 JHP	34	872	12	5	36	872	12	4	
W-W	3.2 BULLSEYE	HORNADY 90 JHP	40	880	16	6	148	952	65	5+	Poor accuracy
W-W	3.2 BULLSEYE	SIERRA 95 FMJ	68	857	24	6+	102	979	29	6+	
W-W	4.0 HERCO	SPEER 88 JHP	89	839	41	3+	121	851	23	2½	
W-W	4.0 HERCO	HORNADY 90 JHP	107	896	43	3½	127	871	51	3	
W-W	4.0 HERCO	SIERRA 95 FMJ	98	884	39	6+	187	869	67	3+	
W-W	3.1 700-X	SPEER 88 JHP	98	949	36	3+	83	929	33	3+	Vertical stringing in both
W-W	3.1 700-X	HORNADY 90 JHP	158	957	62	3½+	89	923	58	3+	
W-W	3.1 700-X	SIERRA 90 JHP	84	965	30	3½	114	934	19	3+	Highest velocity

FOOTNOTES:

1) All groups reported are six-shot groups, fired at 25 yards in the Ransom Rest.
2) Velocity is measured with Oehler 33 chronographs; Sky Screen IIIs spaced 36 inches apart.
3) Unless otherwise noted, group size is measured with a ruler, to the nearest quarter inch.
4) All cases primed with CCI 500 primer

9mmP

THIS IS A cartridge that was seldom loaded for accuracy before the U.S. adopted the round as service pistol fodder. It was long troubled by a reputation for inherent problems for the handloader. The reputation is partially deserved, since there are so many different chamber dimensions used by the various manufacturers. The case has more taper than we are accustomed to and dies must match the chamber rather closely. The dies used here were RCBS with a Lyman expander plug.

Gun #1 — Beretta Model 92F

This handgun is a commercial version of the M9 service pistol, a Beretta 92F. It has been fired extensively, used for the ammo comparison in *The Digest Book of 9mm Handguns*. There were no problems with functioning in the course of our tests. Although we have a Bar-Sto barrel for the pistol, the data was produced using the standard factory tube.

Gun #2 — Browning HiPower

Sometimes called the P35 or GP, this is a recent commercial version of the Browning HiPower. It is well broken in, with several hundred rounds through the gun. There was but a single malfunction with the autoloader and that was traced to a round of ammunition that was loaded a little too long.

Since the United States adopted the Beretta as the new service pistol, there has been a resurgence of interest in the 9mmP round amongst the target shooting fraternity. In a relatively short period of time, we will certainly be shooting the 9mmP in the National Matches at Camp Perry. And that means that a great many people will be loading lots of practice ammo, much of it to be fired in the M9 service handgun, the Beretta Model 92F.

Firing produced some distinct and pleasant surprises. The standard Beretta seems capable of fine accuracy. Three powders and several bullets were excellent: Accurate Arms #7, Du Pont PB and SR7625 with either the 115 Sierra FMJ, 115 Hornady JHP or Hornady 124 FMJ.

Accurate Arms #7 powder is the propellent used by the Israelis in loading the 9mmP ammo that they have used so widely in the past few years. It was designed and developed with this purpose in mind and produced relatively high velocities with no adverse pressure indications. The most accurate load combinations came from the use of IMR powder company's PB. PB is not commonly associated with accuracy loading of the 9mmP and I used it at the urging of Dean Grennell. As is so often the case when it comes to matters of this sort, my colleague is correct — PB is an absolutely first-rate powder for the job at hand.

Both the Sierra 115-grain FMJ and the Hornady 124-grain FMJ, the flat point "Air Force" bullet, proved to be accurate bullets.

9mmP LOAD DATA

BRASS	POWDER CHARGE (grains)	BULLET MAKE/WEIGHT/TYPE (grains)	ES	GUN #1 AVG	SD	GROUP (inches)	ES	GUN #2 AVG	SD	GROUP (inches)	COMMENTS
S&W	8.0 AA-7	SIERRA 115 FMJ	25	1159	11	2½	56	1172	22	2½	
S&W	8.3 AA-7	SIERRA 115 FMJ	26	1148	10	1½	60	1207	20	2	Gun 2 had horizontal spread
S&W	8.5 AA-7	SIERRA 115 FMJ	75	1185	30	2	97	1196	43	2½	
S&W	8.0 AA-7	HORNADY 115 JHP	54	1168	20	1¾	28	1186	10	2+	
S&W	8.3 AA-7	HORNADY 115 JHP	59	1187	19	1¾	96	1179	43	2½	
S&W	8.0 AA-7	SPEER 115 FMJ	162	1137	59	3+	55	1158	24	3+	Note #1 large ES — Neither gun likes this load
S&W	8.3 AA-7	SPEER 115 FMJ	81	1130	50	3½+	74	1197	29	4+	Neither gun likes this load
S&W	7.8 AA-7	HORNADY 124 FMJ	37	1102	16	1¾	74	1148	29	3+	
S&W	8.0 AA-7	HORNADY 124 FMJ	12	1124	4	1¾	51	1167	19	2+	Gun #1 =Tight Stats = Tight Group
S&W	4.5 PB	SIERRA 115 FMJ	24	1079	8	1+	65	1110	24	2	Gun #1 group measured 1.230" — Second most accurate
S&W	4.8 PB	SIERRA 115 FMJ	21	1137	9	3	44	1161	16	3+	(Accuracy decreases
S&W	5.0 PB	SIERRA 115 FMJ	82	1163	34	3	44	1193	17	3+	with more than 4.5 grains)
S&W	4.5 PB	HORNADY 115 JHP	35	1113	13	2+	28	1123	11	2+	
S&W	4.8 PB	HORNADY 115 JHP	44	1174	14	2½	48	1162	22	2½+	
S&W	5.0 PB	HORNADY 115 JHP	50	1208	19	5+	34	1206	14	4+	Vertical strings — both guns
S&W	4.5 PB	SPEER 115 FMJ	61	1091	25	3+	39	1121	15	3+	
S&W	4.8 PB	SPEER 115 FMJ	93	1156	29	2½+	24	1183	9	2+	
S&W	4.5 PB	HORNADY 124 FMJ	42	1082	18	1+	71	1107	25	2+	Gun #1 measured 1.090" — Most accurate load
S&W	6.0 HERCO	SIERRA 115 FMJ	46	1159	16	3+	131	1175	34	3+	Compressed load
S&W	6.0 HERCO	HORNADY 115 JHP	92	1165	36	1¾	95	1199	32	2½	Best load with Herco — Gun #2 had 5 in 1¾"
S&W	6.0 HERCO	SPEER 115 FMJ	57	1176	30	3+	132	1165	45	3+	ES on Gun #2 is excessive
S&W	4.5 SR7625	SIERRA 115 JHP	30	1149	10	3+	21	1179	8	3+	
S&W	4.7 SR7625	SIERRA 115 JHP	34	1146	14	1¾	25	1157	11	2+	
S&W	4.9 SR7625	SIERRA 115 FMJ	51	1157	16	3+	29	1179	11	3+	Identical velocities
S&W	4.5 SR7625	HORNADY 115 JHP	44	1186	15	3+	26	1186	10	3+	Inaccurate in Browning
S&W	4.7 SR7625	HORNADY 115 JHP	66	1124	23	2½+	29	1164	10	6	Poor accuracy in both guns
S&W	4.9 SR7625	SPEER 115 FMJ	63	1137	20	5+	117	1186	36	6	
S&W	4.5 SR7625	SPEER 124 FMJ	69	1125	26	2½	42	1157	30	2½	
S&W	4.5 SR7625	HORNADY 124 FMJ	24	1121	10	3	34	1144	11	3+	Gun #1 — Last five in 1¾"
S&W	4.7 SR7625	HORNADY 124 FMJ	8	1136	3	2+	27	1149	9	2+	Gun #2 — Five in one hole
S&W	4.9 SR7625	HORNADY 124 FMJ	34	1155	15	2½	23	1172	7	3	Gun #1 — Five in 1"

FOOTNOTES:

1) All groups reported are six-shot groups, fired at 25 yards in the Ransom Rest.
2) Velocity is measured with Oehler 33 chronographs; Sky Screen IIIs spaced 36 inches apart.
3) Unless otherwise noted, group size is measured with a ruler, to the nearest quarter inch.
4) All cases primed with CCI 500 primer

.38 SUPER AUTO

SELDOM HAS an old cartridge come roaring back in popularity as has the .38 Super. The reason for the resurgent interest is two-fold: increased use of the round in IPSC shooting and the wide acceptance of the superb Bar-Sto barrel as a cure for the legendary headspacing problem of the original Colt barrels.

Gun #1 — Colt Government Model, with Bar-Sto barrel

This particular pistol is Dean Grennell's and one that has fired some spectacular groups. It has been gussied up with some cosmetics, like a pair of sexy grips from Dean's workshop, but it is fundamentally a plain Government Model with Irv Stone's barrel installed. It out-shot the companion gun with almost every load.

Gun #2 — Astra Model A-80

This is the only high-capacity .38 Super made and a gun that I regard as something of a sleeper. It handles a wide variety of factory and handloaded ammunition without problems. It is also acceptably accurate, pleasant to shoot and handle.

Accurate Arms #7 was probably the best fuel for our test guns, with Hercules Blue Dot in second place. The Super can handle even heavier bullets, as the IPSC shooters are finding out. It's only a matter of time until the bulletmakers produce heavier jacketed bullets in the correct diameter for the old guns. Our tests were confined to the heavier bullets currently available.

In a well-fitted barrel, such as the one in the Colt, the .38 Super is a fine cartridge. The problem arises from the fact that the cartridge is technically termed *semi-rimmed*, meaning that there was a vestigial rim extending beyond the dimensions of the case body. For many years, Colt chose to headspace on the rim. People who handloaded the cartridge routinely had the devil's own time getting the system to work well.

Irv Stone, of Bar-Sto Precision, took the bull by the horns and produced a barrel that ignored the rim and headspaced on the case mouth, like the .45 ACP. The result is spectacular increases in accuracy in almost all guns fitted with the new barrel. The Astra A-80 is fitted with a barrel using the case-mouth-headspacing system, right from the factory.

All of this concern with the vagaries of a sixty-year-old cartridge that was not sufficiently popular for any guns to be produced in the U.S. by any manufacturer except Colt, has had its effect. Colt has begun to produce both their Commander and Government Model .38 Supers with a barrel which headspaces on the case mouth. The accuracy has to improve.

.38 SUPER AUTO LOAD DATA

BRASS	POWDER CHARGE (grains)	BULLET MAKE/WEIGHT/TYPE (grains)	ES	GUN #1 AVG	SD	GROUP (inches)	ES	GUN #2 AVG	SD	GROUP (inches)	COMMENTS
RP (+P)	5.9 UNIQUE	HORNADY 115 JHP	127	1090	50	3+	66	1038	27	3+	
RP (+P)	6.2 UNIQUE	HORNADY 115 JHP	96	1136	43	2½	118	1075	84	4+	
RP (+P)	6.4 UNIQUE	HORNADY 115 JHP	132	1209	48	1½+	136	1108	57	3+	Good group in Colt — Note ES — (?)
RP (+P)	5.9 UNIQUE	SPEER 125 FMJ	44	1073	18	3+	31	993	11	5+	
RP (+P)	6.2 UNIQUE	SPEER 125 FMJ	84	1134	37	4	162	990	64	4+	
RP (+P)	6.4 UNIQUE	SPEER 125 FMJ	68	1165	27	3½	67	1075	28	4+	
RP (+P)	5.9 UNIQUE	SIERRA 130 FMJ	63	1064	25	6+	43	987	15	5+	
RP (+P)	6.2 UNIQUE	SIERRA 130 FMJ	35	1122	13	3+	44	1031	20	4+	
RP (+P)	6.4 UNIQUE	SIERRA 130 FMJ	114	1186	43	2+	37	1074	15	3+	
RP (+P)	7.4 800-X	HORNADY 115 JHP	53	1265	20	2+	92	1170	34	4+	
RP (+P)	7.6 800-X	HORNADY 115 JHP	55	1271	21	2+	63	1187	23	3	Gun #2 — vertical string
RP (+P)	7.8 800-X	HORNADY 115 JHP	95	1347	35	2	162	1175	62	4+	Highest velocity in Colt +good group
RP (+P)	7.0 800-X	HORNADY 124 FMJ	39	1243	14	2+	44	1159	18	2½	
RP (+P)	7.2 800-X	HORNADY 124 FMJ	22	1241	8	3+	42	1142	17	3+	More powder, but less velocity
RP (+P)	7.0 800-X	SPEER 124 FMJ	43	1193	15	4+	29	1098	12	6+	
RP (+P)	7.2 800-X	SPEER 124 FMJ	60	1218	22	4+	56	1148	28	5+	
RP (+P)	7.8 AA-7	HORNADY 124 FMJ	24	1011	9	2	63	951	22	2½	Tight round group in #1, good group in #2
RP (+P)	8.1 AA-7	HORNADY 124 FMJ	30	1062	10	6+	65	985	30	4+	
RP (+P)	8.4 AA-7	HORNADY 124 FMJ	34	1094	12	2+	121	999	50	6+	
RP (+P)	7.8 AA-7	SPEER 124 FMJ	66	978	26	3+	133	911	55	4+	
RP (+P)	8.1 AA-7	SPEER 124 FMJ	45	1063	16	2	59	975	25	6	Vertical stringing in #2
RP (+P)	8.4 AA-7	SPEER 124 FMJ	18	1097	7	1½	67	1009	26	3+	Best group in either gun, 1.610" in Colt
RP (+P)	7.8 AA-7	SIERRA 130 FMJ	65	1021	21	4	62	924	23	3	
RP (+P)	8.1 AA-7	SIERRA 130 FMJ	25	1054	9	2½	31	995	14	2	Best Astra group
RP (+P)	8.4 AA-7	SIERRA 130 FMJ	13	1101	4	3	43	1013	19	6+	
RP (+P)	9.0 BLUE DOT	HORNADY 124 FMJ	33	1298	14	1¾	68	1172	24	4+	
RP (+P)	9.5 BLUE DOT	HORNADY 124 FMJ	77	1309	28	1¾	24	1212	9	4+	
RP (+P)	9.0 BLUE DOT	SIERRA 130 FMJ	37	1282	14	1¾	58	1171	22	3+	
RP (+P)	9.5 BLUE DOT	SIERRA 130 FMJ	57	1332	23	2½	38	1220	14	3½	Primers flat — hot

FOOTNOTES:

1) All groups reported are six-shot groups, fired at 25 yards in the Ransom Rest.
2) Velocity is measured with Oehler 33 chronographs; Sky Screen IIIs spaced 36 inches apart.
3) Unless otherwise noted, group size is measured with a ruler, to the nearest quarter inch.
4) All cases primed with CCI 500 primer

.38 SPECIAL

THE .38 SPECIAL has been around since 1899 and shows no signs of diminishing in popularity. Even in the magnum era, the bulk of the nation's policemen go to work with a reliable .38 Special revolver on their hip. Aside from its use as a defensive or combat round, the .38 still enjoys popularity as a target cartridge. Moreover, in a discussion of handloading handgun ammunition, you have to give full credence to the fact that most handloaders begin with this one.

Gun #1 — Colt Detective Special

Colt has been producing double-action revolvers with exceptionally short barrels for many years. The Detective Special pictured here is a recent vintage specimen, with two-inch barrel and the later enclosed ejector rod housing. Contrary to what often is said about short barrels lacking accuracy, this one produced a number of groups near two inches.

Gun #2 — Smith & Wesson Model 15

Smith & Wesson introduced the gun that they called the Combat Masterpiece in the early Fifties. It was one of the first new models in the firm's postwar modernization program. With a ribbed barrel and target sights, the gun became most popular with legions of policemen. Essentially, the Model 15 is an updated Military and Police Model, the bread-and-butter gun of the Smith & Wesson line. They still make them and they still shoot.

There are a few load combinations in handloading lore that are used so often as to warrant the term "classic." One such load is 2.7 grains of Hercules Bullseye and a 148-grain wadcutter bullet. That was the second load that we tried and it proved to be an excellent choice. Nevertheless, a slightly stiffer charge of Bullseye (3.0 grains) tightened the groups of both of our test guns perceptibly.

Of the powders we used, the highest velocity came from Bullseye and the 110-grain Sierra JHP bullet. Five grains of Bullseye drove that bullet to the velocity level where consistent expansion is likely to occur.

Approach the use of Accurate Arms #7 with all due caution. It appears that pressure rises sharply in this application. The accuracy delivered by the Israeli-produced powder is excellent, but small changes in charge weights can increase velocities (and pressures) a bit more quickly than in a number of other powders.

None of the loads listed are particularly hot and that is in deference to the older and lighter-weight guns that are often encountered. If a shooter wants accuracy above all other considerations, he could spend a lot of time working with the .38 Special.

.38 SPECIAL LOAD DATA

BRASS	POWDER CHARGE (grains)	BULLET MAKE/WEIGHT/TYPE (grains)	ES	GUN #1 AVG	GUN #1 SD	GROUP (inches)	ES	GUN #2 AVG	GUN #2 SD	GROUP (inches)	COMMENTS
WCC-GI	2.5 BULLSEYE	HORNADY 148 BBWC	26	635	9	3+	49	681	15	2½	
WCC-GI	2.7 BULLSEYE	HORNADY 148 BBWC	41	656	12	2+	43	691	14	1¾	The classic load — good accuracy
WCC-GI	3.0 BULLSEYE	HORNADY 148 BBWC	43	714	26	2	41	756	17	1½	Most accurate load in both guns
WCC-GI	3.0 BULLSEYE	HORNADY 158 LSWC	29	629	11	4+	45	671	17	1¾	
WCC-GI	4.3 BULLSEYE	HORNADY 125 JHP	22	808	9	2½	53	845	20	2	Vertical stringing in #2
WCC-GI	5.0 BULLSEYE	SIERRA 110 JHP	60	934	25	2+	62	964	23	2½	Defense load — highest velocity in both guns
WCC-GI	2.7 BULLSEYE	SPEER 148 HBWC	59	685	20	3	66	697	24	1¾	Bullet is oversize for both guns — shaving
WCC-GI	4.1 BULLSEYE	HORNADY 158 JHP	62	782	27	3+	69	801	29	2½	
WCC-GI	4.5 AA-7	HORNADY 148 BBWC	80	680	30	4+	73	739	28	2½+	
WCC-GI	4.8 AA-7	HORNADY 148 HBWC	51	850	17	3+	90	905	36	2	Sharp velocity increase
WCC-GI	5.5 AA-7	HORNADY 158 LSWC	39	741	18	2¾	36	786	17	1¾	
WCC-GI	6.0 AA-7	HORNADY 158 LSWC	78	891	26	3	69	954	24	2	Sharp velocity increase
WCC-GI	6.0 AA-7	HORNADY 158 JHP	72	801	27	3+	64	845	21	1½	Good load in #2
WCC-GI	3.3 231	HORNADY 148 HBWC	48	682	17	2½	70	728	31	1¾	
WCC-GI	3.6 231	HORNADY 148 HBWC	31	740	10	4+	51	792	22	2½	Vertical stringing
WCC-GI	3.8 231	HORNADY 148 HBWC	69	805	25	3+	39	843	15	3	
WCC-GI	3.8 231	HORNADY 158 LSWC	37	684	14	3+	32	730	11	2½	
WCC-GI	4.4 231	HORNADY 158 LSWC	55	755	20	2¾	47	806	16	2	
WCC-GI	2.8 700-X	SPEER 148 HBWC	78	739	29	3+	96	753	40	2	
WCC-GI	3.0 700-X	SPEER 148 HBWC	85	753	33	3+	43	787	43	2½	Second most accurate load
WCC-GI	3.0 700-X	HORNADY 158 LSWC	93	674	31	2+	123	700	33	1½	
WCC-GI	3.3 700-X	HORNADY 158 LSWC	92	731	30	3+	79	765	33	2+	
WCC-GI	4.5 UNIQUE	HORNADY 158 JHP	56	68	24	2½	80	613	29	2	
WCC-GI	5.5 UNIQUE	HORNADY 158 JHP	56	766	63	3	85	760	37	2	
WCC-GI	5.5 UNIQUE	HORNADY 158 LSWC	56	776	20	6+	105	831	43	3+	
WCC-GI	5.5 UNIQUE	LYMAN CAST 168 LSWC	56	831	19	3+	47	865	23	3+	
WCC-GI	5.5 UNIQUE	HORNADY 125 JHP	71	848	31	3½	73	898	27	3+	
WCC-GI	5.5 UNIQUE	HORNADY 110 JHP	73	922	37	3+	78	944	36	2+	Second highest velocity

FOOTNOTES:

1) All groups reported are six-shot groups, fired at 25 yards in the Ransom Rest.
2) Velocity is measured with Oehler 33 chronographs; Sky Screen IIIs spaced 36 inches apart.
3) Unless otherwise noted, group size is measured with a ruler, to the nearest quarter inch.
4) All cases primed with CCI 500 primer

GUN DIGEST BOOK OF HANDGUN RELOADING

.38 SPECIAL

THE TARGET shooters used to be just about evenly divided between the Colt men and the S&W fans. Each company produced fine target-grade revolvers for the NRA Outdoor Pistol Centerfire Matches. Use of those guns in timed and rapid-fire matches meant the shooter had to master the art of single-action cocking and firing.

Gun #1 — Colt Officer's Model Match

Built on the company's medium or .41 frame, the OMM was produced in several versions, including the rare Mark III. It is a hefty but well balanced revolver with a rapid-taper heavy barrel and smooth action. Lots of the OMMs were used by target shooters in the Fifties and Sixties. This particular one is a recent acquisition that was being neglected in the back corner of the handgun case of a local emporium.

Gun #2 — Smith & Wesson Model 14

Smith & Wesson also called this gun the K-38 Masterpiece or, in its earlier versions, the Heavy Masterpiece. It is the competition for the Colt OMM. Like the Colt, the S&W M14 has a heavy barrel and target sights. The immediately-noticed difference in the two is the drastic difference in butt shape and the action. Model 14s feel completely different in handling and firing. This particular gun was my late father's. When I began to use the gun in bullseye matches, I had George Mathews change the hammer to the "cockeyed" type for single-action work.

Several interesting facts come to light when you examine the accompanying load data. First, it seems the loads that were accurate in the other two .38s were likewise accurate in these guns. Also note that the Colt OMM delivered higher velocities than the Smith & Wesson, sometimes by a considerable margin. On the possibility that one of the chronographs was off, I changed them around and got identical results.

The reason is that the Colt not only has a tighter bore, but one that is rifled with a 1:14 twist. The S&W spins bullets at the rate of 1/18¾. The faster spin and tighter bore of the Colt mean that the pressures are a bit higher and the velocities are up.

The faster twist causes Colt revolvers to shoot a tiny bit better than their contemporary Smiths. This was not the case with this particular pair of guns, as the S&W had a small edge almost in every case.

The brass used throughout our .38 Special testing was a large batch of GI stuff headstamped WCC68. Made by Winchester-Western in 1968 according to GI specifications, this is a heavy, thick-walled cartridge case, well-suited to our use.

.38 SPECIAL LOAD DATA

BRASS	POWDER CHARGE (grains)	BULLET MAKE/WEIGHT/TYPE (grains)	GUN #1 ES	GUN #1 AVG	GUN #1 SD	GUN #1 GROUP (inches)	GUN #2 ES	GUN #2 AVG	GUN #2 SD	GUN #2 GROUP (inches)	COMMENTS
WCC-GI	2.5 BULLSEYE	HORNADY 148 BBWC	41	744	14	2	64	675	25	2+	
WCC-GI	2.7 BULLSEYE	HORNADY 148 BBWC	40	765	17	1¾	62	700	21	2+	The classic load — good accuracy
WCC-GI	3.0 BULLSEYE	HORNADY 148 BBWC	33	841	14	1½	30	780	12	1½	Most accurate load in both guns (see page 157)
WCC-GI	3.0 BULLSEYE	HORNADY 158 LSWC	20	761	7	1¾	66	679	23	1¾	Good accuracy
WCC-GI	4.3 BULLSEYE	HORNADY 125 JHP	59	937	21	1¾	154	764	59	1½	Good accuracy
WCC-GI	5.0 BULLSEYE	SIERRA 110 JHP	83	1085	34	1¾	143	924	49	1¾	Highest velocity in #1
WCC-GI	2.7 BULLSEYE	SPEER 148 HBWC	28	787	10	2	88	721	34	1¾	
WCC-GI	4.1 BULLSEYE	HORNADY 158 JHP	70	846	31	2½	91	801	39	2+	
WCC-GI	4.5 AA-7	HORNADY 148 BBWC	103	820	38	1¾	60	776	21	1+	1.037" in #2
WCC-GI	4.8 AA-7	HORNADY 148 HBWC	57	1005	20	2½	52	953	20	2½	Sharp velocity increase (see page 157)
WCC-GI	5.5 AA-7	HORNADY 158 LSWC	49	937	17	2½	52	901	19	1¾	
WCC-GI	6.0 AA-7	HORNADY 158 LSWC	53	1053	20	3+	16	952	4	1¼	
WCC-GI	6.0 AA-7	HORNADY 158 JHP	47	964	21	2+	98	783	32	2½	
WCC-GI	3.3 231	HORNADY 148 HBWC	30	798	10	1¾	30	727	14	1½	1.299" in #2
WCC-GI	3.6 231	HORNADY 148 HBWC	60	860	23	2+	67	800	27	1½	
WCC-GI	3.8 231	HORNADY 148 HBWC	57	929	21	2+	58	860	25	2½	
WCC-GI	3.8 231	HORNADY 158 LSWC	52	789	22	2¼	49	746	17	2+	
WCC-GI	4.4 231	HORNADY 158 LSWC	52	781	20	3¼	81	824	42	2½	
WCC-GI	2.8 700-X	SPEER 148 HBWC	79	801	22	3+	90	777	21	3+	
WCC-GI	3.0 700-X	SPEER 148 HBWC	79	822	30	3¼	92	801	22	3+	
WCC-GI	3.0 700-X	HORNADY 158 LSWC	89	731	32	3+	90	720	29	3+	Not as accurate in 6" barrels (see page 157)
WCC-GI	3.3 700-X	HORNADY 158 LSWC	94	751	38	3+	89	740	30	3+	
WCC-GI	4.5 UNIQUE	HORNADY 158 JHP	56	688	32	2½	80	675	33	2¼	
WCC-GI	5.5 UNIQUE	HORNADY 158 JHP	50	801	21	2½	57	772	30	2+	
WCC-GI	5.5 UNIQUE	HORNADY 158 LSWC	56	803	22	2¾	98	861	32	3+	
WCC-GI	5.5 UNIQUE	LYMAN CAST 168 LSWC	63	852	23	2¾	99	849	47	3+	
WCC-GI	5.5 UNIQUE	HORNADY 125 JHP	71	907	33	3½	73	905	22	3+	
WCC-GI	5.5 UNIQUE	HORNADY 110 JHP	73	988	30	3+	62	962	22	3+	Highest velocity in #2

FOOTNOTES:

1) All groups reported are six-shot groups, fired at 25 yards in the Ransom Rest.
2) Velocity is measured with Oehler 33 chronographs; Sky Screen IIIs spaced 36 inches apart.
3) Unless otherwise noted, group size is measured with a ruler, to the nearest quarter inch.
4) All cases primed with CCI 500 primer

.357 MAGNUM

FEW CARTRIDGES ever caught and fired the imagination of the shooting public as did the .357 magnum when introduced in 1935. Smith & Wesson built the first guns for the round, but Colt quickly followed suit. The gunwriters of the late Thirties — except Elmer Keith — were super-cautious about loads for the new round, but in time, it has come to be accepted as a useful and versatile cartridge.

Gun #1 — Ruger GP-100

Ruger made the Security Six series of double-action revolvers for a number of years before they came out with this new gun recently. The particular gun in question is one that was shipped to us in the normal course of business by Ruger for test and evaluation. With factory ammunition, this GP-100 quickly established itself as one of the most accurate handguns we have ever fired. The accuracy of the gun is vindicated with a wide variety of handloads. I am perfectly willing to accept that this gun may not be typical, but it is a superb shooter.

Gun #2 — Ruger Blackhawk, Old Model

Ruger's first centerfire firearm was the famous Blackhawk and this specimen is entirely typical. Styled after the Colt Frontier Model at a time when Colt was not making that gun, the Blackhawk came along at the peak of the TV Western craze and sold well. It features a modernized action with coil springs, as well as useful adjustable sights. Other barrel lengths were, and are, available; this one has a 6½-inch tube. This is shooting partner Stan Waugh's gun and I don't think you will get it away from him now that he knows how well it shoots.

The .357 magnum was not my favorite cartridge at the beginning of this effort, but the events of two exciting days of shooting could change all that. In these two test guns and the pair that follow on successive pages, we had several groups under an inch and enough close to that level to make it clear that the accuracy potential of the round is quite high.

Two powders were particularly good in these .357s: old reliable 2400 and the newer H110. The best bullets for accuracy purposes are the relatively new ones in the 140-grain weight. These jacketed bullets can be driven to velocities over 1500 feet per second in some guns and still produce target accuracy. We also had some superb results with several 158-grain jacketed bullets.

When I started the shooting program that produced this chapter, I would have bet that the smallest group would be produced by anything but what it turned out to be. Our smallest group measured .472 inches. It was from the GP-100 with 19 grains of H110 and the Hornady 140-grain JHP.

.357 MAGNUM LOAD DATA

BRASS	POWDER CHARGE (grains)	BULLET MAKE/WEIGHT/TYPE (grains)	GUN #1 ES	GUN #1 AVG	GUN #1 SD	GUN #1 GROUP (inches)	GUN #2 ES	GUN #2 AVG	GUN #2 SD	GUN #2 GROUP (inches)	COMMENTS
R-P	17.5 2400	HORNADY 125 JHP	82	1521	29	1¾	108	1642	39	2	
R-P	20.0 2400	HORNADY 110 JHP	37	1693	12	1+	75	1806	29	2½	Gun #1: 1.222", Highest velocity for both guns
R-P	16.5 2400	SPEER 140 JHP	49	1522	21	2	50	1547	20	2½	
R-P	16.8 2400	SPEER 140 JHP	36	1470	13	2+	77	1543	33	2+	Vertical strings both guns
R-P	16.5 2400	HORNADY 140 JHP	31	1464	14	1	63	1533	20	2½+	Gun #1: .793"
R-P	16.8 2400	HORNADY 140 JHP	33	1473	12	1¾	37	1565	14	2	Gun #1: 1.245" — Hot — Flat primers
R-P	16.5 2400	SIERRA 140 JHP	36	1462	13	2+	85	1529	29	2½	
R-P	16.8 2400	SIERRA 140 JHP	66	1459	25	2+	39	1561	13	2½	Hot load, flat primers
R-P	14.8 2400	SPEER 158 JHP	46	1285	18	2	46	1369	20	1¾	
R-P	14.8 2400	HORNADY 158 JHP	59	1317	22	1+	64	1398	26	1¾	Gun #1: 1.000"
R-P	14.8 2400	SIERRA 158 JSP	36	1291	13	1¾	71	1385	26	1	Gun #2: .692", best load this gun
R-P	15.0 2400	HORNADY 158 JHP	56	1337	24	1	42	1399	18	1	Gun #1: .965", Gun #2: .748"
R-P	15.0 2400	SPEER 158 JSP	81	1295	30	1½	73	1341	20	1¾	
R-P	19.8 H110	HORNADY 125 JHP	63	1354	26	1½+	74	1485	29	1½+	
R-P	19.8 H110	SIERRA 125 JSP	53	1341	22	1¾	97	1461	34	2	
FEDERAL	18.5 H110	SPEER 140 JHP	35	1316	12	2	84	1392	30	2+	
FEDERAL	19.0 H110	SPEER 140 JHP	69	1352	25	1+	35	1466	14	1¾	
FEDERAL	18.5 H110	HORNADY 140 JHP	37	1296	13	1+	55	1434	20	1½	
FEDERAL	19.0 H110	HORNADY 140 JHP	43	1334	15	1–	69	1424	26	1¼	Gun #1: .472" — Best group, any gun or load
FEDERAL	18.5 H110	SIERRA 140 JHP	51	1317	19	1–	42	1426	26	1¾	Gun #1: .543" — Second best group
FEDERAL	19.0 H110	SIERRA 140 JHP	55	1385	21	1¾	74	1449	26	1¾	
FEDERAL	16.0 H110	SPEER 158 JHP	59	1138	21	1¾	49	1240	17	2¾	
FEDERAL	16.0 H110	HORNADY 158 JHP	44	1191	17	1+	93	1258	34	2½	
FEDERAL	21.0 296	HORNADY 125 JHP	53	1366	21	1+	61	1497	23	2+	Best load with 296
FEDERAL	21.0 296	SIERRA 125 JSP	92	1345	34	1½+	78	1485	26	2½+	
FEDERAL	19.0 296	SPEER 140 JHP	48	1302	19	1¼	87	1389	83	2½	
FEDERAL	19.0 296	HORNADY 140 JHP	89	1312	30	1+	60	1421	24	2+	
FEDERAL	19.0 296	SIERRA 140 JSP	69	1344	31	2	61	1381	20	2¼	
FEDERAL	14.5 AA-9	HORNADY 140 JHP	119	1377	43	1+	30	1460	12	2+	Best load with AA-9
FEDERAL	15.0 AA-9	HORNADY 140 JHP	46	1415	16	1+	103	1488	40	2+	
FEDERAL	15.0 AA-9	SPEER 140 JHP	143	1373	55	2½	85	1452	31	2½	

FOOTNOTES:

1) All groups reported are six-shot groups, fired at 25 yards in the Ransom Rest.
2) Velocity is measured with Oehler 33 chronographs; Sky Screen IIIs spaced 36 inches apart.
3) Unless otherwise noted, group size is measured with a ruler, to the nearest quarter inch.
4) All cases primed with CCI 550 primers

.357 MAGNUM

AMMUNITION for the .357 has changed enormously, since it was introduced and the handloading procedures also have changed a bit. In the Thirties, there were no available jacketed bullets with which to load the .357. Everyone used hard cast, sometimes gas-checked bullets and, more often than not, those bullets went to the heavier side of the scale. The pendulum swung back in the post-war era, when jacketed bullets as light as 100 grains appeared on the market. Medium-weight guns also appeared, like the two we used to run the same ammo through.

Gun #1 — Colt Python

The Python was introduced in the mid-Fifties and has earned a reputation for quality and performance. The bores of the Colt .357s are a little tighter than on other guns and early Pythons were rumored to have bores that actually had a tiny amount of taper. Our test Python is a recent commercial specimen, with the famous dark blue finish and the ribbed barrel with the attractive underlug.

Gun #2 — Smith & Wesson Model 586

S&W pioneered the cartridge and guns in the Thirties, but this latest .357 is clearly intended to replace the heavier, N-framed Models 27 and 28. It is built on a completely new frame size, intermediate between K and N, and therefore easier to carry and handle. It is interesting to note that the 586 has an underlugged barrel which bears a striking resemblance to the Python's.

Both of these six-inch revolvers were more than acceptably accurate, but neither could match the phenomenal GP-100 discussed on the previous pages. After loading for four guns with the same data, it seems pretty clear that the most accurate bullet is going to weigh 140 grains.

With all of the guns tried, it was a 140-grain bullet atop a healthy charge of slow-burning pistol powder that turned in the best peformance. Since the long-barreled Python and 586 get those bullets up to velocities in excess of 1400 feet per second, it seems that most .357 magnum loading needs would be met with one of these combinations.

Depending on the gun, the best powders were 2400 and H110, followed closely by 296 and AA-9. We didn't try either 4227 and didn't experiment with a great many variations in 296 or AA-9.

In loading each of these rounds of test ammo, one thing was done a little differently. The final step in loading on our press setup is crimping. I personally believe magnum revolver rounds need to be firmly, even heavily, crimped, For this operation we used a Redding Profile Crimp Die, which combines the best of the taper and roll crimps.

The idea is to hold the bullet in place until the powder has raised pressures to the point that velocities will be consistent. I believe this die contributes to better ammunition.

.357 MAGNUM LOAD DATA

BRASS	POWDER CHARGE (grains)	BULLET MAKE/WEIGHT/TYPE (grains)	GUN #1 ES	GUN #1 AVG	GUN #1 SD	GUN #1 GROUP (inches)	GUN #2 ES	GUN #2 AVG	GUN #2 SD	GUN #2 GROUP (inches)	COMMENTS
R-P	17.5 2400	HORNADY 125 JHP	81	1601	33	2½	39	1617	13	3½	
R-P	20.0 2400	HORNADY 110 JHP	50	1761	17	5+	100	1770	39	2½	
R-P	16.5 2400	SPEER 140 JHP	33	1456	11	1¾	63	1478	23	1	Gun #2: .550", Gun #1: 1.677"
R-P	16.8 2400	SPEER 140 JHP	37	1481	16	2+	39	1492	22	2½	
R-P	16.5 2400	HORNADY 140 JHP	65	1435	26	1¾	67	1480	21	1	Gun #2: .840" (see page 161)
R-P	16.8 2400	HORNADY 140 JHP	69	1477	21	2¼	68	1500	23	2+	
R-P	16.5 2400	SIERRA 140 JHP	88	1412	27	2¾	66	1444	24	3+	
R-P	16.8 2400	SIERRA 140 JHP	109	1430	42	5+	156	1459	50	2+	Hot load
R-P	14.8 2400	SPEER 158 JHP	72	1244	36	3½	191	1302	57	3¾	
R-P	14.8 2400	HORNADY 158 JHP	85	1293	28	2+	66	1323	26	2+	
R-P	14.8 2400	SIERRA 158 JSP	63	1276	26	2½	124	1312	46	2¼	
R-P	15.0 2400	HORNADY 158 JHP	50	1321	16	3+	99	1341	36	2+	
R-P	15.0 2400	SPEER 158 JSP	57	1329	18	2¾	88	1358	56	2½	
R-P	19.8 H110	HORNADY 125 JHP	151	1426	53	1½	99	1461	43	2½	
R-P	19.8 H110	SIERRA 125 JSP	106	1393	36	1¾	111	1431	39	2½	
R-P	18.5 H110	SPEER 140 JHP	71	1292	18	2¾	78	1312	39	3½	
R-P	19.0 H110	SPEER 140 JHP	67	1299	38	2¼	81	1313	37	3½	
R-P	18.5 H110	HORNADY 140 JHP	89	1290	31	1¼	109	1312	36	2+	
R-P	19.0 H110	HORNADY 140 JHP	103	1323	38	1¾	80	1363	32	1	Gun #2: .525" — Best group from Smith (see page 161)
R-P	18.5 H110	SIERRA 140 JHP	29	1317	14	2	18	1331	17	2	
R-P	19.0 H110	SIERRA 140 JHP	33	1340	18	2½	22	1344	18	2+	
R-P	16.0 H110	SPEER 158 JHP	87	1087	19	2½	23	1099	24	2+	
R-P	16.0 H110	HORNADY 158 JHP	99	1113	38	2½	88	1168	39	2	
R-P	21.0 296	HORNADY 125 JHP	101	1285	56	2¾	111	1389	33	2½	
R-P	21.0 296	SIERRA 125 JSP	142	1372	55	2½	127	1395	48	2½	
R-P	19.0 296	SPEER 140 JHP	99	1244	37	3+	88	1199	44	3+	
R-P	19.0 296	HORNADY 140 JHP	130	1262	43	3¼	108	1315	60	1½	
R-P	19.0 296	SIERRA 140 JSP	87	1266	44	2½	44	1412	66	1¾	
R-P	14.5 AA-9	HORNADY 140 JHP	84	1333	46	2½	84	1329	40	2½	Best load w/AA-9
R-P	15.0 AA-9	HORNADY 140 JHP	72	1333	24	1¾	81	1390	29	2½	
R-P	15.0 AA-9	SPEER 140 JHP	55	1348	19	3	57	1371	24	1½	

FOOTNOTES:

1) All groups reported are six-shot groups, fired at 25 yards in the Ransom Rest.
2) Velocity is measured with Oehler 33 chronographs; Sky Screen IIIs spaced 36 inches apart.
3) Unless otherwise noted, group size is measured with a ruler, to the nearest quarter inch.
4) All cases primed with CCI 550 primers

.41 MAGNUM

OF ALL the cartridges for which we loaded ammunition in the course of compiling this chapter, the .41 magnum is easily the newest. Most of the others were at least several decades old when the .41 arrived and only the .44 magnum shares post-World War II origins with it. First conceived as a policeman's round, the .41 never really found its way into the holsters of all that many of the men on the beat. But it surely caught the imagination of the more discriminating of the nation's outdoorsmen and, more recently, the silhouette shooters. Our first pair of .41s are typical of the shorter-barreled versions of .41 revolvers.

Gun #1 — Smith & Wesson Model 57

The very first .41 magnums were Model 57s and their fixed-sight cousins, the Model 58s. Built around the durable N-frame, the 58s are heavy and that is one of the major reasons they had a rough time breaking into the police market. But the hunters and frontiersmen, who carry a gun on or around them most of the time, bought the revolver and handloaded for it. This particular gun is an older one with a four-inch barrel and many thousands of rounds through it — well broken in.

Gun #2 — Ruger Blackhawk

Almost as soon as Smith & Wesson announced the .41, Ruger wrapped their durable Blackhawk single-action revolver around the cartridge. In the western states and Alaska, where single-actions sell as well as they did the day after the Little Big Horn, the Blackhawks were extremely popular. More than a few were bought because they were a good bit less expensive than the Smith & Wesson or because the buyer couldn't find a .44. More often than not, the guy who took a .41 as second-best ended up loving the thing, as does the owner of this 4⅝-inch, stag-handled Ruger.

The author is a shooter who has no prejudices at all — except where .41 magnums are concerned. This is my all-time favorite handgun cartridge. I see the .41 as a near perfect blend of power and accuracy, all wrapped up in a handgun that is more durable than a comparable revolver chambered for the .44 magnum. All three of the major manufacturers produce brass and ammo for the cartridge.

Bullets are available in the .410-inch diameter from Speer, Sierra, Hornady and, more recently, Nosler. There are persistent rumors that Speer may broaden their line of superb TMJ (Totally Metal Jacketed) bullets in the next year or so. All sorts of bullet molds are available for the purpose of home casting bullets for the .41.

Since the .41 is truly a magnum cartridge, it well deserves a respectable powder charge. The slow-burning pistol powders are most commonly used for this purpose. We tried three of them — H110, 296 and 4227 — as well as Blue Dot. Accuracy was excellent with all of them.

.41 MAGNUM LOAD DATA

BRASS	POWDER CHARGE (grains)	BULLET MAKE/WEIGHT/TYPE (grains)	GUN #1 ES	GUN #1 AVG	GUN #1 SD	GUN #1 GROUP (inches)	GUN #2 ES	GUN #2 AVG	GUN #2 SD	GUN #2 GROUP (inches)	COMMENTS
FEDERAL	12.5 BLUE DOT	HORNADY 210 JHP	64	1045	23	1¾	63	1103	24	2+	
FEDERAL	13.0 BLUE DOT	HORNADY 210 JHP	67	1089	25	3	41	1154	13	3	Vertical stringing, both guns
FEDERAL	12.5 BLUE DOT	SPEER 220 JSP	48	1112	17	2½	89	1176	33	2½	
FEDERAL	13.0 BLUE DOT	SPEER 220 JSP	56	1104	21	3+	58	1183	22	3½	
FEDERAL	14.0 BLUE DOT	HORNADY 210 JTC	48	1197	17	1½	21	1284	7	1½	Best load with Blue Dot
FEDERAL	14.5 BLUE DOT	HORNADY 210 JTC	40	1220	14	1¾	51	1300	19	1¾	
FEDERAL	19.0 296	HORNADY 210 JTC	101	995	38	3½	82	1089	30	4	
FEDERAL	20.0 296	HORNADY 210 JTC	33	1062	12	2½	38	1137	18	2+	
FEDERAL	20.5 296	HORNADY 210 JTC	56	1116	19	3+	20	1210	9	2¾+	Vertical stringing, both guns
FEDERAL	20.5 296	HORNADY 210 JHP	62	1132	22	2¾	22	1241	8	3+	Vertical stringing, both guns
FEDERAL	19.0 296	SIERRA 220 JTC	75	998	32	2½	44	1085	16	3	More vertical stringing
FEDERAL	20.0 296	SIERRA 220 JTC	36	1089	14	3	22	1157	8	3¼	
FEDERAL	20.5 296	SIERRA 220 JTC	16	1166	6	2½	31	1246	12	2+	
FEDERAL	18.5 IMR4227	HORNADY 210 JHP	57	1031	21	2½	22	1101	8	1¾	
FEDERAL	19.0 IMR4227	HORNADY 210 JHP	71	1095	26	2½	20	1128	7	2+	
FEDERAL	19.5 IMR4227	HORNADY 210 JHP	26	1263	9	1¾	21	1265	8	1¾	Round groups, both guns
FEDERAL	18.5 IMR4227	SIERRA 220 JTC	94	1042	34	1½	54	1085	20	1½	
FEDERAL	19.0 IMR4227	SIERRA 220 JTC	53	1293	19	2¾	8	1284	3	2½	
FEDERAL	19.5 IMR4227	SIERRA 220 JTC	26	1274	9	3+	22	1281	11	3+	
R-P	19.0 H110	HORNADY 210 JTC	13	1262	5	1¾	5	1257	1	2+	Gun #1 had 4 shots @ 1257 fps; Gun #2 had 3
R-P	19.5 H110	HORNADY 210 JTC	7	1214	2	1¾	5	1230	2	1	Gun #2: .819" Most accurate load
R-P	20.0 H110	HORNADY 210 JTC	17	1225	10	3	80	1229	32	2½	
R-P	20.0 H110	SPEER 220 JSP	33	1297	10	3	5	1210	2	3½	
R-P	19.5 H110	HORNADY 210 JHP	14	1133	9	1½	53	1229	18	1¾	
R-P	19.0 H110	SIERRA 220 JTC	9	1261	3	1½	8	1255	2	1½	Low stats
R-P	19.5 H110	SIERRA 220 JTC	25	1247	10	2	15	1242	7	2+	
R-P	20.0 H110	SIERRA 220 JTC	5	1218	2		22	1220	11	2½	
R-P	19.5 H110	LYMAN CAST 210 SWC	40	1189	13	1½	48	1283	16	2½	

FOOTNOTES:

1) All groups reported are six-shot groups, fired at 25 yards in the Ransom Rest.
2) Velocity is measured with Oehler 33 chronographs; Sky Screen IIIs spaced 36 inches apart.
3) Unless otherwise noted, group size is measured with a ruler, to the nearest quarter inch.
4) All cases primed with Federal 155 primer.

.41 MAGNUM

.41 FOLKS are a small but growing group, typically vocal in their enthusiasm for the cartridge. But most of them who use the gun for silhouette or hunting purposes will prefer a longer barrel. Typical velocities can increase by as much as 200 fps when the barrel is four inches or so longer.

Gun #1 — Smith & Wesson Model 657

This particular revolver almost didn't make it into the test battery. It is an entirely typical 657 with an 8⅜-inch barrel, obtained in early 1986 when the factory announced the stainless steel version of the gun. I have another long-barreled .41, a blue Model 57 that is exceptionally accurate. Since I wanted to find out how this particular revolver behaved with handloads, I included it in the test. I'm not sorry — it's at least as good as its blued cousins.

Gun #2 — Ruger Bisley Model

The Bisley Ruger is a variation of the Blackhawk. The major difference is the grip frame, which is shaped somewhat like the one on the original Colt revolvers. The Bisley name is for the target range in Great Britain where the big matches were once held. As the use of the name suggests, the first Bisleys were intended for target work. The Ruger Bisley may end up with the same distinction, in the sense that the grip shape is far easier to handle with heavy-recoiling ammo than the usual frame. This particular specimen is nearly new, with only a hundred or so rounds through it.

With Blue Dot powder, these two revolvers produced velocities nearly 1400 fps. But accuracy deteriorated to the point that it was a waste of components. With other powders, such as H110, 296 and 4227, velocities are somewhat lower but the accuracy level is much higher. And that, after all, has as much to do with performance as extreme velocity.

Any number of fine bullets proved to be accurate in our testing, but three stood out. They were the 210-grain Hornady JHP, the 210-grain Hornady JTC/Sil bullet and the Sierra 220-grain Flat Point.

The two latter bullets are reflective of a shooting phenomenon. They are projectiles primarily intended for the silhouette shooter, who faces one of the damnedest challenges ever offered a pistolero. He has to knock down the heavy steel cut-outs of game animals at ranges as great as 200 meters. That's a long shot, so he needs a lot of inherent accuracy. But the distant targets can't be powder-puffed over, they have to be *driven* from their stands. And that means that he has to be using a powerful six-gun. Properly loaded, the .41 magnum is equal to the challenge.

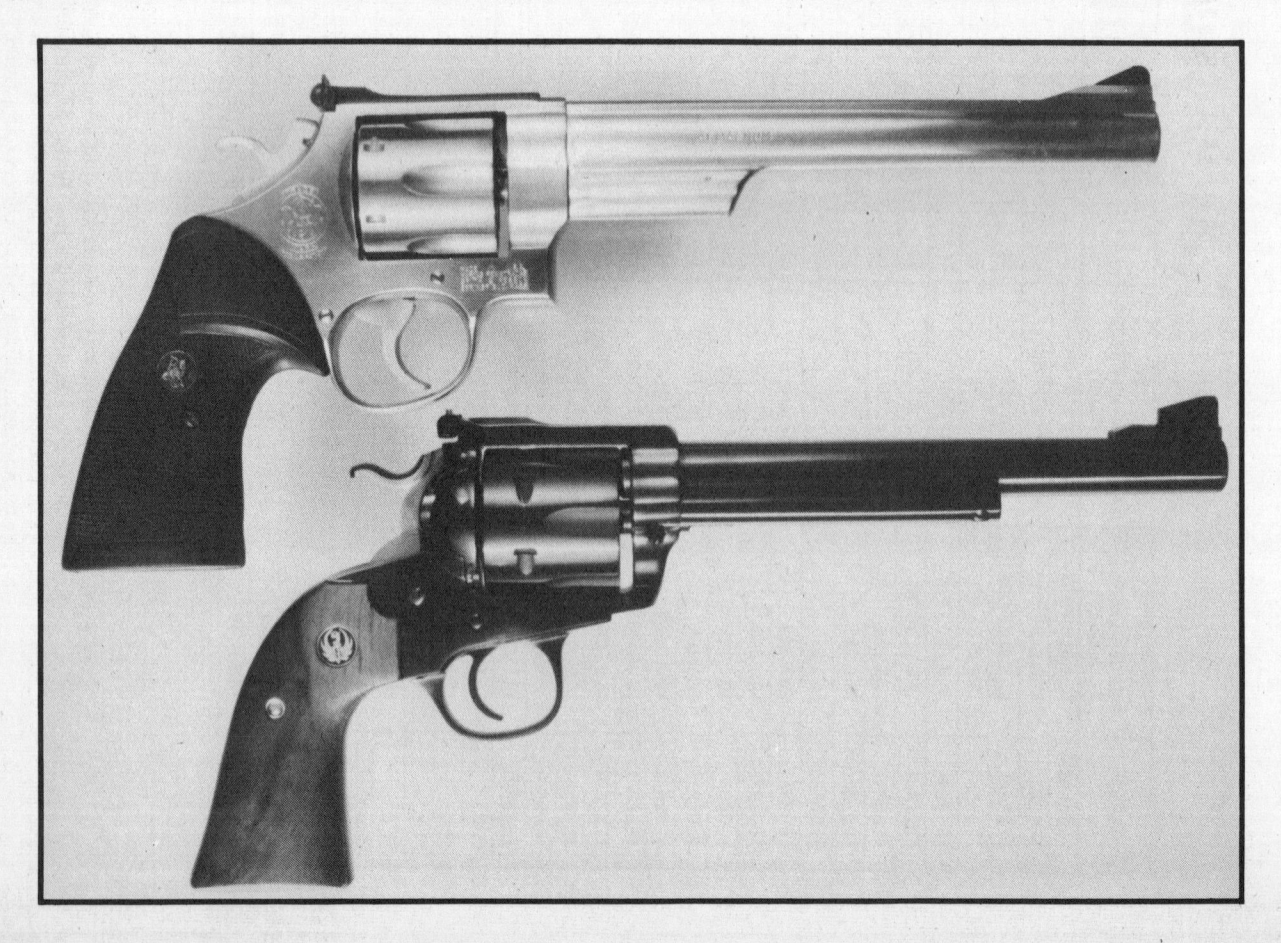

.41 MAGNUM LOAD DATA

BRASS	POWDER CHARGE (grains)	BULLET MAKE/WEIGHT/TYPE (grains)	GUN #1 ES	GUN #1 AVG	GUN #1 SD	GUN #1 GROUP (inches)	GUN #2 ES	GUN #2 AVG	GUN #2 SD	GUN #2 GROUP (inches)	COMMENTS
FEDERAL	12.5 BLUE DOT	HORNADY 210 JHP	49	1216	20	3	51	1255	18	1¾	
FEDERAL	13.0 BLUE DOT	HORNADY 210 JHP	51	1222	19	2¾	54	1259	22	1¾	
FEDERAL	12.5 BLUE DOT	SPEER 220 JSP	48	1212	21	3	52	1231	22	1¾	
FEDERAL	13.0 BLUE DOT	SPEER 220 JSP	50	1244	21	2½	56	1252	33	2+	
FEDERAL	14.0 BLUE DOT	HORNADY 210 JTC	89	1330	32	1½	91	1369	33	2+	Best load with Blue Dot
FEDERAL	14.5 BLUE DOT	HORNADY 210 JTC	103	1355	40	2½	53	1396	19	2¾	
FEDERAL	19.0 296	HORNADY 210 JTC	57	1231	22	1¾	33	1281	11	2+	
FEDERAL	20.0 296	HORNADY 210 JTC	61	1250	21	2+	56	1301	18	2+	
FEDERAL	20.5 296	HORNADY 210 JTC	46	1290	16	1½	38	1345	13	1¾	Best load w/296, both guns around 1¼"
FEDERAL	20.5 296	HORNADY 210 JHP	59	1284	21	1¼	42	1316	13	1¼	
FEDERAL	19.0 296	SIERRA 220 JTC	39	1251	22	1¾	52	1313	18	1½	
FEDERAL	20.0 296	SIERRA 220 JTC	69	1277	24	2¼	57	1328	20	2	
FEDERAL	20.5 296	SIERRA 220 JTC	34	1299	13	1¾	56	1350	22	2	Highest velocity load, both guns
FEDERAL	18.5 IMR4227	HORNADY 210 JHP	24	1140	9	2+	50	1165	18	2¼	
FEDERAL	19.0 IMR4227	HORNADY 210 JHP	32	1122	9	2+	40	1175	18	2¼	
FEDERAL	19.5 IMR4227	HORNADY 210 JHP	38	1140	15	1¼	41	1190	14	1½	
FEDERAL	18.5 IMR4227	SIERRA 220 JTC	33	1109	13	1+	16	1122	15	1¼	Gun #1: 1.031"
FEDERAL	19.0 IMR4227	SIERRA 220 JTC	41	1121	20	1¼	15	1140	16	1½	Gun #2: .981", second most accurate load
FEDERAL	19.5 IMR4227	SIERRA 220 JTC	55	1155	17	1¼	31	1193	15	1	Gun #1: 1.201"
R-P	19.0 H110	HORNADY 210 JTC	56	1254	21	1¼	30	1295	11	1½	
R-P	19.5 H110	HORNADY 210 JTC	45	1261	21	1¼	31	1308	12	1½	
R-P	20.0 H110	HORNADY 210 JTC	34	1279	23	1½	40	1321	15	1¾	
R-P	20.0 H110	SPEER 220 JSP	31	1240	21	2+	50	1269	24	3+	
R-P	19.5 H110	HORNADY 210 JHP	33	1211	22	2+	50	1222	20	1¾	
R-P	19.0 H110	SIERRA 220 JTC	36	1222	19	1½	41	1256	24	1½	
R-P	19.5 H110	SIERRA 220 JTC	39	1239	16	1¾	51	1284	18	1	Gun #2: .839", most accurate load
R-P	20.0 H110	SIERRA 220 JTC	50	1252	16	1¾	66	1292	26	2	
R-P	19.5 H110	LYMAN CAST 210 SWC	56	1200	19	2+	63	1211	29	2¼	

FOOTNOTES:

1) All groups reported are six-shot groups, fired at 25 yards in the Ransom Rest.
2) Velocity is measured with Oehler 33 chronographs; Sky Screen IIIs spaced 36 inches apart.
3) Unless otherwise noted, group size is measured with a ruler, to the nearest quarter inch.
4) All cases primed with Federal 155 primer.

.44 SPECIAL

AS AN updated and improved version of the earlier and shorter .44 Russian cartridge, the .44 Special (often called the .44 S&W Special) was introduced in the S&W Hand Ejector New Century Model in the first decade of this century. Collectors and shooters alike have come to use the same name for that handgun — the Triple Lock. The .44 Special is a round that is much beloved of revolvermen and one that has returned in new guns in modern times because of popular demand.

Gun #1 — Smith & Wesson Model 624

The Model 624 is the true lineal descendant of the original Triple Lock. It is built on the company's large N-frame and the "6" prefix means that the revolver is crafted of stainless steel. There is a lot of logic to the use of this handgun in the hands of a handloading outdoor type. For one thing, it is a considerably lighter gun than the .44 magnum. Carefully handloaded, the 624 is a fine shooter.

Gun #2 — Colt New Frontier Single Action

Colt has made many more single-action revolvers in the .44 Special persuasion in the last decade than they ever did in the days before WWII. Gunwriters such as Skeeter Skelton nagged the Colt firm into producing a number of variations of the single-action in .44. Our test gun is one variation of the basic single-action. It is the New Frontier model, with adjustable sights and 7½-inch barrel.

There is no shortage of load data for the .44 Special cartridge and some of it dates to a time when there was no .44 magnum. The .44 Special is not a magnum and shouldn't be loaded in that fashion.

Many of the fast-burning pistol powders perform nicely in the .44 Special. 231, HP38 and Bullseye are all good performers. If you need a little more velocity, try Unique. For many shooters, the powder of choice is 2400. We were able to get the famous Elmer Keith bullet, Lyman's #429421, up to around 1100 fps with the popular 2400.

The legendary accuracy of the .44 Special did not materialize in either gun. The best groups were around two inches and any number of other revolvers used in these tests in other calibers will beat that. I firmly believe either of our test guns will do a good bit better with more tinkering and load development.

If you are using an older .44 Special, be watchfully cautious with your handloading, staying alert for signs of excessive pressure. The quoted loads are for the guns listed only.

.44 SPECIAL LOAD DATA

BRASS	POWDER CHARGE (grains)	BULLET MAKE/WEIGHT/TYPE (grains)	GUN #1 ES	GUN #1 AVG	GUN #1 SD	GUN #1 GROUP (inches)	GUN #2 ES	GUN #2 AVG	GUN #2 SD	GUN #2 GROUP (inches)	COMMENTS
R-P	4.5 HP38	HORNADY 240 LSWC	52	614	16	3+	19	663	16	2¾	
R-P	5.0 HP38	HORNADY 240 LSWC	34	669	12	2¾	36	726	14	2	Most accurate load in Colt: 1.881"
R-P	4.5 HP38	SIERRA 240 JHP	38	534	13	4+	44	625	15	3+	
R-P	5.0 HP38	SIERRA 240 JHP	26	594	10	3½+	58	675	18	3+	
R-P	5.0 HP38	LYMAN CAST #429421	25	661	8	2+	14	716	4	2+	Group size almost identical in both guns, best S&W load
R-P	9.5 AA-7	LYMAN CAST #429421	29	551	28	3+	112	697	53	2½	Velocity is too low for accuracy
R-P	7.5 UNIQUE	ALBERTS(swaged) 240 LSWC	19	792	10	4½	17	833	7	4+	
R-P	8.0 UNIQUE	ALBERTS(swaged) 240 LSWC	44	819	17	5	64	870	21	3½	
R-P	8.5 UNIQUE	ALBERTS(swaged) 240 LSWC	65	946	56	3¼	59	962	23	3	
R-P	8.5 UNIQUE	HORNADY 180 JHP	96	783	32	2½	162	956	59	3+	
R-P	9.0 UNIQUE	HORNADY 180 JHP	77	919	30	3+	110	1010	43	2½	
R-P	10.0 2400	ALBERTS(swaged) 240 LSWC	60	658	27	3½	40	699	40	3	Vertical stringing in Colt
R-P	11.0 2400	ALBERTS(swaged) 240 LSWC	62	717	20	3+	85	763	26	2½	
R-P	11.0 2400	SPEER 240 TMJ	50	659	51	3½	48	707	16	2¼	
R-P	12.0 2400	SPEER 240 TMJ	77	709	28	3+	57	809	21	2+	
R-P	13.0 2400	SPEER 240 TMJ	111	824	51	3½	89	906	41	3+	
R-P	13.5 2400	SPEER 240 TMJ	62	844	23	2½	96	951	31	2	
R-P	14.0 2400	SPEER 240 TMJ	61	860	24	3+	35	963	12	2¾	
R-P	16.0 2400	SPEER 240 TMJ	46	1036	17	3¼	40	1127	16	2¾	
R-P	16.0 2400	LYMAN CAST #429421	36	1073	13	3½+	83	1173	27	2½	Highest velocity load in both guns. Use only in strong, modern guns
R-P	5.2 231	HORNADY 240 LSWC	44	721	22	3+	45	760	28	3¼	
R-P	5.5 231	HORNADY 240 LSWC	60	747	29	2¾	59	788	27	2¾	
R-P	6.0 231	HORNADY 240 LSWC	39	812	22	3+	46	844	22	2¾	
R-P	6.5 231	HORNADY 240 LSWC	37	868	32	2¾	48	901	40	2½	

FOOTNOTES:

1) All groups reported are six-shot groups, fired at 25 yards in the Ransom Rest.
2) Velocity is measured with Oehler 33 chronographs; Sky Screen IIIs spaced 36 inches apart.
3) Unless otherwise noted, group size is measured with a ruler, to the nearest quarter inch.
4) All cases primed with Federal #150.

.44 MAGNUM

MAYBE IT'S because the number just trips lightly off the tongue and maybe it's because movie actors made so much pretend use of their power, but the .44 is here to stay. Originally produced at the insistence of the late Elmer Keith, the .44 magnum has been around for better than thirty years and shows no signs of diminishing popularity. While it once could claim to be the most powerful of handgun cartridges, the .44 may now be eclipsed by a couple of limited production numbers. It is just about all that the average fellow is likely to need.

Gun #1 — Smith & Wesson Model 29

The first .44 magnums were like this one, built on the ever-present N-frame. While most .44 users would opt for a longer barrel to develop a bit more velocity, some still cling to the four-inch version that we have at hand. Guns like these are often fired with .44 Special ammo. It is the most carryable of the .44 magnums and developed speeds in excess of 1300 fps with some loads. When the stiffer loads were touched off with the gun securely bolted into the Ransom Rest, the muzzle pivoted toward the sky — quickly.

Gun #2 — Ruger Blackhawk

In the early .44 magnum era, Ruger produced a revolver much like this one, chambered for the cartridge. In short order, the gun was changed to a heavier version called the Super Blackhawk. In the past few years, Ruger introduced the compromise gun that you see here. It has the Super Blackhawk mainframe, but the butt is the plain Blackhawk type and the cylinder is fluted. The result is an attractive and easy-to-pack six-gun that sports a 5½-inch barrel.

It was another pleasant surprise to find a pair of handguns that shot as well as these two did. While the best .44 magnum group came from the Ruger, the S&W was consistently accurate with nearly anything that was put through it.

Despite their relatively short barrels, these two guns are still magnums, with internal ballistic characteristics that work best with healthy powder charges. We used the classic slow-burning powders in loading the rounds. H110, IMR4227, 296 and 2400 worked well with a variety of bullets. With each of these powders, the test guns were each capable of twenty-five-yard groups in the area of an inch-and-a-half. That is fine accuracy for a magnum revolver.

The hotter loads with each powder were perfectly safe, showing only minor signs of primer flattening and not a bit of sticky extraction. Nevertheless, these and any other really hot loads should be reserved for something less than constant use in the guns as they can be pretty hard on action timing.

.44 MAGNUM LOAD DATA

BRASS	POWDER CHARGE (grains)	BULLET MAKE/WEIGHT/TYPE (grains)	ES	GUN #1 AVG	GUN #1 SD	GROUP (inches)	ES	GUN #2 AVG	GUN #2 SD	GROUP (inches)	COMMENTS
FEDERAL	18.0 AA-9	HORNADY 240 JTC	26	1107	11	2½	33	1228	12	2	
FEDERAL	18.4 AA-9	HORNADY 240 JTC	33	1122	14	2+	34	1250	14	2½	
FEDERAL	19.0 AA-9	HORNADY 240 JTC	40	1161	14	2¼	33	1285	13	3	
FEDERAL	19.0 AA-9	HORNADY 240 JHP	31	1166	10	1½	39	1267	14	1½	
FEDERAL	19.0 AA-9	HORNADY 285 JSP	28	1133	10	1½	21	1257	8	1+	Most accurate load for both guns
FEDERAL	19.0 AA-9	BARNES 275 JSP	27	1138	9	1¾	27	1243	9	1¾	
FEDERAL	24.0 IMR4227	HORNADY 180 JHP	48	1201	22	2+	64	1293	23	1¾	
FEDERAL	24.0 IMR4227	HORNADY 240 JTC	49	1238	15	2+	19	1335	7	2½	
FEDERAL	24.0 IMR4227	SIERRA 240 FMJ/FP	33	1282	17	2+	47	1359	19	1½	Compressed
FEDERAL	24.0 IMR4227	SPEER 240 TMJ	23	1196	8	2½	28	1275	9	1¾	Compressed
FEDERAL	24.0 IMR4227	SPEER 240 JHP	37	1187	12	3+	48	1312	16	2½	Compressed
FEDERAL	24.5 IMR4227	HORNADY 240 JTC	39	1204	14	2½	35	1305	15	1¾	Compressed
FEDERAL	23.0 296	HORNADY 240 JHP	27	1175	9	1+	32	1272	11	1¾	Most accurate load in #1: 1.211"
FEDERAL	23.5 296	HORNADY 240 JHP	33	1212	10	1¾	30	1292	10	2½	
FEDERAL	23.5 296	SIERRA 240 JHP	21	1222	9	1+	49	1343	17	2	Second most accurate load in #1: 1.244
FEDERAL	23.5 296	HORNADY 265 JSP	42	1152	20	2+	37	1205	13	2½	
FEDERAL	23.5 296	SIERRA 250 FMJ	36	1159	17	1½	24	1238	12	1½	
FEDERAL	21.0 H110	SIERRA 240 JHP	56	1089	22	2	31	1134	10	1¾	
FEDERAL	22.0 H110	SIERRA 240 JHP	47	1088	23	3	114	1153	33	3	
FEDERAL	22.0 H110	SIERRA 265 JSP	101	1136	40	1¾	26	1133	9	2	
FEDERAL	20.0 2400	SIERRA 240 JHP	26	1250	8	2+	55	1305	21	1¾	
FEDERAL	21.0 2400	SIERRA 240 JHP	22	1281	9	2	31	1350	14	1½	
FEDERAL	21.0 2400	HORNADY 265 JSP	27	1143	10	2½	25	1301	8	1½	
FEDERAL	21.5 2400	HORNADY 265 JSP	27	1178	11	1¾	29	1317	11	2	
FEDERAL	22.0 2400	HORNADY 240 JTC	33	1240	14	3½	24	1381	9	3	
FEDERAL	22.0 2400	HORNADY 240 JHP	33	1299	14	3+	56	1374	20	3½	
FEDERAL	16.5 BLUE DOT	HORNADY 240 JHP	39	1301	19	2+	42	1345	15	1¾	
FEDERAL	17.5 BLUE DOT	HORNADY 240 JHP	42	1316	14	1¾	45	1380	11	2½	Highest velocity in Gun #1 & #2
FEDERAL	17.5 BLUE DOT	HORNADY 265 JSP	36	1249	22	2+	68	1341	22	2	

FOOTNOTES:

1) All groups reported are six-shot groups, fired at 25 yards in the Ransom Rest.
2) Velocity is measured with Oehler 33 chronographs; Sky Screen IIIs spaced 36 inches apart.
3) Unless otherwise noted, group size is measured with a ruler, to the nearest quarter inch.
4) All cases primed with Federal #155 primers.

.44 MAGNUM

LIKE ITS smaller brother in .41 size, the .44 magnum needs a bit more barrel to really shine. All makers of the guns know that and can provide barrels in excess of ten inches. When the propelling powder is as slow-burning as the ones that we have chosen, the greater length allows the powder to burn more efficiently and down-range performance is better.

Gun #1 — Smith & Wesson Model 29

Until a few years ago, this was the longest barrel you could get on a 29, but Smith & Wesson has introduced an even longer barreled .44 in a special Silhouette version. About 8⅜ inches of barrel is all that most of us will need. This particular gun had only a few shots through it before the tests began. It proved to be quite accurate, close to the performance of its short-barreled brother.

Gun #2 — Ruger Super Redhawk

This is a brand-new gun on the market and the first example of the model that I have seen. The Super Redhawk is clearly intended to be the ultimate big-bore double-action revolver. If the rest of the breed turn out to be the performer that this one is, the Super Redhawk will be a winner. Although it is heavy, the revolver is still a usable handgun. The basic frame is the heaviest ever made for a DA revolver, with an extension at the forward end that provides a seat for the barrel. On this revolver, the scope mount clamps securely to the frame.

With both of the long-barreled .44 magnums, there was a noticeable difference in the arc of recoil in the Ransom Rest; they are far less vigorous in their recoil. We fired some hand-held shots and either one would be pleasant to shoot.

In assembling the handloads with bullets ranging from 180 to 275 grains in weight, it became perfectly clear that the guns would produce tight groups with at least one combination of the bullets and powders chosen. Going through the same batch of load combinations used in the guns on the previous page, the groups would fluctuate a little, but the norm was accuracy. Many combinations were under or near two inches.

There are plenty of bullets available for the .44 and we used them from the little 180-grain pill just introduced by Hornady up to the hefty 265-grain Hornady bullet originally made for the .444 Marlin. This was a pleasing cartridge and pair of guns to load, one of the more rewarding efforts in the entire series.

But it was not without its inexplicable curiosities. Look at the velocities produced with the Ruger Blackhawk on the preceeding page, then at the ones delivered by the Smith & Wesson on this one. The Ruger velocities were higher in the 5½-inch barrel than they were in the longer 8⅜-inch S&W!

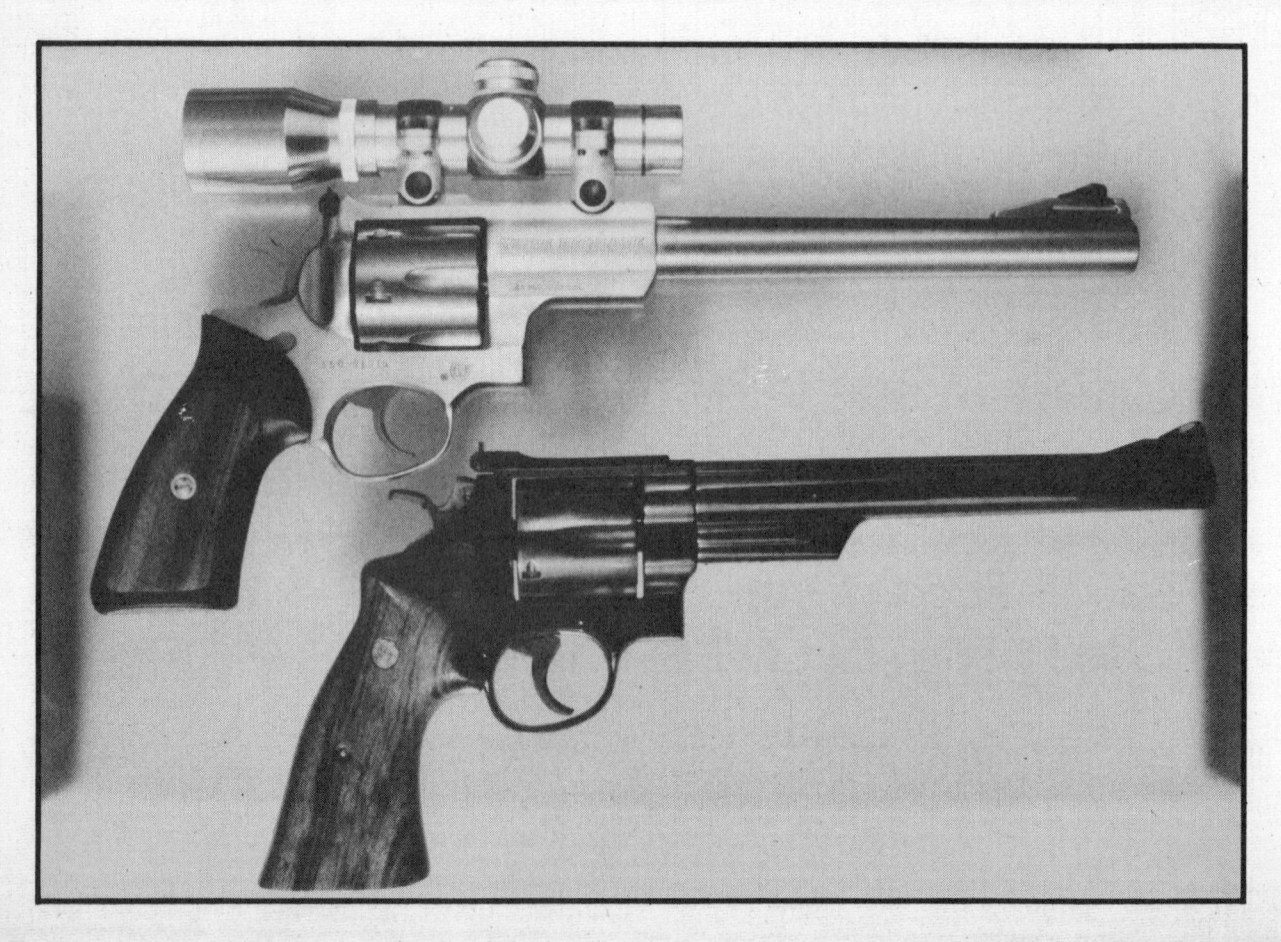

.44 MAGNUM LOAD DATA

BRASS	POWDER CHARGE (grains)	BULLET MAKE/WEIGHT/TYPE (grains)	GUN #1 ES	GUN #1 AVG	GUN #1 SD	GUN #1 GROUP (inches)	GUN #2 ES	GUN #2 AVG	GUN #2 SD	GUN #2 GROUP (inches)	COMMENTS
FEDERAL	18.0 AA-9	HORNADY 240 JTC	44	1247	22	2	56	1600	29	2+	
FEDERAL	18.4 AA-9	HORNADY 240 JTC	15	1268	6	2½	58	1608	23	2+	Highest velocity in #2
FEDERAL	19.0 AA-9	HORNADY 240 JTC	78	1282	30	3+	72	1300	29	3¼	
FEDERAL	19.0 AA-9	HORNADY 240 JHP	53	1258	19	1½	59	1335	22	1½	Both guns just over 1¼"
FEDERAL	19.0 AA-9	HORNADY 285 JSP	25	1229	8	2	68	1277	24	1¾	
FEDERAL	19.0 AA-9	BARNES 275 JSP	29	1234	10	3+	32	1259	12	2½	
FEDERAL	24.0 IMR4227	HORNADY 180 JHP	82	1315	29	2½	77	1359	31	2¾	
FEDERAL	24.0 IMR4227	HORNADY 240 JTC	62	1318	24	2¾	37	1409	12	2¾	
FEDERAL	24.0 IMR4227	SIERRA 240 FMJ/FP	30	1332	11	1½	32	1401	19	1¾	Compressed
FEDERAL	24.0 IMR4227	SPEER 240 TMJ	69	1243	29	2½	39	1380	12	2¾	Compressed
FEDERAL	24.0 IMR4227	SPEER 240 JHP	51	1282	23	2	49	1410	23	2+	Compressed
FEDERAL	24.5 IMR4227	HORNADY 240 JTC	51	1278	9	3½	60	1420	21	2+	Compressed
FEDERAL	23.0 296	HORNADY 240 JHP	42	1268	14	1½	47	1390	17	1+	Second most accurate load
FEDERAL	23.5 296	HORNADY 240 JHP	83	1287	26	2+	77	1420	33	1+	Most accurate load in #2
FEDERAL	23.5 296	SIERRA 240 JHP	68	1347	24	1½	54	1424	21	1¾	
FEDERAL	23.5 296	HORNADY 265 JSP	77	1211	27	4+	60	1428	30	3+	
FEDERAL	23.5 296	SIERRA 250 FMJ	38	1220	15	3+	56	1399	30	3+	
FEDERAL	21.0 H110	SIERRA 240 JHP	114	1126	38	2½	50	1289	29	3+	
FEDERAL	22.0 H110	SIERRA 240 JHP	89	1115	32	3	88	1190	36	3+	
FEDERAL	22.0 H110	SIERRA 265 JSP	41	1113	18	2½	112	1308	39	1½	
FEDERAL	20.0 2400	SIERRA 240 JHP	57	1288	20	1¾	66	1366	28	1½	
FEDERAL	21.0 2400	SIERRA 240 JHP	90	1331	36	2¼	88	1452	33	2+	
FEDERAL	21.0 2400	HORNADY 265 JSP	68	1269	26	1½	27	1350	9	2	
FEDERAL	21.5 2400	HORNADY 265 JSP	43	1306	17	2¼	77	1373	76	2¾	
FEDERAL	22.0 2400	HORNADY 240 JTC	48	1330	18	3+	49	1447	18	2¾	
FEDERAL	22.0 2400	HORNADY 240 JHP	48	1371	18	1¾	47	1499	23	2½	
FEDERAL	16.5 BLUE DOT	HORNADY 240 JHP	50	1333	19	2	56	1500	23	2+	
FEDERAL	17.5 BLUE DOT	HORNADY 240 JHP	78	1383	29	1½	77	1512	29	1¾	Highest velocity in #1
FEDERAL	17.5 BLUE DOT	HORNADY 265 JSP	61	1313	24	2	80	1483	30	1¾	

FOOTNOTES:

1) All groups reported are six-shot groups, fired at 25 yards in the Ransom Rest.
2) Velocity is measured with Oehler 33 chronographs; Sky Screen IIIs spaced 36 inches apart.
3) Unless otherwise noted, group size is measured with a ruler, to the nearest quarter inch.
4) All cases primed with Federal #155 primers.

.45 ACP

THE .45 has been the U.S. service cartridge for the vast majority of this troubled century. Beyond its service in uniform, the round is exceptionally popular with the civilians. It has, in recent years, attracted a following as a law enforcement round. In competition, the .45 ACP is used for everything from bullseyes to bowling pins. Despite the popularity of the .38 Super on the IPSC circuit, the .45 is probably still the more widely used cartridge. And out in Arizona, Jeff Cooper continues to advocate (with irrefutable logic) the .45 as the premier defense round. This is a *popular* cartridge.

Gun #1 — Smith & Wesson Model 645

The S&W factory fooled with a .45 auto for the best part of two decades before they produced the 645 just a couple of years ago. What they offered in the new gun for the old round was a fine handgun indeed. It is made of stainless steel and has the double-action trigger that is so popular. Beyond that, the 645 is a rugged and dependable automatic pistol that shoots with commendable accuracy. Many police departments have adopted the pistol. The test gun has fired about eight hundred shots, enough to break it in satisfactorily.

Gun #2 — Colt Gold Cup

It is nearly impossible to describe the effect of the Colt .45 auto in a paragraph. With all due deference to the single-action fanciers and the fans of K-frame Smiths, the .45 is America's favorite handgun. It has been made by countless factories other than the original one in Hartford and is the favorite gun of all kinds of competitors. It has fought several wars and seems to go on forever. Our test gun is one of the flossiest of the breed ever to leave the factory — a Gold Cup. This is a late version, with the heavier slide. It has fine adjustable sights, deep blue finish and an excellent trigger pull. Except for the trigger pull, the gun is "stock."

.45 ammunition is handloaded for all sorts of purposes and we weren't able to explore all of the variations of the many themes. All .45 ammunition — actually, all automatic pistol ammo — has a limitation that revolver fodder doesn't. Autoloading pistol ammunition must be contoured to move easily and reliably through a feeding/chambering cycle that revolver ammo doesn't have to match. The shooter manually inserts the round in the chamber of the revolver, a luxury he doesn't have with a .45 auto.

This means that such matters as overall cartridge length and degree of crimp are important. The bulletmakers are aware of this and all three (Speer, Sierra and Hornady) produce bullets that functioned without fail in our test guns, when those bullets were correctly seated in good RCBS dies.

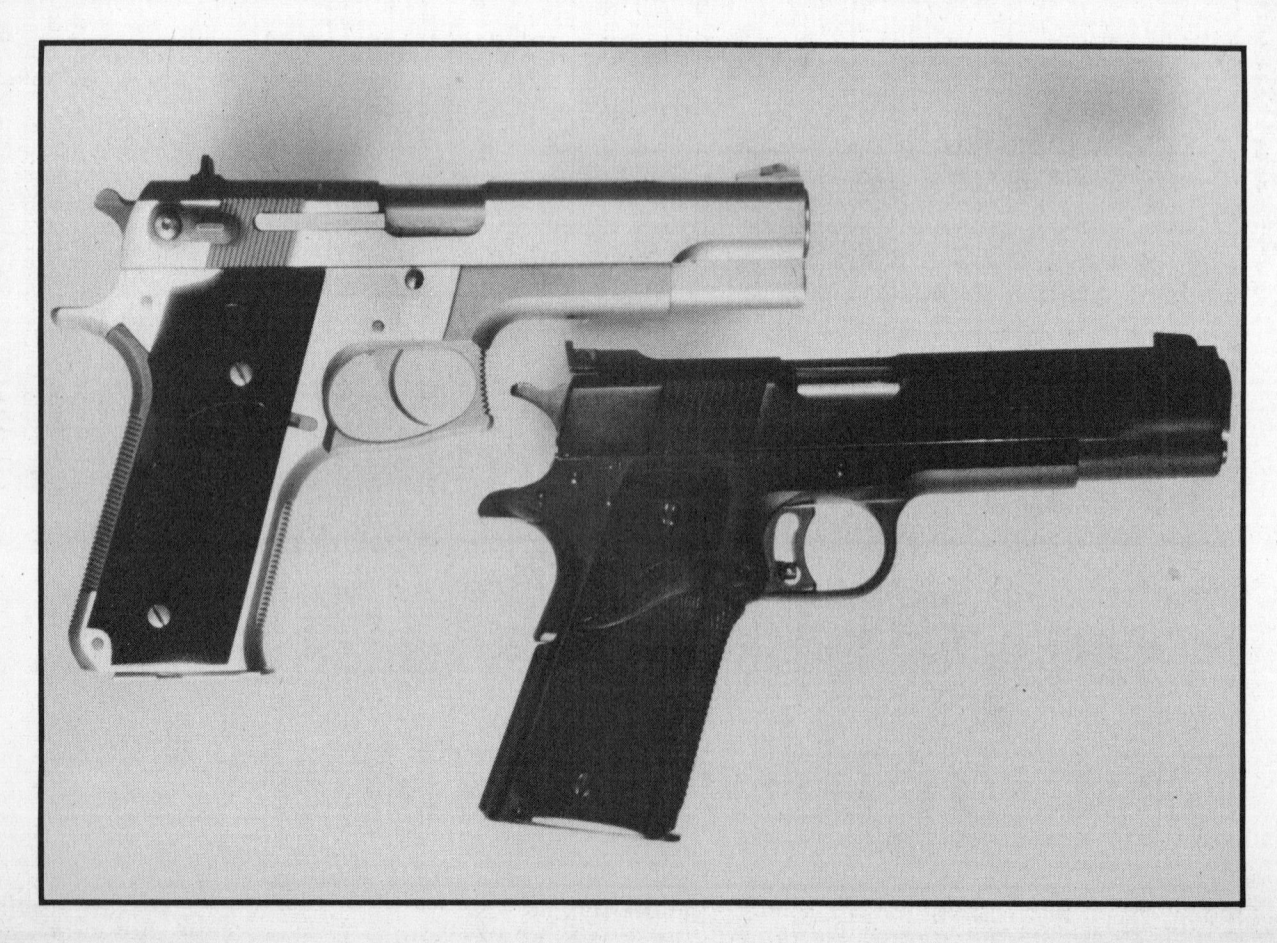

.45 ACP LOAD DATA

BRASS	POWDER CHARGE (grains)	BULLET MAKE/WEIGHT/TYPE (grains)	ES	GUN #1 AVG	GUN #1 SD	GROUP (inches)	ES	GUN #2 AVG	GUN #2 SD	GROUP (inches)	COMMENTS
W-W	4.1 BULLSEYE	HORNADY 185 FMJ	42	741	15	2+	48	739	16	2	
W-W	4.4 BULLSEYE	HORNADY 185 FMJ	31	750	12	2+	50	759	21	3	Horizontal stringing
W-W	4.7 BULLSEYE	HORNADY 185 FMJ	54	795	17	3+	34	810	13	4	
W-W	4.1 BULLSEYE	SIERRA 185 MATCH	56	695	20	2½	46	706	15	3+	
W-W	4.4 BULLSEYE	SIERRA 185 MATCH	43	700	15	3	61	719	19	3+	
W-W	4.1 BULLSEYE	SPEER 185 MATCH	55	698	20	2½	60	690	24	3	
W-W	4.7 BULLSEYE	HORNADY 200 FMJ C/T	57	779	20	1½	45	776	15	2¾	Best load w/Bullseye
W-W	4.7 BULLSEYE	SPEER 200 TMJ	38	761	12	3+	62	771	20	2½	Five in one hole, Gun #2
W-W	4.7 BULLSEYE	SIERRA 200 MATCH	48	756	19	3+	19	766	8	2+	
W-W	4.7 BULLSEYE	HORNADY 230 FMJ/FP	27	777	9	2+	58	768	18	4+	
W-W	4.7 700-X	HORNADY 185 MATCH	64	845	26	1¾	99	850	36	2	
W-W	4.7 700-X	SPEER 185 TMJ	120	715	48	1¾	69	779	78	4	
W-W	4.7 700-X	HORNADY 200 FMJ C/T	57	806	21	2	99	818	40	2	
W-W	4.7 700-X	HORNADY 230 FMJ/FP	101	786	38	3+	94	792	35	3+	
W-W	4.7 700-X	HORNADY 185 MATCH	127	825	40	3+	93	882	30	1¾	High ES
W-W	5.0 700-X	SIERRA 185 JHP	50	858	18	2+	45	848	16	6+	
W-W	5.0 700-X	HORNADY 200 FMJ C/T	55	851	21	2+	60	884	21	2½	
W-W	5.0 700-X	HORNADY 230 FMJ/FP	74	844	24	1¾	30	845	11	1¾	Most accurate load, Gun #1: 1.689"
W-W	5.0 700-X	HORNADY 185 MATCH	29	818	11	1¾	40	815	18	1¾	Most accurate load, Gun #2: 1.778"
W-W	5.5 231	SIERRA 185 JHP	62	753	21	2¾	67	756	25	5+	Horizontal strings, Gun #2
W-W	5.5 231	HORNADY 200 FMJ C/T	32	789	12	2½	38	774	16	4	
W-W	5.5 231	HORNADY 230 FMJ F/P	46	765	15	3	22	755	10	5	
W-W	4.6 RED DOT	SPEER 185 TMJ	67	763	31	3+	87	747	41	3+	
W-W	5.4 RED DOT	HORNADY 200 C/T	79	829	29	2½	89	890	32	3½	Highest velocity in Gun #2
W-W	5.4 RED DOT	HORNADY 230 FMJ/FP	79	838	41	2+	81	860	29	2½	

FOOTNOTES:

1) All groups reported are six-shot groups, fired at 25 yards in the Ransom Rest.
2) Velocity is measured with Oehler 33 chronographs; Sky Screen IIIs spaced 36 inches apart.
3) Unless otherwise noted, group size is measured with a ruler, to the nearest quarter inch.
4) All cases primed with Federal #150 primers

.45 ACP

WHEN THE U.S. Army went off to World War I, it didn't have enough 1911 Colts with which to equip the troops. The large frame revolvers already in production at both the S&W and Colt factories were pressed into service with cylinders correctly chambered to take the .45 Auto round. With this began a long tradition of use of the .45 ACP in revolvers. That's one aspect of the never-ending story of .45s in the American consciousness. Another is the vast number of autos that have been customized in one way or another for a wide variety of purposes.

Gun #1 — Colt Commander

This particular Commander has been lightly customized by West Virginia gunsmith Wayne Novak. Most of the subtle little things that Novak did to the gun aren't visible in the photographs, but the fancy Craig Spegal checkered grips are. So are Novak's superb carry gun sights. What is more significant is the installation of a Bar-Sto barrel, well before Novak did his carry gun package. The barrel and bushing were installed by Irv Stone himself. He has done this on many thousands of .45s over the years and it always improves the gun. Across the range of loads tried in the four .45 handguns, this short-barreled little auto is clearly the most accurate.

Gun #2 — Smith & Wesson Model 25-2

This same basic revolver is available from Smith & Wesson in a great many variations and calibers, but the .45 ACP gun has been just recently discontinued. That's a shame, because these big, beefy Smith & Wessons are fine, accurate revolvers. Many have been shortened to four- or five-inch length for carry gun purposes. One of the most worthwhile accessories for the 25-2 is the Ranch Products full moon clip. This little chunk of sheet metal clips six rounds of the ammo into a cluster that drops neatly into the cylinder, extracting easily when the rounds are fired. We just gave up and did it the old way, picking the rounds out with our fingers.

Testing of the .45 ammo was uneventful and interesting. We used bullets in three different weights: 185, 200 and 230 grains. Across the board, all bullets shot very well in all the guns. There was little in the way of abrupt changes in group size as usually happens with other calibers. It is interesting to run a finger down the column showing group size and thereby note the consistent way all of our four test guns handled the variety of ammunition which we loaded and fired in them. The .45 ACP is clearly an accurate round of ammunition.

Best results in all guns came with moderate charges of Bullseye or 700-X.

.45 ACP LOAD DATA

BRASS	POWDER CHARGE (grains)	BULLET MAKE/WEIGHT/TYPE (grains)	GUN #1 ES	GUN #1 AVG	GUN #1 SD	GUN #1 GROUP (inches)	GUN #2 ES	GUN #2 AVG	GUN #2 SD	GUN #2 GROUP (inches)	COMMENTS
W-W	4.1 BULLSEYE	HORNADY 185 FMJ	35	699	19	1½	35	721	27	1½	Most accurate load for both guns
W-W	4.4 BULLSEYE	HORNADY 185 FMJ	66	655	25	1½	47	702	32	1¾	Second most accurate load for both
W-W	4.7 BULLSEYE	HORNADY 185 FMJ	84	652	30	2	89	708	22	2	
W-W	4.1 BULLSEYE	SIERRA 185 MATCH	57	732	20	3+	42	740	16	2	
W-W	4.4 BULLSEYE	SIERRA 185 MATCH	118	668	37	3+	103	699	35	3¼	
W-W	4.1 BULLSEYE	SPEER 185 MATCH	61	792	21	2	38	804	27	2	
W-W	4.7 BULLSEYE	HORNADY 200 FMJ C/T	43	756	15	1½	26	789	23	2	
W-W	4.7 BULLSEYE	SPEER 200 TMJ	58	754	27	1¾	79	735	44	2+	
W-W	4.7 BULLSEYE	SIERRA 200 MATCH	92	761	29	2¼	23	723	10	2+	
W-W	4.7 BULLSEYE	HORNADY 230 FMJ/FP	45	743	17	3+	37	730	15	1¾	
W-W	4.7 700-X	HORNADY 185 MATCH	94	801	36	2½	65	803	26	2½	
W-W	4.7 700-X	SPEER 185 TMJ	103	736	37	3½	95	762	32	1½	
W-W	4.7 700-X	HORNADY 200 FMJ C/T	87	776	33	1¾	78	784	33	1¾	
W-W	4.7 700-X	HORNADY 230 FMJ/FP	55	761	19	1¾	63	747	19	1¾	Best load w/700-X
W-W	5.0 700-X	HORNADY 185 MATCH	71	828	27	3	79	841	27	2	
W-W	5.0 700-X	SIERRA 185 JHP	115	800	44	2+	110	830	1¾	2½	
W-W	5.0 700-X	HORNADY 200 FMJ C/T	77	824	33	2½	78	845	37	1¾	
W-W	5.0 700-X	HORNADY 230 FMJ/FP	43	816	19	3	77	833	18	1¼	Horizontal strings in #1
W-W	5.5 231	HORNADY 185 MATCH	56	770	20	1¾	29	789	12	1+	Most accurate load, Gun #2: 1.089"
W-W	5.5 231	SIERRA 185 JHP	46	721	18	4	34	734	13	2½	
W-W	5.5 231	HORNADY 200 FMJ C/T	64	757	25	2+	54	756	17	2	
W-W	5.5 231	HORNADY 230 FMJ/FP	48	756	19	1¾	49	736	18	2+	
W-W	4.6 RED DOT	SPEER 185 TMJ	68	718	24	3+	59	720	24	2½	
W-W	5.4 RED DOT	HORNADY 200 FMJ C/T	59	826	23	2	71	834	25	1¾	
W-W	5.4 RED DOT	HORNADY 230 FMJ/FP	102	858	59	3+	108	863	68	3+	Highest velocity; note high ES

FOOTNOTES:

1) All groups reported are six-shot groups, fired at 25 yards in the Ransom Rest.
2) Velocity is measured with Oehler 33 chronographs; Sky Screen IIIs spaced 36 inches apart.
3) Unless otherwise noted, group size is measured with a ruler, to the nearest quarter inch.
4) All cases primed with Federal #150 primers

.45 COLT

AFTER HAVING been rather sharply corrected on several occasions, I have abandoned the use of the term .45 "Long" Colt, but that's just what we are talking about here. The cartridge was originally chambered in the Colt Frontier revolver for Army use in the early 1870s. That it has survived to the present day is tribute to the American shooter's affinity for big, venerable cartridges.

Gun #1 — Smith & Wesson Model 25-5

After a considerable clamor for such a gun, Smith & Wesson began to produce their massive N-framed revolver in .45 Colt in the mid-1970s. The revolver has all of the features associated with top-of-the-line revolvers from S&W: adjustable sights with insert, smooth and silky action, high polish blue finish and target type grips. Despite its similarity to the .41 and .44 magnums, neither gun nor cartridge is a magnum and should not be loaded to magnum velocity.

Gun #2 — Ruger Blackhawk

While Ruger once produced the popular Blackhawk with an auxiliary cylinder for .45 ACP, current guns are available with only the .45 Colt cylinder. They are rugged, heavily built revolvers which raise the traditional Western single-action revolver to the highest point of its development. The .45 Blackhawk was unavailable for a few years and rumor held that it was because intrepid reloaders were firewalling their loads.

At the beginning of the load session with the Blackhawk and S&W 25-5, I would have bet the most desirable powder would have been my long-time personal favorite, Unique. That is the propellant that I have used for a number of years in this cartridge. Unique did okay and the quoted loads are acceptable in either gun.

At the urging of my friend and co-author Mike Venturino, who is constantly fiddling with those old-timey guns and cartridges, I tried his pet .45 Colt powder. That's Red Dot and I'm sure the Hercules people won't mind if I transfer allegiance to another powder in their line. Red Dot was clearly superior in the small range of powders that we loaded and fired through the two guns.

The guns were two of the more frustrating ones we fired in the course of assembling data for this chapter. Both of them had "the revolver syndrome" in which five out of six chambers are aligned with the bore of the gun. But the sixth was either misaligned or had an erratic throat dimension or other malady which caused one shot in each group to wander from its mates.

If a shooter can live with a five-shot revolver, these two are perfect candidates for the title. With a little precise testing, you can isolate the bad chamber and mark same, aligning it under the hammer as the sixth one to be fired. Accuracy is then excellent.

.45 COLT LOAD DATA

BRASS	POWDER CHARGE (grains)	BULLET MAKE/WEIGHT/TYPE (grains)	ES	GUN #1 AVG	SD	GROUP (inches)	ES	GUN #2 AVG	SD	GROUP (inches)	COMMENTS
FEDERAL	8.0 UNIQUE	SIERRA 240 JHP	25	743	9	3½	72	778	24	3½	Round group in both guns
FEDERAL	8.4 UNIQUE	SIERRA 240 JHP	44	801	15	1¾	42	834	14	3	
FEDERAL	8.6 UNIQUE	SIERRA 240 JHP	30	841	12	3½	36	876	14	3	Gun #2 had five shots in 1½" — possible misaligned chamber
FEDERAL	8.8 UNIQUE	SIERRA 240 JHP	46	885	20	5+	53	927	18	4+	Sharp decrease in accuracy
FEDERAL	8.0 UNIQUE	HORNADY 250 JHP	81	776	29	3½	13	797	5	3+	Gun #2 had five in 2"
FEDERAL	8.4 UNIQUE	HORNADY 250 JHP	92	862	31	2½	46	830	16	5	Gun #2 — horizontal stringing
FEDERAL	8.6 UNIQUE	HORNADY 250 JHP	91	812	33	3+	31	859	10	2½	
FEDERAL	8.8 UNIQUE	HORNADY 250 JHP	45	841	15	2½	67	891	24	3½	
FEDERAL	8.0 UNIQUE	SPEER 260 JHP	94	770	35	2½	65	783	20	2¾	Gun #2 — 5 in 1½"
FEDERAL	8.4 UNIQUE	SPEER 260 JHP	65	816	24	2¾	84	853	30	3+	Vertical stringing both guns
FEDERAL	8.6 UNIQUE	SPEER 260 JHP	33	830	16	3+	43	853	17	3+	Same velocity in #2 as previous load
FEDERAL	8.8 UNIQUE	SPEER 260 JHP	20	838	17	3+	27	877	8	2¾	
FEDERAL	6.3 RED DOT	SIERRA 240 JHP	39	719	15	1¾	45	759	18	1½	Most accurate load in both guns
FEDERAL	6.6 RED DOT	SIERRA 240 JHP	18	760	8	3	39	799	15	4	Both guns had 5 shots in 1¾"
FEDERAL	6.8 RED DOT	SIERRA 240 JHP	55	797	19	4	45	826	16	4	Gun #1 had 5 in 2"
FEDERAL	6.3 RED DOT	HORNADY 250 JHP	69	709	22	4+	28	746	12	3	
FEDERAL	6.6 RED DOT	HORNADY 250 JHP	99	752	34	2½	79	757	25	3½	
FEDERAL	6.8 RED DOT	HORNADY 250 JHP	42	774	15	2½	44	804	17	4	Gun #1 had 5 in 1¾"
FEDERAL	6.3 RED DOT	SPEER 260 JHP	60	714	21	3+	46	746	16	2	Gun #1 had 5 in 1½"
FEDERAL	6.6 RED DOT	SPEER 260 JHP	39	722	22	3½	36	767	12	3	
FEDERAL	6.8 RED DOT	SPEER 260 JHP	51	759	20	3+	45	790	16	3+	
FEDERAL	6.3 RED DOT	SPEER 250 LSWC	56	735	24	3+	30	776	11	3+	
R-P	8.4 HERCO	SIERRA 240 JHP	88	768	28	2	72	833	25	2½	
R-P	8.8 HERCO	SIERRA 240 JHP	47	831	19	2½	41	877	14	4	
R-P	8.4 HERCO	HORNADY 250 JHP	48	769	17	3	47	833	18	3+	Gun #2 had five in 1"
R-P	8.8 HERCO	HORNADY 250 JHP	71	785	24	1½	56	847	20	2½	
R-P	8.4 HERCO	SPEER 260 JHP	62	774	24	2½	43	813	18	1¾	
R-P	8.8 HERCO	SPEER 260 JHP	48	821	16	2½	56	853	20	2½	
R-P	6.1 700-X	SIERRA 240 JHP	73	787	24	3+	50	796	18	3+	
R-P	6.1 700-X	HORNADY 250 JHP	51	772	26	3+	96	794	32	3+	

FOOTNOTES:

1) All groups reported are six-shot groups, fired at 25 yards in the Ransom Rest.
2) Velocity is measured with Oehler 33 chronographs; Sky Screen IIIs spaced 36 inches apart.
3) Unless otherwise noted, group size is measured with a ruler, to the nearest quarter inch.
4) All cases primed with Federal #150 primer.

CHAPTER 13
RELOADING FOR THOMPSON/CENTER'S CONTENDER

On the facing page, we have some — but by no means all — of the cartridges for which Contender barrels have been chambered. Above are some T/C catalogs.

It Has The Most Omnivorous Appetite For Ammunition Of All The World's Firearms!

THE THOMPSON/CENTER *Contender* appeared upon the handgun scene in early 1967 or thereabouts. It did not win instant acceptance and roaring popularity overnight, but it was able to hang in there until the shooters in this country and the rest of the world came to realize what a uniquely useful design it really is. As of early 1987, about 300,000 Contender receivers have been produced. According to personnel at the factory, the typical Contender owner winds up with about five barrels for the basic receiver. Discounting the fact that a lot of those barrels are of the Super-14 design — fourteen inches in length — and taking each hypothetical barrel at the usual minimum of ten inches, that extrapolates to a trifle over 236 miles, if all the barrels were laid end-to-end.

The really special charm of the Contender is the ability of the given owner to change barrels in a minute or so, with the resulting capability of firing a fairly incredible variety of ammunition, depending solely upon the caliber for the particular barrel installed at the time.

The Contender handles rimfire ammunition, as well as centerfire. Earlier models employed a small insert in the face of the hammer that could be rotated half a turn with a screwdriver to make the shift from rimfire to centerfire or vice versa. A fairly recent modification puts a small lever atop the hammer spur. In center position, the gun won't fire, even if cocked and snapped with a live round in the chamber. Move the lever to the right and it fires rimfire ammo such as the .22 long rifle or .22 magnum; move it to the left and it fires any centerfire cartridge for which a barrel can be had. There are two firing pins in the receiver: the upper for rimfire, the lower for centerfire.

For the calendar 1985 sales year, buyers of guns and/or barrels cast their votes as to cartridge popularity and the polls showed the following ranking, in descending order of popularity:

.22 long rifle
.223 Remington
.44 magnum
.30-30 Winchester
7mm T/CU
.45 Colt/.410-bore
.357 Remington Maximum
.357 magnum
.35 Remington
.222 Remington
.22 Hornet
.30 Herrett
.32-20 Winchester
.22 WRM
.357 Herrett
6.5mm T/CU
.41 magnum
.32 H&R magnum

When the *Thompson/Center Catalog No. 14* appeared early in 1987, the listing of cartridges for the Contender carried a few surprises. Back, for the first time in many years, were the 9mm Parabellum (Luger) and the .30 M1 Carbine cartridges, both offered only in the ten-inch bull barrels. The only listed octagonal barrels were in the two rimfire calibers, .22 LR and .22 WRM, with ten-inch bull barrels available for each and the .22 LR available in the Super-14, as well.

The 7-30 Waters cartridge, initially offered as a barrel for the Contender carbine, was available in ten-inch and fourteen-inch length as bull barrels for the Contender pistol. Unlike the highly popular 7mm T/CU that retains its original wildcat status year after year, the 7-30 Waters started out as a factory load, available from dealers in the Federal cartridge line.

Listed briefly, the No. 14 catalog offered these barrels for the Contender:

    .22 LR 10 octagon, bull, Super-14
    .22 WRM 10 octagon, bull
    .22 Hornet 10 bull
    .222 Rem. 10 bull, Super-14
    .223 Rem. 10 bull, Super-14
    6mm T/CU Super-14
    6.5mm T/CU 10 bull, Super-14
    7mm T/CU 10 bull, Super-14
    7-30 Waters 10 bull, Super-14
    .32 H&R magnum 10 bull
    .30 M1 Carbine 10 bull
    .32-20 WCF 10 bull
    .30 Herrett 10 bull, Super-14
    .30-30 WCF 10 bull, Super-14
    9mm Parabellum (Luger) 10 bull
    .357 magnum 10 bull
    .357 Rem. Maximum 10 bull, Super-14
    .357 Herrett Super-14
    .35 Rem. Super-14
    .41 magnum 10 bull
    .44 magnum 10 bull, Super-14
    .45 Colt/.410-bore 10 bull, 10 vent-rib

A great many other cartridges have been offered as Contender barrels in the years since the gun was first introduced. In a lot of instances, sales figures simply did not warrant keeping them in the line and they were dropped when the initial stock was exhausted. Initially, all of the barrels were of the tapered octagonal configuration and the forend was carried by a snap fastener. As burlier calibers were added, a problem reared its revolting head: Given enough recoil, the forend would come loose and drop to the ground with each shot.

The first of the ten-inch bull barrels appeared in early 1974, with the introduction of the .30 Herrett and the No.

What we might call the 9mm family — from left, the 9mm Short or .380 ACP; the 9x18mm Ultra; the 9mm Luger or Parabellum; the .38 ACP or .38 Colt Super and the 9mm Winchester magnum. Nos. 3 and 4 are the only ones that have been chambered for Contender.

1 edition of the T/C catalog appeared about the same time. The forends for the bull barrels were secured by a slotted-head screw, giving improved structural integrity at some minor expense of barrel-changing convenience.

At the early outset, the octagonal barrels were offered in a choice of 8.75-inch or ten-inch length and a few were made up in six-inch length: very few, indeed. The six-inch or snubnose contender barrels usually were in .22 LR, but I once had one in .45 Colt. They had an intriguing appearance, with the barrel chopped off just ahead of the forend. Saleswise, they proved to rank in popularity at the approximate level one might predict for chocolate-covered chicken feet at a gourmet convention and the concept did not survive through many catalogs. The 8.75-inch barrels were an early casualty, too.

Contender barrels were produced in at least two calibers

This gives you a fairly graphic illustration of the fuss and commotion that results from setting off a round of 7mm T/CU in the Super-14 barrel. Photo was taken near sundown at a 1-second exposure, with flash and you may be able to see the ghost image of the scoped barrel after recoil. This combo gave my first groups below one minute of angle (MOA) out of a Contender, using a 4X Redfield scope and firing off the portable benchrest.

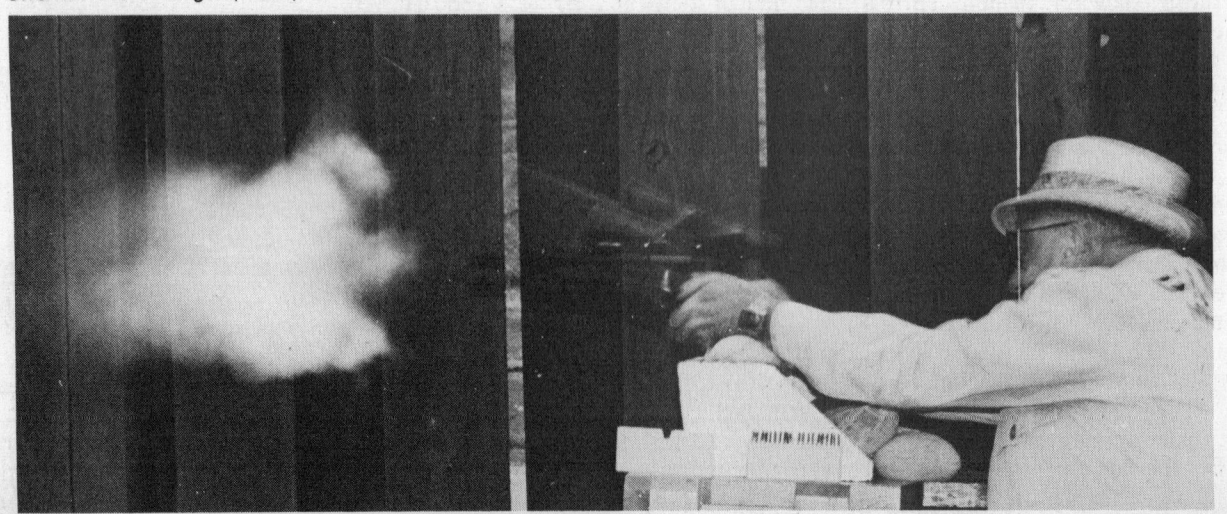

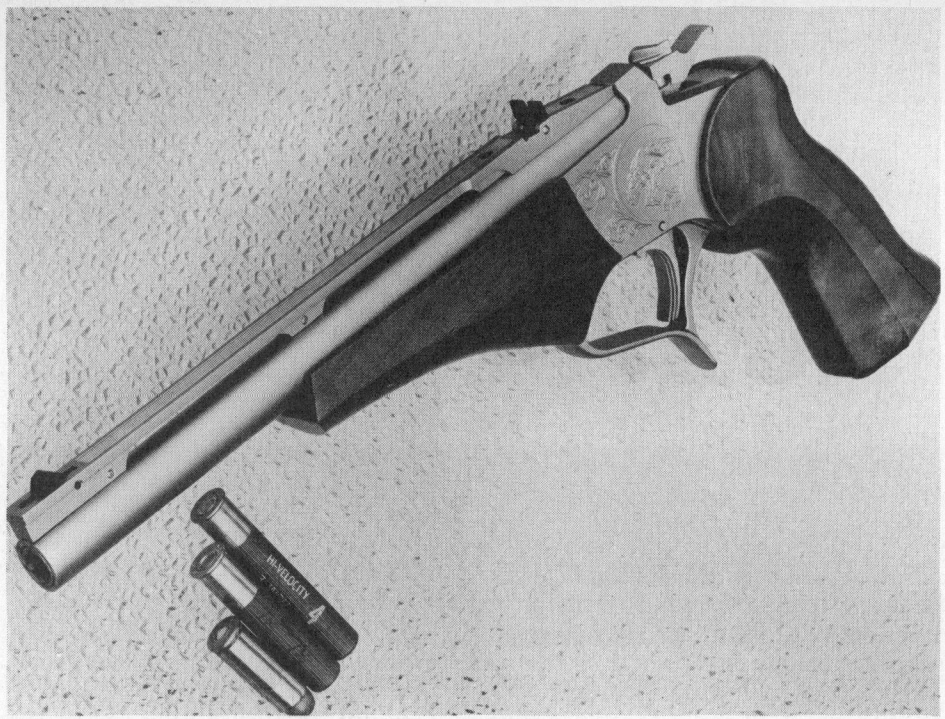

Here's one of the new Armour Alloy II Contenders, fitted with a vent-ribbed barrel that takes either the .45 Long Colt or, with detachable choke installed, the two sizes of .410-bore shotshells. Rear sight folds down.

that I do not have. One was the .25-35 Winchester Center Fire (WCF), a fairly *rara avis* cartridge in its own obsolescent right. Having had gratifying results with other unlikely caliber .25 cartridges such as the .256 Winchester magnum, I had occasional twinges of regret at having missed getting a .25-35 WCF Contender barrel during its brief moment on stage. Joe Wright, president of the Thompson/Center Association, had one of the .25-35 WCF barrels and he assures me I didn't really miss out on anything all that intrinsically mournworthy; he spent a lot of research time on his .25-35 WCF barrel and never succeeded in getting it to shoot for rancid owl-sweat. I can relate to that, having had comparable experiences too many times.

The other item of exotica I don't have is a Contender barrel in 5mm Remington. That was a cartridge Remington introduced in the early Seventies: a rimfire bottleneck round that carried a caliber .20 bullet about the same diameter as a pellet for a Sheridan air rifle.

I do, however, have some thoroughly improbable and unlikely Contender barrels. One is a ten-inch bull for the .38-45 Clerke: a wildcat cartridge designed by John A. "Bo" Clerke (pronounced Clark) for use in the M1911-pattern .45 ACP auto pistol. Warren Center made it up personally, at my request, long ago and made another for himself, as long as he was at it. You could regard it as a sort of teenybopper .357-44 Bain & Davis barrel and I have

As it was in the beginning: Receiver No. 2516 carries an octagonal barrel in 9mm Luger with snap-held forend. Scrollwork behind the puma was omitted from production shortly after this receiver was made.

GUN DIGEST BOOK OF HANDGUN RELOADING

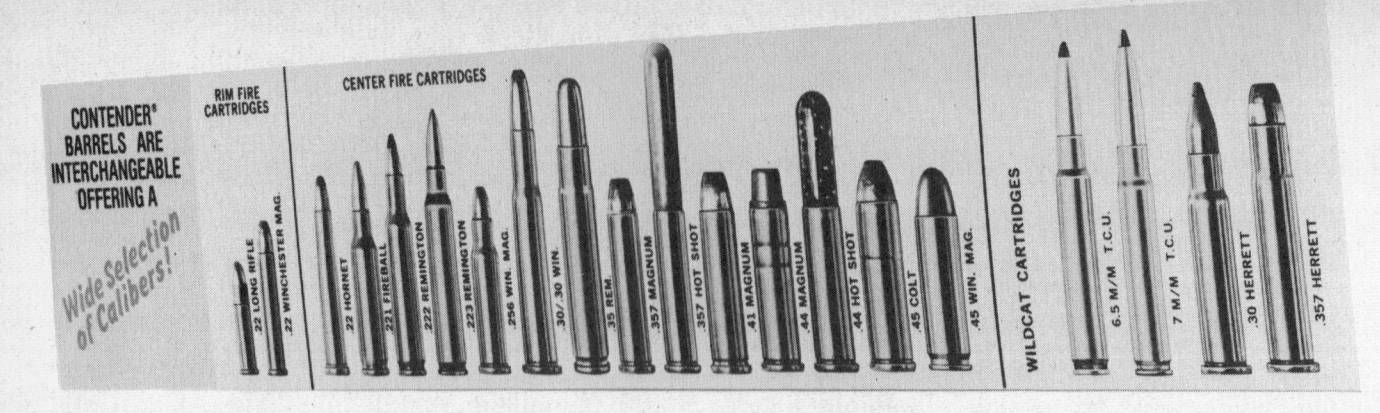

For a time, the barrel cartons carried this illustration of the available Contender chamberings. Since that time, some of these have been dropped and various other cartridges have been added to the T/C catalog.

one of the ten-inch octagonals once available for that cartridge. When I came down with the pulsating hots to get a ten-inch bull barrel in .45 Auto Rim, friend Center humored my whim and made one; exactly one.

"By the time I got that damnable extractor properly fitted," he confided, "there was nothing on earth that would've induced me to make another!"

Yes, Virginia, it is possible to get reasonably hefty bullets out of the .45 Auto Rim cartridge at velocities up around 1500 fps, but if you think I'm going to offer load data on the particulars, think again!

For a short time in the late Sixties, T/C marketed Contender barrels for that truly immortal evergreen cartridge, the .38 Special. It was a state of affairs that was heavily redolent with shimmering joys and, at the same time, quease-making pitfalls.

In the Real World, the .38 Special is tethered and hamstrung to rather modest pressure ceilings. The basic Contender action was entirely capable of working with the cartridge to the levels of the .357 magnum and perhaps just a smidge beyond. The upshot of the situation was that it led the owner of a .38 Special Contender barrel to make up reloads that were quite capable of converting good .38 Special revolvers into rapidly expanding clouds of high-velocity steel fragments. The lurking peril of allowing such cartridges to lie about is painfully obvious, I think.

The .38 Special Contender barrels didn't sell well and were dropped early in the game. I feel that was just as well.

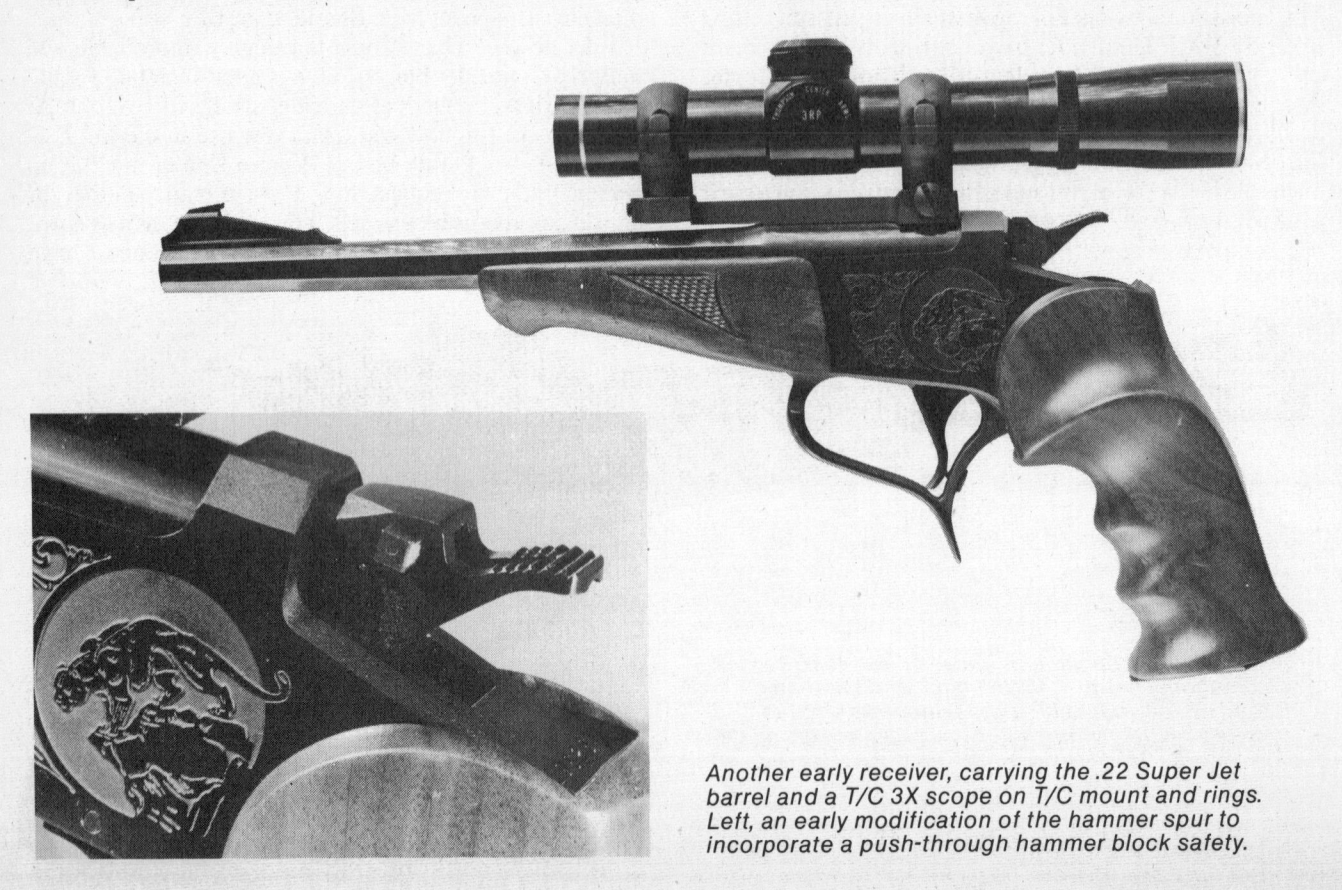

Another early receiver, carrying the .22 Super Jet barrel and a T/C 3X scope on T/C mount and rings. Left, an early modification of the hammer spur to incorporate a push-through hammer block safety.

184   GUN DIGEST BOOK OF HANDGUN RELOADING

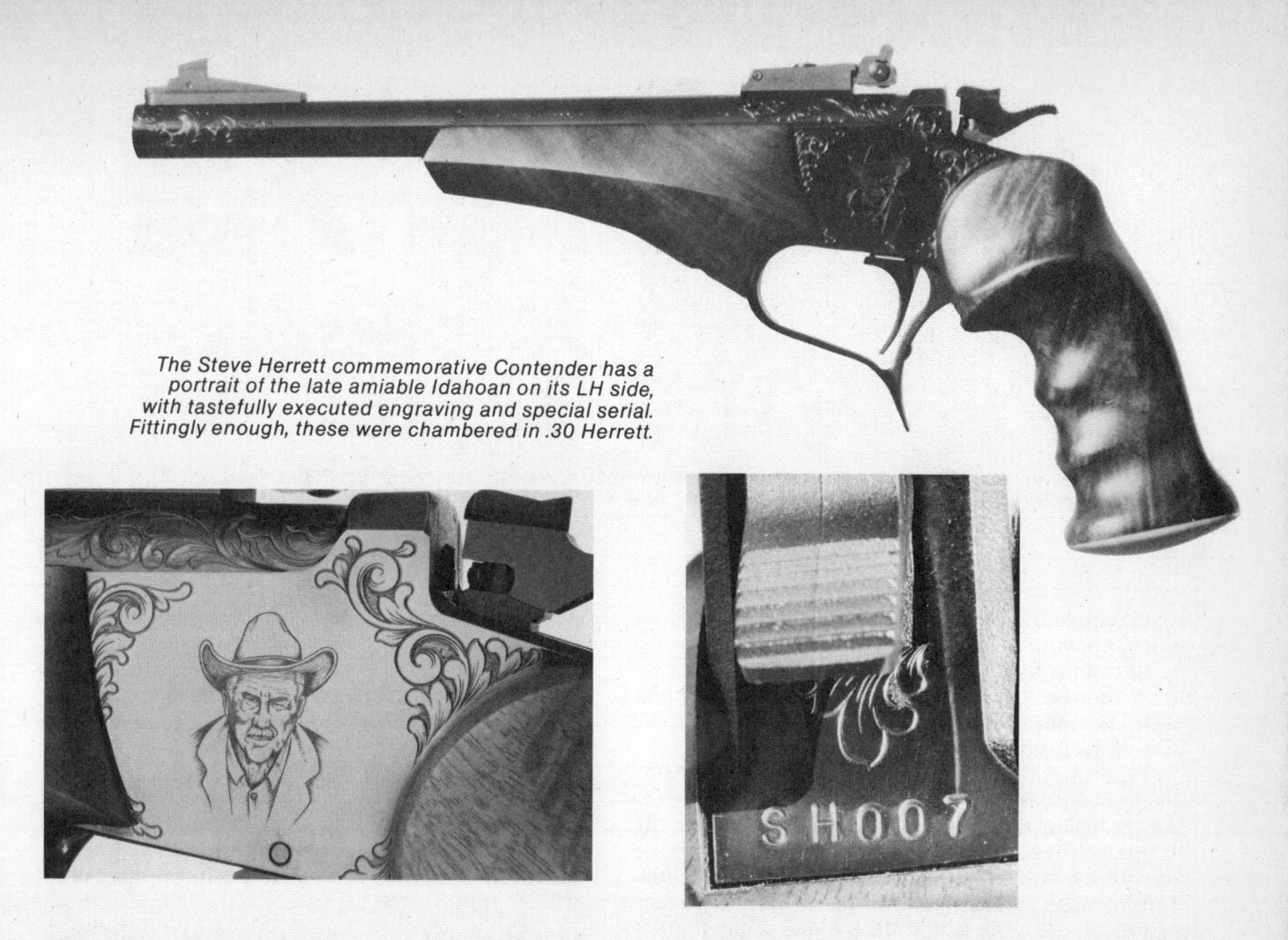

The Steve Herrett commemorative Contender has a portrait of the late amiable Idahoan on its LH side, with tastefully executed engraving and special serial. Fittingly enough, these were chambered in .30 Herrett.

If you happen to own one of those early .38 Special barrels, I heartily recommend you load for it by readily available book data for the cartridge and do not venture beyond that point.

Much the same applies to the Contender barrels once available in .45 ACP. Unlike nearly all autoloaders, the .45 ACP Contender barrel offered total support for the case and, like the .45 Auto Rim counterpart, it could carry the funny, fat little cartridge to undreamed-of heights of performance. Again, one had to exercise every possible precaution to keep such loads from straying into any other gun for which the cartridge was chambered.

There are other instances in which the Contender is much stronger than other guns ordinarily chambered for the cartridge. The .32 H&R magnum and .32-20 WCF are two random examples. If you happen to own and operate guns in those calibers, other than the Contender, you have a choice of two approaches: You can load down to pressure levels that are safe in any gun or you can brand the Contender loads in some distinctive manner, to keep them from ending up in the wrong chambers. The former alternative is probably the better of the two.

In the Contender, the upper level of tolerable pressure depends to a large extent upon the diameter of the given case heads. Ponder the matter and it's pretty obvious: Given some amount of pressure in pounds per square inch (psi), the larger case heads will deliver more straining back-thrust to the action, in direct proportion.

Thus, the Contender action will endure some pretty steep peak pressures in cartridges such as the .22 Hornet or .223 Remington but, when you install a barrel for a cartridge such as the .444 Marlin or .45-70 Government, the peak pressures need to be lower: a *lot* lower.

Apart from the contemporary offering of Thompson/Center Arms, barrels for the Contender can be had from other sources, often in calibers that T/C no longer offers — or never offered at all. The obvious example of after-market barrel suppliers is SSK Industries, (Rte. 1, Della Dr., Bloomingdale, OH 43910). The operation is run by J. D. Jones, who will send you the brochure of current SSK offerings in return for enough postage stamps to carry a two-ounce letter. The list is pretty copious, consisting of seventy-some different calibers and combinations. The list keeps getting bigger.

If you hanker for handguns with more pizzazz on tap than the Contender can handle, SSK can provide wildcats on the Remington Model XP-100 action that will make your cup run over. What they will do to your shooting hand is a disturbing conjecture. There is, for but one lurid example, the .284 Winchester case with its neck opened to accept bullets of .375-inch diameter. SSK can also supply the basic XP-100 action with a system of interchangeable barrels. If you write for a quote on one of those, plan to be seated when you open the envelope. They are not cheap but, as I've sometimes noted, they don't shoot cheap, either.

GUN DIGEST BOOK OF HANDGUN RELOADING

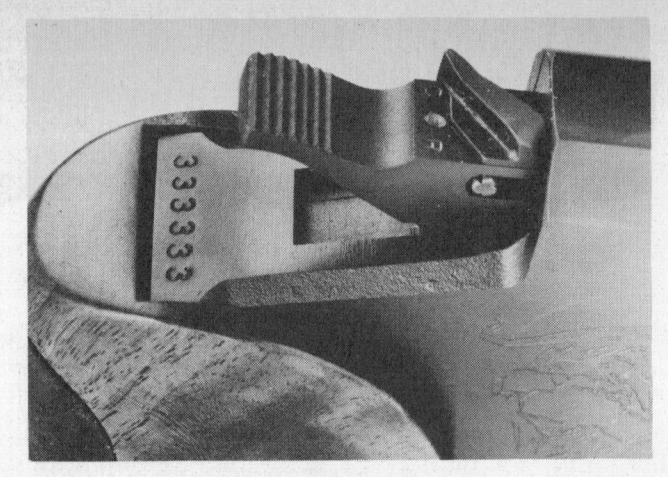

Left, receiver No. 299999, with its selector lever in Safe position. Warren Center got No. 300000, lavishly engraved and decorated, with a .22 Hornet barrel. Right is No. 333333, with its selector lever positioned for firing centerfire cartridges. This is the current style of safety and it is extremely convenient to use.

At present, I own three of the SSK Contender barrels and I would not hesitate for a moment to say that the quality of the SSK barrels is fully on a level with barrels from Thompson/Center; even a trifle better, if that's possible. In order of acquisition, I have the .411 JDJ, the 6.5mm JDJ and the .444 Marlin.

The .411 JDJ can be regarded as the ultimate .41 Maximum, being based upon the .444 Marlin case, necked to accept bullets of .410-inch diameter, same as the .41 magnum. The 6.5mm JDJ uses the .225 Winchester case as parent brass, opened to take bullets of .264-inch diameter, with a shoulder blown out fairly square at the time of fire-forming. The 6.5mm JDJ is one of the flattest-shooting handguns I can call to mind just offhand.

How flat is that? The load in question puts 32.5 grains of IMR-4320 Du Pont powder behind the 120-grain Speer No. 1435 spitzer soft point bullet for ballistics of 2185/1272 and I've had five-shot groups with that load at one hundred yards that got down to a center-spread of but .983-inch: comfortably under one minute of angle (MOA). If you zero that particular load to be dead on the money at 175 yards, its maximum rise above line of sight will be 2.56 inches at one hundred yards and at two hundred yards it will be 2.06 inches low. Time of flight to two hundred yards is .299-second and its remaining ballistics at that distance figure out to 1835/897. In other words, it gets to two hundred yards still packing more punch than a long-barreled .357 magnum does at the muzzle, with sub-MOA accuracy and the sort of string-taut trajectory we hardly dared to dream of only a few years ago.

For good measure, your typical 6.5mm bullet — definitely including the Speer No. 1435 — provides put-down-*right-now* capabilities in excess of reasonable expectations. The reason is that a number of the 6.5mm cartridges — such as the 6.5x55mm Swedish Mauser — deliver relatively modest muzzle velocities, leading the bulletmaker to use slightly thinner jackets on their 6.5mm offerings. That, in turn, means highly effective bullets for use at handgun velocities.

As for the .444 Marlin SSK barrel, a local friend and research associate has been workig with it and he swears that it drives the Remington factory load to group under one inch at one hundred yards. Like you, I find that difficult to gulp down just offhand, even though I regard it as intriguing to contemplate. To date, I've not done any hundred-yard work with Remington factory fodder and, for that reason, can offer no first-hand testimony.

Before passing on to other topics, SSK also offers the best scope mount of which I'm aware for the T/CC. J.D. Jones refers to it as his T'SOB mount and he isn't kidding. The recoil of the really *berserker* cartridges may convert the lenses of the scope to glittering fragments, but the

The hammer on one of the Armour Alloy II Contenders. Note the A-prefix of the serial number. The AA-II receivers and barrels do not interchange with the blued receivers and barrels, are highly rustproof.

Here's how the new selector system works. At left, the lever is positioned to fire centerfire, as indicated by the letter C. Red spot in center indicates Safe position and, if the lever is moved to point to the R, it will fire rimfire cartridges. Photo at right shows the centerfire striker positioned for firing those rounds.

mounts won't be fazed in the slightest and the scope tube, even though eviscerated of its optics, will remain steadfastly in place.

Another after-market supplier of Contender barrels is Bullberry Barrel Works, Ltd., otherwise Woodsmith/Bullberry, 6939 Montcliff Ave., Las Vegas, NV 89117. A bulletin on their caliber offering appeared in a recent mailing from the Thompson/Center Association and I would regard that as a favorable omen. Joe Wright, the current high honcho of T/CA, is a knowledgeable gent, in my book. Twenty bucks, sent to The Thompson/Center Association, Box 792, Northboro, MA 01532, will get one year's membership in the organization and, if you've any interest in Contenders — as I incline to assume you must,

having read this far — that could prove to be a gratifying investment. An earlier T/C group faltered and fell by the wayside. Under the leadership of Mr. and Mrs. Joe Wright the reincarnated T/CA is doing good work and putting out a really worthwhile bulletin, every month or so. The dues are $20 per year and, in my sincere opinion, well worth it.

The Bullberry offering appears to span from five offerings in caliber .17, up to and including the .17 Remington, to the .45 Long Colt, without any of the blustering, blistering behemoths available from the SSK stables. They offer pistol barrels to fifteen inches and carbine barrels from sixteen to twenty-two inches in all of the listed numbers in thirty-four calibers, including the .218 Bee, glory be!

At right below is the original hammer insert, here turned to fire centerfire cartridges. It could be rotated one-half turn to handle rimfire cartridges. At left, we have a close look at the dual firing pins.

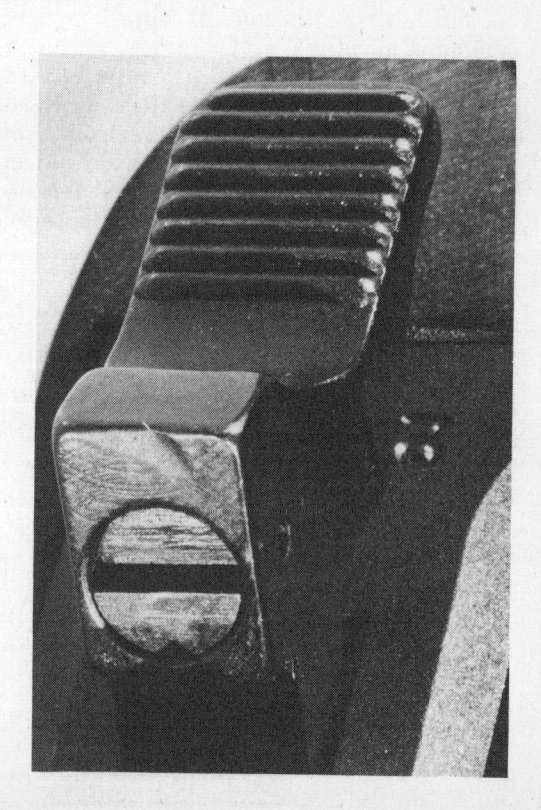

GUN DIGEST BOOK OF HANDGUN RELOADING

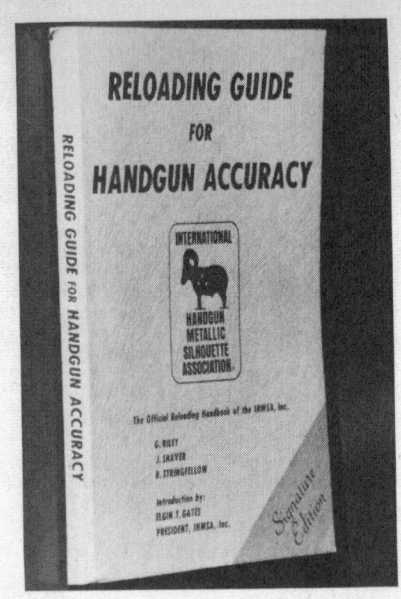

Published by IHMSA, Box 1609, Idaho Falls, ID 83401, this book carries a great deal of excellent load data for use with T/C Contender pistols.

Candidly, I regard the assignment at hand with mixed emotions. For a great many years, I have itched, twitched and faunched to put together a really comprehensive compiliation of Contender load data. Suddenly, I was commanded to do it against a deadline only a few months downstream. The legendary Tantalus hardly had it any worse.

Several manuals and handbooks offer quantities of load data for use in the Contender and/or for the cartridges in which the Contender is chambered. I see neither motive nor justification for quoting data from such sources. For one thing, it would shatter and sprain copyright laws in a deplorable manner. What I propose to do in such instances is to mention that good listings can be found in sources such as the Speer Manual, the Hornady Handbook, the Sierra Manual, the Hodgdon Manual, the Lyman Handbook, the guides from powdermakers such as Hercules, Du Pont, Winchester, Accurate Arms, et al. It is my firm conviction and contention that, if you propose to dabble in reloading cartridges, the bedrock minimum is to have all of the better load data sources on hand for ready reference and I've just listed practically all of them, except for the Nosler book, which didn't mention much dope for handguns the last time I saw a copy of it.

I have some amount of Contender load data that I have worked up on my own initiative at one time or the other. This I plan to quote and offer. I have some reported data from other grape-stompers in the same basic vineyard and I plan to quote that, with the stipulation that it's someone else's, not my own. In the meantime, I plan to cite sources from the manuals and handbooks. No, if you were about to ask, I've never been able to isolate the basic difference between a manual and a handbook. Between you and me, I think the distinction is pretty ephemeral. It's about on a par with the fine semantic separation between reloading and handloading.

Within recent times, I encountered a minor performance problem with a T/CC barrel in a particular frame and shipped it back to the attention of Tim Pancurak, at the T/CA works for service and rectification. I noted in passing that the serial number of the receiver was in the low 20,000 brackets and realized I must've had that receiver on hand for a long while. My oldest receiver is No. 1147. It carries the .22 Hornet barrel as a matter of normal routine. They started at about serial number 1000, so that is only about the 147th receiver ever made. I think I have another up in the 22## brackets, but would not bet upon it heavily. After that, just about all my receivers have serials with five digits or more.

Just about all, if not absolutely all, of the nice, round-number serials repose in the collection of Warren Center. I think it was about the time I figured that No. 50000 had to be coming down the pike that I asked if I could get it and was told that it was already spoken for.

I did a spot of furious fast-think and asked if anyone had put their arm on number 55555. As it turned out, no one had and I filed a requisition for it, receiving it in the fullness of time. It was the first of my Significant-Number Contender receivers. As time shambled forward, I corralled 66666, 77777, 88888, 99999, 111111, 222222 and, at the moment, I'm engaged in endeavors to snare 333333 as well as 299999. Along the way, Warren Center checked and found that 11111 serial had never been used, so he had that

Tuning the post-234256 triggers is an easy snap. Use an .050-inch hex wrench or a homemade tool such as this to set sear engagement and stop screw.

made up and sent to me. Thus, I'm marginally in the market for the missing pieces: 22222, 33333 and 44444. If you happen to have one of those, I'd be pleased to negotiate but, if you're suffused with dreams of paying off the mortgage on the proceeds, be duly advised that I've been getting along without them for quite a while.

The serial number of your Contender reveals quite a bit about the gun, as they have made a few modifications down the years. One of the first was the hammer with the crossbolt safety lock, cranked into production at some point a little ahead of serial 88000. My number 11111, having been made since that time, incorporates the same safety which could confound Contender students in the years ahead.

At or near serial 195000 — the usual commas are not included in Contender serials — they moved the pivot pin for the action opening lever somewhat to the rear, relocating it almost directly above the trigger. This made it substantially easier to get the gun open after firing. At the same time, the adjustment screw for regulating sear engagement was moved from inside the action to a location on the front upper surface of the trigger. Prior to that, you had to remove the barrel to adjust the sear engagement screw.

At serial 234256, they added a combination of firing pin selector and safety as a small lever on the hammer, just ahead of the hammer spur. This, I think, constitutes improvement over previous production. When the lever is in center position, the gun will not fire either a rimfire or centerfire cartridge in the chamber, even if cocked and snapped. If you propose to verify this, point the muzzle in a safe direction, just in case, mindful that all mechanical devices are more or less subject to malfunction. Move the lever to the right for firing rimfire ammo, with the suitable barrel installed, or to the left to fire centerfire cartridges from the appropriate barrels.

Within even more recent times, the Competitor grip or stock has been made the standard item of woodwork sup-

When the selector lever was added to the hammer spur, at No. 234256, it required a special type of hammer extension to fit the spur. These are a great convenience when a scope is installed.

plied with all Contenders. It has a rubber insert at the rear surface, with a sealed chamber of air to cushion the shock of heavy loads. The Competitor is fully ambidextrous the way a handgun handle really should be. I have never cared much for thumb rests, because they make the gun awkward to fire in the so-called weak hand.

People often asked the late Steve Herrett which hand he shot with and his routine response was, "It all depends on whether the critter is to the right or left of the pickup!" I'm a natural northpaw, but I made USAAF expert rating with the .45 auto pistol in 1944, using my left hand. This was about two weeks after I'd earned the same qualification out of my right hand. I feel awfully uneasy about guns I couldn't shoot with my left hand, if I wanted to. Within recent years, I've drifted over to aiming handguns with my left eye, as it's a little sharper on close subjects than the right one.

Among its other manifold capabilities, the Contender

A close look at the detachable choke on a .44 Hot Shot barrel. It must be used when firing shot loads but it's mandatory to remove it before firing .44 magnum cartridges that carry solid bullets!

A fairly intimate view of the barrel marking on a .218 Bee bull barrel, one of my favorite tubes!

GUN DIGEST BOOK OF HANDGUN RELOADING

An early caliber offering was the .22 K-Hornet, but sales were quite lackadaisical, prompting the dropping of that cartridge after a few years.

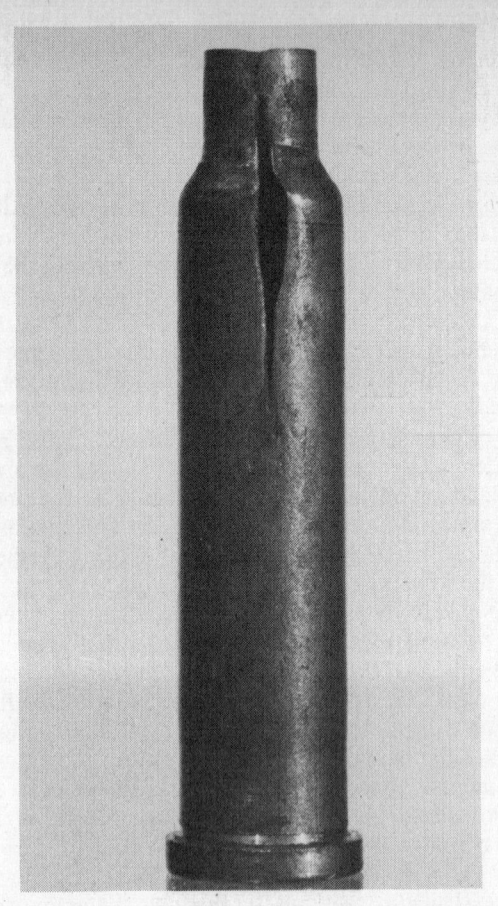

A .22 K-Hornet case that went into unplanned retirement from a split neck and shoulder. If cases were not annealed, this was all too common.

functions as a surprisingly competent one-hand shotgun. In the beginning, they came up with a combo barrel that would handle the .45 Colt or Long Colt cartridge. If you installed the supplied choke tube, it would handle .410 shotshells in the standard 2.5-inch length or three-inch magnum. The Firearms Act of 1934 forbade smoothbore handguns, among several other things such as shoulder-stocked handguns. If you fire a shot charge from a rifled barrel, the charge takes on a swirling motion from the lands and grooves. When it emerges from the muzzle, centrifugal force causes the pellets to depart abruptly from the axis of the bore. Only a short distance from the muzzle, the pattern has a large hole in the center.

The detachable choke tube carried straight ribs that arrested the swirling motion, resulting in a thoroughly creditable pattern, out to respectable distances. The BATF originally gave its blessing to the barrel, then, once it was in production, voiced a yen to have it discontinued. Warren Center proceeded to design what came to be known as a .44 Hot Shot barrel that essentially duplicated the performance of the .410-bore barrel. The cartridge was the familiar .44 magnum, carrying a plastic capsule filled with shot pellets. The same basic design of detachable choke was furnished and, when installed, it arrested the swirl, meanwhile shredding the tough plastic of the capsule to release the pellets.

In June 1970, using the .44 Hot Shot barrel, I bagged an Idaho jackrabbit on the run at a distance we paced off as forty yards. The late Steve Herrett and his friend Hugh Farmer were witnesses to the shot. Herrett snapped a photo of me and my trophy jackbunny. We sent it to Warren Center and he passed it along to Jim Sheridan, who does the advertising for T/C. Sheridan put the photo in the catalogs, captioning it, "Running jackrabbits at thirty yards!" Pained, I pointed out to Sheridan that it was a full, fat forty yards. He responded that no one would ever believe that. We wrangled back and forth about the matter for a few years and he resorted to a procrustean solution by dropping the photo from the catalog!

The fact remains that the .44 Hot Shot barrel was a remarkable short-range game getter. The great virtue was the fact that, beyond some nominal distance, there was no hazard from fallout or ricochets. I used my .44 Hot Shot barrel to thin-down a colony of California ground squirrels in a pasture where a lot of cattle were grazing, without traumatizing so much as an ounce of beef. It was and still is an efficient firearms system.

T/C went on to introduce a comparable Hot Shot barrel for the .357 magnum cartridge, with long capsules of shot for use in it. Within more recent times, they have put the original .45 LC/.410-bore barrels back into production and dropped both the .44 and .357 Hot Shot barrels, meanwhile continuing to offer capsules and loaded ammo in .44 size. I have one of the new Armour Alloy II Contenders in .45 Colt/.410 with a ten-inch vent-rib barrel and it does a fine job with either size of .410 shell. I blush to confess that I have never tried any of these with the .45 Colt cartridge. Perhaps I should.

By the early Seventies, a few custom makers were offering long barrels and shoulder stocks to convert the Contender into a carbine, resulting in a remarkably nice system. T/C did not offer that option until the mid-Eighties. The T/

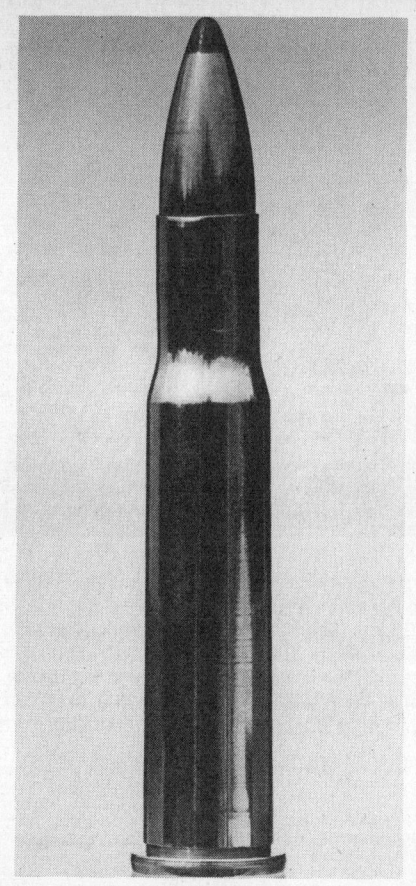

This is the standard .22 Hornet, still a popular Contender caliber and capable of good accuracy.

that effect appears on the butt pad.

As with many other contemporary armsmakers, T/C is putting a lot of stress upon making sure that every owner and operator of any gun they make has been exposed to a current operator's manual. If you have acquired any manner of Thompson/Center firearm, including the various muzzleloaders, either new or used, without getting an operators manual with it, you are invited to request the pertinent manual by writing to Thompson/Center Arms, Box 5002, Rochester, NH 03867. Specify the model, caliber and serial number of the gun you have, when requesting the manual.

It is not a state of affairs widely known, but Thompson/Center maintains some manner of custom shop and can furnish a number of items not in the regular catalog. T/C will not make and market a combination they regard as unsafe. For example, they won't sell you a Contender barrel in .22-250 Remington no matter how you beg, plead, kick or scream. They view it — and rightly — as too much cartridge for the gun. They feel that, if you want to fire .22-250 Rem, you can always get one of their TCR'83 rifles, for barrels of that caliber are available.

A .22 Hornet, left, with a .17 K-Hornet before fire-forming and a fire-formed .17 K-Hornet.

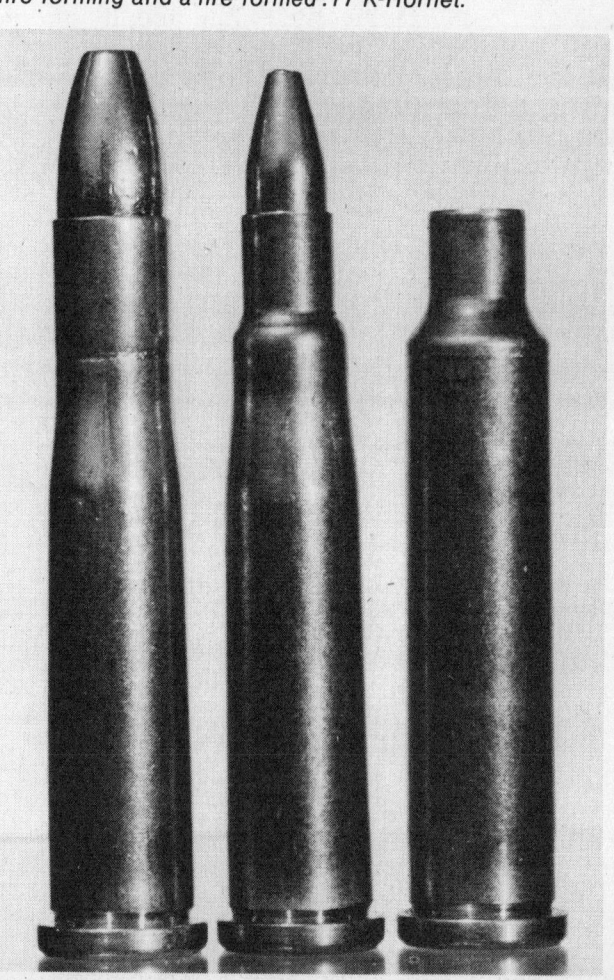

C carbine barrels are twenty-one inches long, currently offered in .22 LR, .22 Hornet, .222 Remington, .223 Remington, 7mm T/CU, 7-30 Waters, .30-30 WCF, .35 Remington, .357 Remington Maximum and .44 magnum. All have open iron sights as well as drilled and tapped holes for scope mounts. In addition, there is the same length in .410-bore by three-inch, in full choke, with a raised ventilated rib. It will, of course, fire the 2.5-inch standard .410 loads equally well. The Contender carbine offers the option of an ultralightweight shoulder gun — typically, not much over five pounds — with superior credentials as to accuracy. It comes up to the shoulder and handles with graceful ease and, with the trigger suitably adjusted, it makes a thoroughly top-drawer one-banger. In the .44 magnum version, you definitely will be able to tell when the cartridge goes off, right now! No left-hand versions are offered, as the carbine is totally ambidextrous.

Nine different states are currently using the Contender carbine for hunter safety instruction as the ideal training rifle for student hunters.

The important thing is not to give in to temptation to hang a ten-inch or fourteen-inch barrel onto the receiver fitted to a carbine stock, as that is a big, rough, federal no-no. Put a shoulder stock on the Contender receiver and the barrel must be at least sixteen inches in length. A notice to

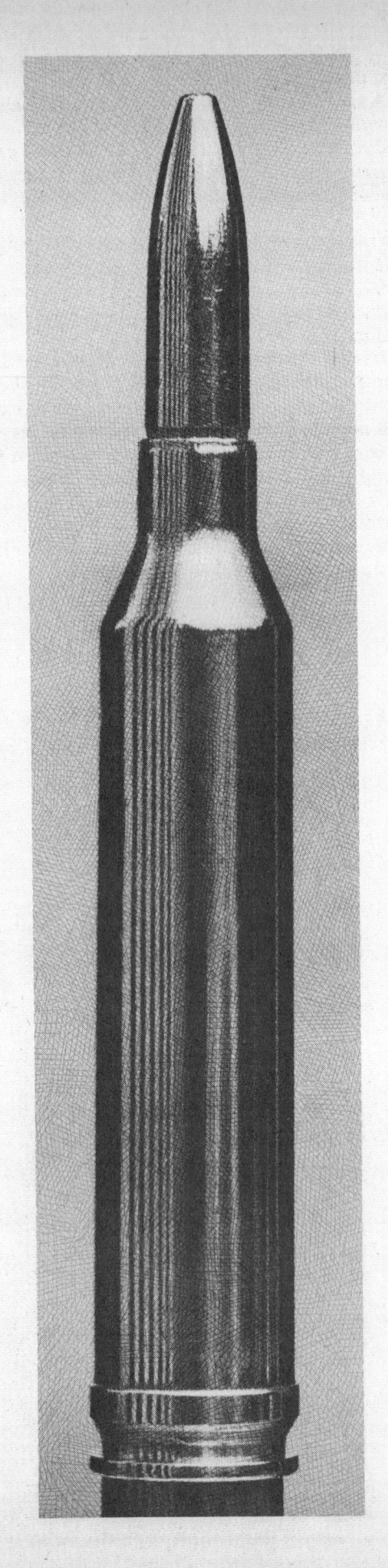

While Contenders have been chambered for a host of unlikely cartridges, the odds are excellent that you'll never see a Contender barrel for the cartridge at left, here. It's the .264 Winchester magnum. Above, headstamp of a .44 Remington magnum load from Federal Cartridge, a highly popular Contender load.

Considerations of liability — a heavy factor, these days — make T/C extremely reluctant to part with unchambered barrels of any bore dimension whatsoever. The centerfire Contender barrels for caliber .22 cartridges have groove diameters of .224- to .225-inch and a 1:14 LH rifling pitch. If they sold an unchambered barrel of those specifications, it could end up as a .22-250 Remington or, worse yet, a .220 Winchester Swift: bad medicine on the Contender action. If they chamber it for .223 Remington before you get your hands on it and if you have it rechambered for the Swift or the .22 CHeetah — somewhat swifter than Swift — and have a mishap because of your intrepid enthusiasm, T/C has a pretty valid defense in their records that the barrel left their plant with a .223 Remington chamber. Put another way, the classic USN defense: "It didn't happen on *my* watch!"

Subject to the restrictions just mentioned, you can get nearly anything within reason from T/C, if you're willing to put your money down and wait your turn. As of quite recent times, for example, I have a grand slam in Hornet barrels for the T/CC: ten-inch octagonal, ten-inch bull, Super-14 and carbine. The ten-inch bull and carbines are current standard listings and both examples I have are excellent shooters, especially the carbine. I also have nonstandard Super-14s in .30 M1 carbine, .32-20WCF and 9mm Luger/Parabellum, with corresponding standard barrels in ten-inch bull configuration. For the project immediately at hand — as well as a few others down the pike a bit — I want to conduct some research into the effect of barrel length upon ballistics and the collection of Contender barrels makes what I consider the ideal test vehicles for the purpose.

GUN DIGEST BOOK OF HANDGUN RELOADING

Ten should not be fed loads beyond certain pressure levels and I don't want to get carried away by the greater action strength of the Contender to the point where I might publish loads safe in it but not so safe in the Bren Ten. The book at hand is being produced at a time when firing production versions of the Colt Delta Elite were not yet available.

When the Contender is barreled up for the lustier cartridges, such as several from SSK Industries, or the Jurras *Howdah* series, installation of the rubber grips from Pachmayr can prove quite helpful in coping with the recoil. Two or even four Mag-na-port vents in the muzzle can cut down drastically on recoil and muzzle-flip. Vito Cellini also has developed compensators that cancel out ninety percent of the recoil and can install these on Contenders and many other guns. Further details from Pachmayr Gun Works, 1220 S. Grand Avenue, Los Angeles, CA 90015; Mag-na-port International, Inc., 41302 Executive Drive, Mount Clemens, MI 48045-3448 or Cellini's, 3115 Old Ranch Road, San Antonio, TX 78217.

Harking back to the interesting concept of the Contender carbines, at the request of the BATF, you cannot buy the butt stock and forend as separate components. They must be sold with a barrel sixteen inches or longer. The

The current triple-deuce family: From left, the .221 Remington Fire Ball; .222 Remington; .223 Remington; and .222 Remington magnum. The last is rarely if ever seen as a Contender round, but others are used.

The .223 Remington family: From left, .223 Remington; 6mm T/CU; 6.5mm T/CU and 7mm T/CU. All three of the Thompson/Center Ugalde cases are made from .223.

I did not go for a Super-14 in .32 H&R magnum. To my thinking, anybody who'd go that far wouldn't stop *there*. I do, however, have a Super-14 barrel in .22 long rifle which is a standard catalog item, as well as a ten-inch bull in .22 WMR.

A further variant, apart from the SSK and Bullberry offerings, is the 7mm International Rimmed, also termed 7 Int-R or 7R. It is based upon the .30-30 WCF case, formed with a rather long neck to accept bullets of .284-inch diameter, but not shortened from the standard .30-30 case length of 2.040 inches. The 7 Int-R was developed by the International Handgun Metallic Silhouette Association (IHMSA), under the directorship of its president, Elgin Gates, who can be reached at IHMSA, Box 1609, Idaho Falls, ID 83401 for further information, details, load data and current prices for both the barrels and loading dies, likewise in stock for the 7 Int-R.

Mike Rock, who holds forth in Albany, Wisconsin, can supply Contender barrels in 10mm Auto, as used in the Bren Ten and, more recently, in Colt's Delta Elite autoloading pistols. I have a Bren Ten and like it quite a bit, but I've resisted the temptation to send for one of the Contender barrels in the same chambering, because I feel the Bren

A quartet of nifty little .22 centerfire rounds, well suited for use in the Contender. From left, .22 Hornet; .218 Bee; .22 Super Jet and .221 Remington Fire Ball.

exclusive use in the revolvers of that chambering, as manufactured by Freedom Arms, Box 1776, Freedom, WY 88120. You may find load data for the .454 Casull, here and there. I've published some of it, myself.

Despite the fact that the cartridges resemble each other closely, except for case length, .454 Casull load data is *not* safe for use in .45 Colt cases to be fired in the Thompson/Center Contender nor, for that matter, in the Ruger Blackhawk, of any variation or model.

The .454 Casull cartridge and the five-shot, single-action Freedom Arms revolver constitute a specialized system starkly incompatible with anything else, whatsoever. The peak pressures of some listed .454 Casull loads may exceed 60,000 copper units of pressure and/or pounds per square inch. Such pressures can and probably will wreck just about any other gun in existence and leave the shooter much the worse for wear.

The .25-35 WCF is at left, with a .30-30 WCF. The .25-35 was offered as a Contender barrel for a brief time and the .30-30 remains a highly popular cartridge.

intent is to avoid the illegal use of barrels shorter than sixteen inches on shoulder stocks.

Let me close with a special word of warning, which I consider important. The .45 Colt cartridge, sometimes termed the .45 Long Colt — to differentiate it from the .45 Automatic Colt Pistol (ACP), which some refer to as the .45 Colt — may be encountered in two levels of load data in some data sources: one for general use, such as in the .45 Colt Single-Action Army Colts and a somewhat more powerful set of data for exclusive use in the Ruger Blackhawk or Thompson/Center Contender. With reasonable prudence, you can use the latter sets of data in your Contender to good effect, with little if any hazard. Always work up to the maximum loads cautiously, if at all.

However, there is another cartridge that resembles the .45 Colt quite closely, except that it is slightly longer in case length. It is called the .454 Casull and it is intended for

Huntington Die Specialties, Box 991, Oroville, CA 95965, can convert .256 Winchester magnum barrels (the cartridge at left) to the .257 Sabre Cat (center). A .25-35 WCF is at right, rounding out the caliber .25 offerings for the Contender. As far as I know, no one has ever used the .25-20 in the Contender.

As these words go onto the paper, the Contender is pretty close to twenty years old and, within the past week, I took delivery of the eagerly awaited receiver number 333333, as well as 299999. That ciphers out to roughly 16,667 guns per year of its career, on the average, or about forty-six per day, weekends included. If you had solicited wagers, back in 1967, that it would last this long and sell that many copies, it's doubtful if you'd have gotten many takers.

Along the way, other manufacturers, their eyes wetly a-gleam, have sought to hitch their corporate bandwagons to the Contender success saga, by offering lookalike designs. Without an exception to the present, the results have been about what I think they should have been. The quasi-Contenders impressed me as painfully mawkish and must have seemed much the same to the gun-buying public. They were speedily yawned into oblivion and, I think, they deserved to be.

An all-too-common house fly perches atop a round of .45 Long Colt and contemplates a round of the .45 Silhouette with understandable awe. Latter is based upon the .45-70 Government case, shortened.

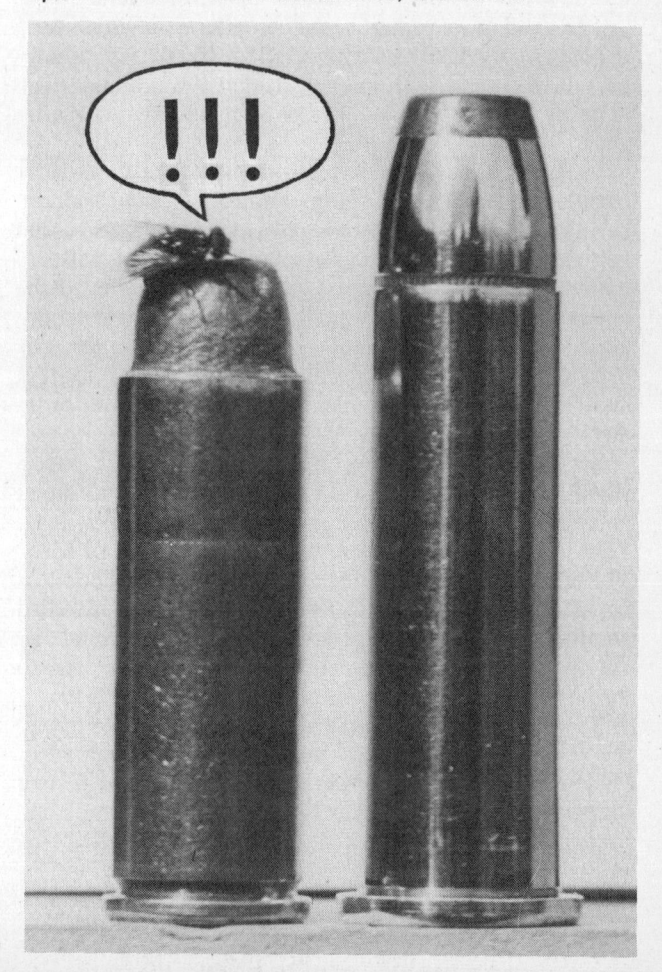

It gets down to the hard fact that any mechanism of exceptional capability is apt to be overtaxed by someone, somewhere, sometime. It lies deep within the innate nature of the beast. In the days of WWII, the General-Purpose ¼-ton truck, much more familiarly known as the "Jeep," was making its first appearance. Being much lower-slung than contemporary vehicles, there was a temptation to bang it around corners at high speed and, rather frequently, the more intrepid did so at a speed that was obviously too high. When that happened, the Jeep rolled and, hardly with any exception, the driver and passengers came out of it with extensive pelvic fractures and assorted other trauma.

My point in bringing all this up is that the Contender — like the WWII Jeep — is a system that is uncommonly capable in comparison to nearly all of its contemporaries. Despite that, the Contender can be overtaxed. It sometimes is overtaxed and the consequences can be downright dire.

GUN DIGEST BOOK OF HANDGUN RELOADING

THE .17 CONTENDERS

Since writing the introductory portion of this section, I've been in conversation with assorted people directly connected with the production and distribution of Contender barrels. As a result, I am now in a position to clarify and perhaps even ramify comments offered earlier.

As noted, the great caliber .17 mania hit the gun world in early 1968 and subsided quite abruptly at some point in 1972. Along the way, there were fanatical splinter groups in ardent pursuit of caliber .14 and even caliber .10; perhaps others, as well.

If anyone knows why T/C stopped making caliber .17 barrels, you'd expect it to be Warren Center, himself. Various theories have been advanced: the high pressures developed and the hazards of liability suits; the berserkly-increased cost of producing barrels with such tiny holes and then going on to rifle them. What was the real reason?

"The damned things simply didn't sell. We made only a few of each caliber and they just sat there, for a long time, soaking up inventory tax. By the time someone finally bought the last of them, we were glad to drop them from the catalog listing."

There you have what I'm tempted to term the full, uncensored, Center-grade explanation. Future firearms historians, kindly make note. The quote, by the way, is not necessarily verbatim, but I think the gist of it is sufficiently close.

Center surprised me by mentioning they'd even made up a few barrels in caliber .17 Remington; a bit of Contender lore that had escaped my interest in such matters.

The good news is that, if you lust after and hanker for a .17 Remington Contender barrel, SSK Industries can supply one of same. J.D. Jones reports that he continues to sell a small but tangible number of such things, year after year.

I noted earlier that I have ten-inch, octagonal Contender barrels on hand in .17 K-Hornet and .17 Mach IV chambering and I knew exactly where to look for the boxes that held the dies for loading both of them. It was just a matter of patient excavation. I dug down and back, retrieved the boxes and, on opening them, found that neither box contained any loading dies. What became of those die sets is as irretrievably lost as the specific details on what King Tut had for breakfast on the morning of his eighteenth birthday.

Surviving evidence suggests that I made up at least a small handful of loads in .17 K-Hornet and fired them. Nearly all of the cases split grandiosely at the shoulders. That is an ongoing problem with K-Hornets of all sorts. The original .22 Hornet load needs to be fired in a conventional .22 Hornet barrel, after which it needs to be annealed carefully and judiciously and reloaded in a .22 Hornet die. When fired in a .22 K-Hornet chamber, it may fire-form to the new contours, although you can expect to lose a few.

If I were to get a fresh set of dies and endeavor to reload the .17 K-Hornet today, I would dispense a charge of H4198 or IMR-4198 powder that came up to the base of the case neck, seat a 25-grain Hornady .17 bullet and hope to get a velocity somewhere in the lower to middle 2000 fps velocity range.

Contemporary editions of the *Hornady Handbook* provide ample load data on the .17 Mach IV cartridge for use

The caliber .17 Contender barrels were produced in the era of the octagonal tubes which, even then, left a lot of "meat" around the hole in the muzzle.

in rifles. It is formed down from the .221 Remington Fire Ball case, then fire-formed to blow the shoulder out to thirty degrees, rather than the twenty-three degrees of the standard .221 RFB case. The Hornady data was worked up in a twenty-five-inch barrel with a 1:10 twist and they show four different charges that top out at 3700 fps. Start at their 3500 fps listings, or even lower, and you may be able to work the ten-inch Contender barrel up into the lower 3000 fps velocities.

The Hornady Handbook takes the .17 Remington to a top of 4100 fps, out of a twenty-four-inch barrel with a 1:10 twist. J.D. Jones reports velocities up to about 3500 fps in his SSK barrels for the .17 Remington. If you buy a barrel from Jones, be sure to request information as to his favorite load data for the combination.

As for the .17 Bumblebee, I wrote that up in 1968, using a borrowed barrel and set of dies. By indulging in a spot of autoplagiarization — one of the more satisfying vices, take my word — I can report performance of the .17 Bumblebee, out of a ten-inch barrel, as follows:

Actually, there were two little caliber .17 cartridges at the start, both developed by the late Frank "Hempy" Hemsted. One was the .17 Hummingbird — based upon the .22 Hornet case and the .17 Bumblebee — or Bumble Bee — was based upon the .218 Bee Case. The first test vehicles were BSA Cadet Martinis, fitted with barrels nearly an inch in diameter.

We were able to get a ten-inch octagonal Contender barrel made up in .17 Bumblebee and ran some tests. At the time, I thought it was the first caliber .17 handgun in

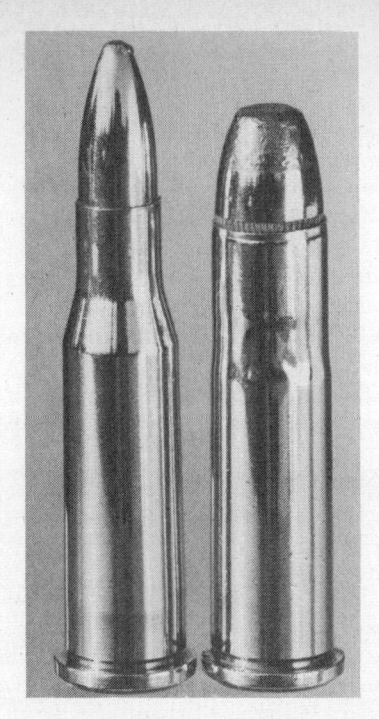

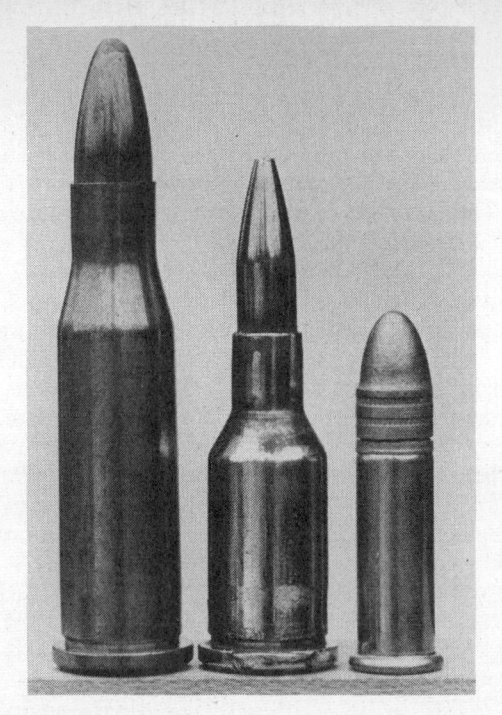

Left, a .218 Bee cartridge with a .32-20 WCF; both have same head dimensions. Center, the .218 Bee, a .17 Bumblebee, based upon the Bee case and a .22 LR round. Right, the .22 Super Jet, with a .218 Bee round.

existence and, to date, no one has ever advised me to the contrary. In point of fact, when we tried to get an existing section of .17 barrel Contenderized, Warren Center advised that it would present great technical difficulties, but they just happened to have a .17 Contender barrel there at the plant, so he rechambered it and sent it back for the envisioned tests. Thus, it appears that Center or someone else at the factory must have pre-dated me on it a little.

Our starting load was 7.0 grains of H4227 behind handmade bullets weighing 13.2 and 18.5 grains. The 13.2-grain JHP went 2583/196 and the 18.5-grain did 2444/245; not bad for a start, we thought.

With 7.0 grains of H110 and the 18.5-grain JSP bullet, we got 2710/302 and 7.5 grains of H110 with the same bullet gave 2801/322. We were inclined to regard that as a maximum load.

Going back to the 13.2-grain JHP bullet, we got:
7.5 gr H110	3215/303
8.0	3496/358
8.3	3546/367

That last load showed slight resistance to opening the breech and I'd be inclined to view 8.0 grains of H110 as maximum. Best velocity for the 18.5-grain JSP was 7.5 grains of 2400, going 2898/345.

Off the sandbagged benchrest, with a Bushnell Phantom scope installed, the load with 7.0 grains of H4227 behind the 18.5-grain JSP bullets put five of them into a center-spread of 1.45 inches at a distance of fifty yards.

All in all, the .17 Bumblebee was a cute, spunky little cartridge, but not the simplest project to form down from the parent .218 Bee brass. For one reason or another, it turned out to be one of those promising wildcats that just sort of fades away from lack of interest on the part of the shooting public. Sometimes things work out that way and, yes, it does seem a pity.

Hornady is the only major bulletmaker ever to offer a caliber .17 bullet commercially. They continue to make a nice little 25-grain jacketed hollow point (JHP), their catalog No. 1719, at a price of $9.45 per box of one hundred, as of early 1987. That's only about a buck a box more than they charge for their caliber .22 bullets.

Hornady is the only major bulletmaker offering .17 bullets to the present and in only one weight. A round of .22 long rifle ammo gives scale here.

The Hornady Handbook, third edition, on pages 58 through 61, offers load data on the .17 MACH IV and .17 Remington. The Hodgdon Manual No. 25, pages 96 through 98, covers the .17 Ackley Bee, .17 MACH IV, .17-222, .17-223 and .17 Remington. The last listing has data for Du Pont (now IMR), Winchester and Hercules powders, in addition to Hodgdon powders. The Accurate Arms loading book has an entry for the .17 Remington, as does the No. 46 Lyman Reloading Handbook. The HM-25 also has a listing of the .17 Bumblebee on page 378, in their pistol section. All of the other data listings were worked up in rifle-length barrels.

If you find yourself with a Contender barrel that is an orphan, datawise, I suggest you request load data for it from Hornady Manufacturing Company, Box 1848, Grand Island, NE 68802, as they would be more apt to have information on it than any other source I can think of.

During the brief but feverish heyday of the caliber .17s, a great many different gunsmiths and wildcatters worked with it and chambering dimensions probably span a broad gamut. Be soberly advised that most of these teeny-bopper cartridges attain some eye-popping peak pressures and exercise all possible caution and prudence in making up loads to try out in them.

Hodgdon's top load for the .17 Bumblebee generates 2455 fps out of a ten-inch Contender barrel and, with the 25-grain bullet, that's good for 335 foot-pounds of energy (fpe): no humungous vast amount, in comparison to any number of other handgun cartridges. At their best, some of the little .17s were respectably accurate and their downrange ballistics were pretty good, due to the small amount of frontal area and generally decent ballistic coefficient. The BC of the Hornady bullet, if you were about to ask, is .190 and, given a muzzle velocity of 3200 fps — the lowest listed in the HH tables — when zeroed for 250 yards, it's within less than five inches from the line of sight, clear to three hundred yards: not too shabby for a feisty little bullet.

.22 HORNET

For various reasons, not all of them explicable, the .22 Hornet remains alive and doing fairly well in the offering of Contender barrels. There is no denying that the cartridge performs quite well in the Contender, although the .218 Bee, as noted later, does even better.

At various times in the distant past, .22 Hornet rifle barrels have been made with a groove diameter of .223-inch and bullets of the same diameter should be used when reloading for such guns. Contender barrels for the .22 Hornet, as well as T/C carbine barrels, are made with a groove diameter of .224-.225-inch and reloads will perform better in them, if bullets of the usual .224-inch diameter are used. The rifling pitch used in T/C barrels is one turn in twelve inches, or 1:12, LH twist.

Load data for the .22 Hornet in Contenders is available in goodly abundance from a number of sources and thus hardly warrants extensive coverage here. Bullets in the weight range of 40 to 45 grains are best suited for the .22 Hornet Contenders and ten inches is the standard barrel length offered; fourteen-inch barrels are available only on a custom basis.

Suitable powders include Winchester 680, Hercules 2400, Du Pont IMR-4227 and IMR-4198 or Hodgdon H4227 and H4198. The top load for the 40-grain bullet in the #10 Speer Manual uses 2400 powder at 8.5 grains for 2122/400 or a maximum of 9.5 grains for 2366/497.

The same source shows IMR-4227 as the best powder for 45-grain bullets, with 10.3 grains delivering 2086/435 and a maximum of 11.3 grains for 2294/526. As always, maximum loads should be approached with caution, if at all.

.22 K-HORNET

This is an "improved" wildcat version of the original .22 Hornet, developed by the late Lysle Kilbourn. It is produced by firing a .22 Hornet round in the K-Hornet chamber and results in a blown-out case with a sharper shoulder and straighter sides in the portion behind the shoulder.

The added case capacity thus obtained is rather slight and so is the potential gain in cartridge performance, when fired in the ten-inch Contender barrel. While exact ballistics vary somewhat, the edge of the K-Hornet over the standard Hornet is hardly apt to be much over fifty feet per second. Few shooters regard this as an advantage over the standard case of sufficient value to justify the additional bother of working with the non-standard case and loading dies. Sales of K-Hornet Contender barrels were extremely slow, resulting in the early discontinuance of barrels for the wildcat.

Current editions of both the Speer Manual and Sierra Manual carry data listings for the .22 K-Hornet in the Contender, to maximum ballistics of 2370/561 and 2500/625 for the 45-grain bullets, respectively.

The rate of case failure, through splitting at the neck or shoulder, is rather high with the K-Hornet. It can be controlled somewhat by first firing factory loads in a standard Hornet chamber, then carefully annealing the case necks before going on to make up fire-forming loads for use in the K-Hornet chamber.

.218 BEE

There are at least two categories of what we'll call Unappreciated Cartridges. First, there are the cartridges that richly-richly *deserve* to be unappreciated for the good

While some early .22 Hornet rifles required a .223-inch bullet, bore dimensions of the Contender take the .224 diameter. Bullets labeled "Hornet" have thinner jackets.

and sufficient reason that there is little or nothing in their basic nature that merits appreciation in the first place. Quite possibly, you can think of some archtypical examples and, as you're probably coming to suspect, I have my own private list of likely nominees.

On the other hand, we have a number of cartridges offering capabilities that should get them a lot more popularity from the shooting public than they've ever been given. Some are blighted by the fact that no one has ever made an especially good gun chambered to handle them. Others, even without that handicap, are appreciated far short of what they should be.

The .218 Bee is a prime and salient example of the latter category. Originally, it was developed as a round for use in lever-action Winchester rifles, in which it offered no more than the casual level of accuracy one might expect. Later, Winchester brought out their Model 43 bolt-action rifle in .22 Hornet and .218 Bee. The accuracy potential showed a useful gain, but the action was rather springy and case separations occurred after no more than a few reloadings.

Rather early in the career of the Contender, the .218 Bee was listed among the available barrels and it seemed like the natural of all time — *and was.* For perhaps the first time ever, the cartridge had a proper gun in which to function. With the resizing die carefully adjusted, the reloads would headspace on the shoulder, rather than on the rim. With its greater powder capacity, the Bee could outpace and outpunch the Hornet quite handily. Its accuracy potential in the Contender proved exceptionally gratifying.

But, as has happened all too often, the word did not get around to the buying public and the public didn't buy the barrels so, as a pretty direct result, T/C finally sold out their first production run, sighed and dropped it from the catalog.

I have one of the ten-inch octagonal barrels from the days when the Bee was in the catalog and a ten-inch bull barrel of the same chambering. As you might expect, the bull barrel is the better of the two; an icily capable shooter. It begs, pleads and whimpers for a scope of at least two-power for the sake of taking advantage of its outer limits of capability.

A few of the more conscientious data sources continue to carry listings for the .218 Bee in rifles. Any such listed loads can be used in the Contender. Despite the caliber designation, it needs to be loaded with bullets of .224-inch diameter for use in the Contender. Not to get tedious about it, small rifle-type primers should be used.

It isn't often that one lucks on to a load that is head, shoulders and most of the thorax above its contemporaries. That happens to be true in the instance of the .218 Bee Contender. I'm pleased to share the recipe with you.

Start with the Sierra 45-grain JSP bullet, either their #1300 or #1310, and charge the load-ready .218 Bee case with 13.0 grains of Hodgdon H4227. Remind yourself that it's bad luck to be superstitious. Seat the Sierra bullet to a length overall (loa) of 1.790 inches, as closely as you can proportion it.

In the ten-inch barrel, this load churns up ballistics of about 2500/625 and, fired carefully with a good scope from a steady rest, it is not uncommon to crack the minute of angle (MOA) barrier with it out to one hundred yards or farther.

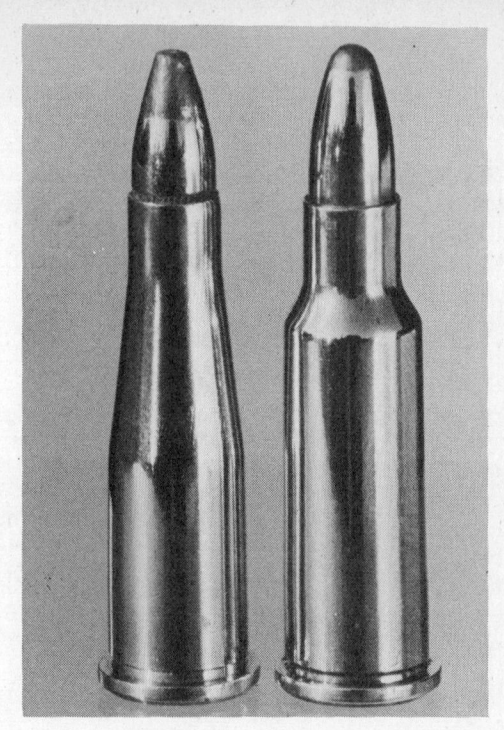

Above, the .22 Remington Jet, left, with a .22 Super Jet. Latter is best made by necking down .256 Win. mag case.

Below, the .357 magnum, right, serves as the parent case for the .22 Jet left, and the .256 Winchester magnum that can be used in making up the .22 Super Jet, center.

Empty brass in .256 Win. magnum is made only by Winchester and is not too easy to find. Huntington Die Specialties, address given earlier, can supply forming dies to produce the .256 cases from .257 mag brass. Right, from left, a .22 Super Jet with a .256 Win. mag.

Credit where credit is due: That load was first discovered by my respected contemporary, Bob Milek, who reported it to his readers. Being one of those, I tried it in my Bee and have never even come close to beating its performance.

This section is cranked in for the benefit of those who had the perception and foresight to get a Bee barrel when the getting was good. It is unlikely that it will ever be a catalog item again.

REMINGTON .22 JET

Originally introduced in 1961 for use in the Smith & Wesson Model 53 revolvers, this cartridge is based upon the .357 magnum case, at a length of 1.288 inches, rather than the 1.290-inch length of the parent case.

The Model 53 S&W revolver offered the option of switching cylinders for use of .22 long rifle ammunition, as well as the Jet loads, and had a hammer that could be changed from rimfire to centerfire configuration. As a result, reloads made for use in the Model 53 required bullets of .222-inch diameter, rather than the usual .224-inch size. Hornady developed a jacketed soft point (JSP) bullet of .222-inch diameter for use in .22 Jet loads to be used in the Model 53 and continues to list it in their catalog.

Put charitably, the .22 Jet was not a popular success in the Model 53 revolver. Its report was singularly ear-piercing and the extravagant taper of the case caused fired cases to back up in the chamber, thus hampering cylinder rotation, unless both the case and chamber were scrupulously grease-free.

Contender barrels for the .22 Jet handled the cartridge without the mentioned problems and had a groove diameter of .224-inch, with a rifling twist of one turn in fourteen inches (1:14). Consequently, reloads made up for use in the Contender tended to perform better with the larger bullet diameter, .224-inch, although loading dies were apt to be dimensioned for use with .222-inch bullets, unless modified by a slightly larger expanding plug.

Both the Hornady Handbook and Sierra Manual carry load data for the .22 Jet and both sources use the Contender, rather than the Model 53 S&W. Sierra takes the 40-grain bullet to a maximum of 2450/533 and the 45-grain to 2200/484. Hornady's listing tops out at 2600/601 for their 40-grain bullet and 2500/625 for their 45-grainer.

Remington catalogs list their 40-grain factory load at 2100/392 in an 8¾-inch barrel. Although still made, the .22 Jet cartridge is no longer stocked by most dealers and, as a Contender cartridge, it offered little or no advantage over the much more popular .22 Hornet round.

.22 SUPERJET

This was an improved version of the basic .22 Remington Jet cartridge and one would have to observe that, if ever a cartridge stood in need of improving, the .22 Jet was one of the likeliest candidates.

The .22 Superjet, or Super-Jet, was conceived by Dan Cotterman, about 1963, when he was the reloading editor of *Gun World* magazine. He was assisted in his endeavors by Bain & Davis Sporting Goods and Keith Davis, then a partner of the firm. Currently, Bain & Davis gets their mail at 307 E. Valley Blvd., San Gabriel, CA 91776 and Jon Harting is the man to address in regard to such matters.

If you happen to have a standard .22 Jet Contender barrel, it seems likely that Bain & Davis still has the necessary chambering reamer to convert it to .22 Superjet.

The .22 Jet was designed as a cartridge for use in revolvers and, as was just discussed, it left rather a lot to be desired in that utilization. The .22 Superjet was not notably more successful as a revolver cartridge, although Cotterman had a BSA Cadet Martini re-chambered for it and found that it performed rather well in its new configuration.

In the early Seventies, when T/C still offered the .22 Jet as an option in their ten-inch octagonal barrels, I ordered two of them and had one converted to .22 Superjet by Bain & Davis.

One might assume that a working stock of cases could be fire-formed by blowing out .22 Jet loads in the modified chamber: not necessarily so. That approach led to a really demoralizing casualty rate in lost cases. The approach that did work ever so much better was to run .256 Winchester magnum cases up into the resizing die for the .22 Superjet — RCBS made the loading dies for .22 Superjet and, presumably still can supply them — and take off from that point.

In terms of powder capacity, the .22 Superjet falls somewhere above the .218 Bee and below the .222 Remington. A few suggested loads are listed here, with their performance in the ten-inch, octagonal Contender barrel, as converted to .22 Superjet by Bain & Davis. The .22 Superjet should be reloaded with small rifle primers and bullets of .224-inch diameter, for use in converted Contender barrels.

Parent case: Winchester .256 Winchester magnum, virgin
Primer: Remington #7½ small rifle

40-grain, .224-inch Sierra #1200 bullet at 1.668 inches loa

17.3 grains Norma N-200	2430/524/49	3.48 MOA
15.2 grains Hodgdon H4198	2181/422/62	9.04
16.4 grains Hodgdon H322	1943/335/63	6.93
19.0 grains Hodgdon H335	2266/456/47	6.14
18.0 grains Winchester 748	1883/315/55	3.51

45-grain, .224-inch Speer #1023 bullet at 1.702 inches loa

17.3 grains Norma N-200	2400/575/29	2.13 MOA
15.2 grains Hodgdon H4198	2208/487/31	3.90
16.4 grains Hodgdon H322	2037/414/34	1.78
19.0 grains Hodgdon H335	2280/519/48	1.75
18.0 grains Winchester 748	1963/385/30	2.07

Both loads with H335 showed large muzzle flashes in bright daylight. Performance figures are the usual velocity/energy/standard deviation. Group spread, in minutes of angle, was between centers for five shots at twenty-five yards. Charges brought the level of powder to the base of the case neck. No obvious indications of excessive pressure were noted in the test gun.

.221 REMINGTON FIRE BALL

This cartridge appeared in 1962, for use in the equally new Remington Model XP-100, bolt-action, single-shot pistol. SAAMI specifies a maximum working pressure of 55,500 c.u.p. for it and, to the best of my knowledge, that gives it the distinction of the highest peak pressure of any handgun cartridge, to date. Within quite recent times, Remington has dropped the .221 RFB for use with the Model XP-100, currently offering the pistol in .223 Remington and 7mm Remington Bench Rest, as well as a few other calibers by way of their custom shop.

T/C made Contender barrels in .221 RFB for a time, usually if not always in the ten-inch length. At present, Contender barrels are not available in the chambering.

Essentially, the .221 RFB is a shortened version of the remarkably popular .222 Remington cartridge and RCBS can furnish an extended shell holder plus a form-and-trim die that make an easy task of producing .221 RFB cases from any other case with the same head dimensions, such as the .222 Remington, .223 Remington or .222 Remington magnum. The excess neck needs to be removed and the case trimmed to the specified length of 1.400 inches. As a usual rule, such conversions do not require inside neck reaming or outside neck turning.

The #10 Speer Manual lists load data for the .221 RFB, out of the Model XP-100 Remington, with bullet weights from 40 to 70 grains and it would be prudent to use the lower quoted loads in the T/C Contender. Suitable powders include Hercules Reloder 7, IMR-4198, H4198, IMR-4227, H4227, Winchester 680, Hercules 2400 and Hodgdon H110, with Winchester 748 and Hodgdon BL-C2 for the 70-grain bullet.

Small rifle primers should be used when reloading the .221 Remington Fire Ball.

.222 REMINGTON

Contender barrels are cataloged currently in both ten-inch and Super-14 for the .222 Remington — often termed the "Triple-Deuce" — and load data listed in virtually any handbook or manual can be used to varying good effect when reloading this cartridge for the Contender. Mindful of the barrel that is substantially shorter than typical rifle barrels used in developing the load data, ballistics from the Super-14 barrel will be somewhat below the listed levels and those from the ten-inch will be still lower. In selecting a powder, the faster-burning numbers will sacrifice less velocity.

In their booklet, "Making Your Contender Perform!" T/C lists one load for the .222 Remington, using 16.5 grains of Hercules 2400 powder behind the 50-grain Sierra Blitz bullet, quoting a velocity of 2643 fps/776 fpe out of the ten-inch bull barrel and noting that, despite the caliber designation, bullets of .224-inch diameter should be used when reloading for the .222 Remington Contender.

.223 REMINGTON or 5.56mm

As noted elsewhere, this is one of the most popular of all

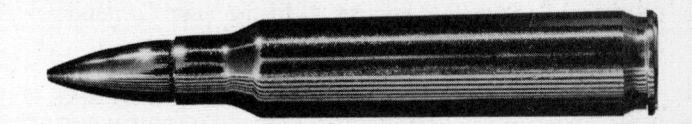

Adoption as a military cartridge has made the .223 Remington extremely popular with shooters everywhere.

centerfire cartridges, in terms of Contender barrel sales. It is hardly surprising, in view of the fact that adoption of the basic cartridge by U.S. armed forces — as well as those of many other countries — make the fired cases interestingly available.

Despite that fact, Thompson/Center counsels against use of military brass when reloading for the Contender and

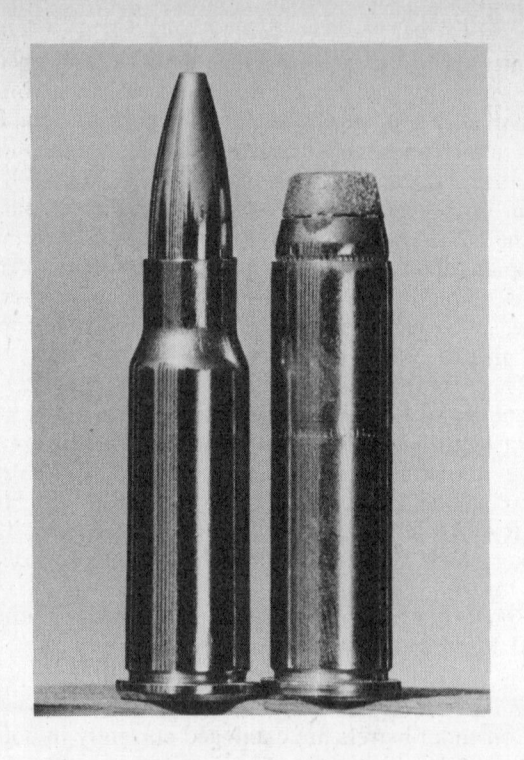

The .256 Win. mag was originally developed for use in some Colt revolvers that, somehow, never got produced. Descended from the .357 magnum, right, it is a superb performer in the Contender. Sierra No. 1600 works well.

especially for use as parent brass when forming cases for the T/CU line of wildcats.

It should be pointed out that contractors producing military ammunition are little, if at all, concerned over the reloading properties of the cases after their initial firing. As a result, some of the military cases can pose problems with brittle brass, case separations and the like. Avoidance of military brass is a good idea for the beginning or inexperienced reloader. A further vexation is the fact that military .223 brass rather frequently has stamp-crimped primer pockets, requiring cutting or swaging the crimp away before a new primer can be seated.

Any of the data widely available in handbooks and manuals for the .223 Remington can be used when reloading for use in the Contender and, as with the .222 Remington, the faster-burning powders will give the most satisfactory results. My own favorite powder for use in the .223 Remington case — in barrels of any length — has long been Hodgdon H335, largely due to the tight groups it is quite apt to deliver. IMR-4198 is another good choice, per data on page 12 of the T/C booklet, "Making Your Contender Perform!"

.22 PPC & 6mm PPC

This pair of cartridges, differing primarily in bullet diameter, has won an enviable reputation in benchrest rifle competition. Each is based upon the .220 Russian case, with Boxer-type small rifle primers. The designation stands for Pindell-Palmisano Cartridge, having been developed by Ferris Pindell and Dr. Lou Palmisano. The cases are not produced domestically at present and obtaining a sufficient supply has been one of the problems. The situation may be improving, as Dr. Palmisano had negotiated with the Finnish firm of Sako for cases, ammunition and rifles in these designations.

A few experimental prototype ten-inch Contender bull barrels have been made up for both PPC calibers and they have demonstrated excellent accuracy. It seems rather unlikely that either cartridge is apt to become a standard catalog barrel for the Contender in the near future, although some custom barrel makers such as SSK or Bullberry might be prevailed upon to produce them.

.256 WINCHESTER MAGNUM

This is truly a fine little cartridge, endlessly in hopeless quest of a suitable gun to handle it. Despite the caliber designation, it handles bullets of .257-inch diameter, in the too-common custom of nomenclating cartridges with casual disregard for actual diameters. In essence, it is the .357 magnum case, necked down to accept bullets of .257-inch diameter, in weights up to 87 grains or so.

Some years back, Ruger introduced their singular Hawkeye single-shot pistol in .256 Win. mag. It looked quite a bit like a single-action revolver, except that the cylinder served as a breech-block and the cartridge was chambered in the rear of the barrel proper.

The Ruger Hawkeye never performed all that impressively well and was dropped from production within only a few years. T/C brought out some ten-inch, octagonal barrels in .256 Win. mag for the Contender and they functioned quite a bit better. Later, a small number of Super-14 bull barrels were chambered in .256 Win. mag for the Contender and they proved to have really exceptional capabilities.

Unfortunately, as often happens, the word did not get spread about; the gun-buying public didn't buy the guns or barrels and the listing was dropped. Today, it is still possible to obtain a Super-14 bull barrel in .256 Win. mag from the T/C Custom Shop, but be advised it may cost a little and delivery may take a while.

The .256 Win. mag is one of those cartridges that make good and grateful use of additional bore length. During its days as a standard catalog offering, it was available only in

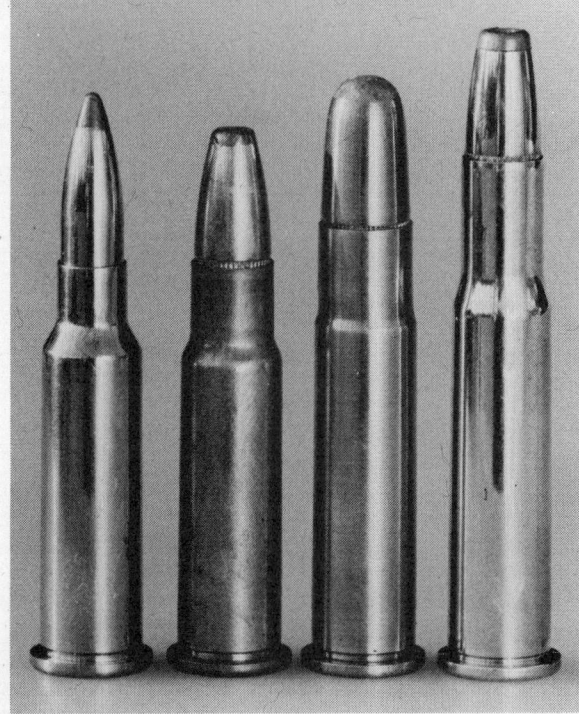

Top, RCBS forming dies for the .257 Sabre Cat; a somewhat more energetic round than the .256 Win. mag. Above, from left, .257 Sabre Cat; .30 Herrett; .357 Herrett and .30-30 WCF. Latter serves as parent brass for the other three.

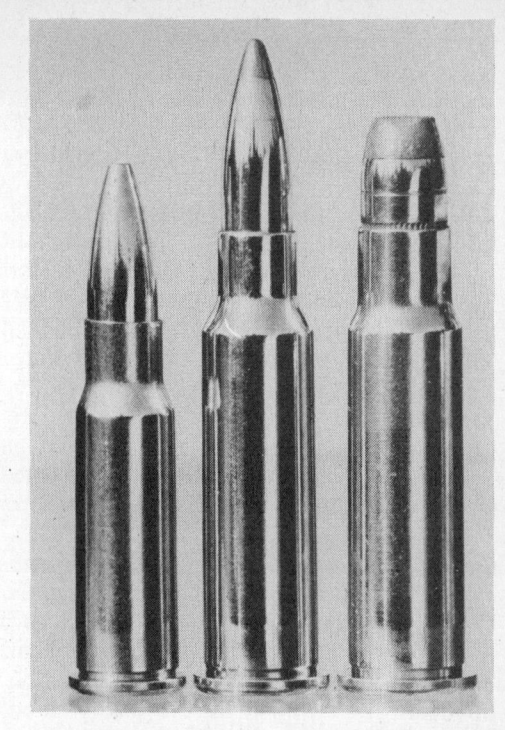

For comparison, here's the .256 Win. mag; .257 Sabre Cat and .30 Herrett. You could think of the .257 Sabre Cat as sort of .25 Herrett, well adapted for varmint duties.

ten-inch, octagonal barrels. It either was dropped or on the way out by the time the first ten-inch bull barrels appeared about early 1974.

Several data sources continue to list loads for the .256 Win. mag and I hope they continue to do so. In the early Eighties, I negotiated for a pair of Super-14 barrels in .256 Win. mag, having an envisioned use for the spare one. When they arrived, I made up a quantity of test loads, installed scopes and test-fired both barrels off the bench. One of the barrels, with one particular load, impressed me rather profoundly, so I left it in its original chambering. I sent the other barrel up to the Huntington Die Specialties, Box 991, Oroville, CA 95965, to be rechambered to .257 Sabre Cat.

The barrel that retained its .25 Win. mag chambering purely doted upon a load consisting of 15.0 grains of Hodgdon H4227, behind a 75-grain Sierra JHP bullet. Average velocity of the load in that barrel was 2295 fps, for 877 fpe and it was not uncommon for fifty-yard groups to stay within half an inch between centers for five shots.

There may be other loads that do better, although not a whole *lot* better, obviously. In the fourteen-inch barrel, the .256 Win. mag has the potential of a truly superb varmint cartridge, with modest recoil and a punch to belittle most if not all .357 magnums.

Winchester still supplies unprimed brass in .256 Win. mag, but it is not too commonly encountered in the shops. The RCBS custom shop can supply forming dies to convert .357 magnum cases and I use such a set to remain in the .256 Win. mag business, these latter days.

I like this funny, spunky little cartridge. I am impressed by what it can do.

.25-35 WINCHESTER CENTERFIRE (WCF)

This is an elderly and little-known variant on the redundantly familiar .30-30 WCF. The hyphenated designation indicates that it dates to the days of black powder and had a caliber .25 bullet — .257-inch, actually — driven by 35.0 grains of black powder.

For a brief while, T/C offered Contender barrels in .25-35 WCF. It is one of their few calibers that I never got around to obtaining, as I mentioned before.

I scanned through my library in quest of some load data for the .25-35 and was surprised to discover how scarce it is. The Lyman Reloading Handbook covers quite a few calibers that hardly any other sources list, but the LRH-46 does not have that particular cartridge. The Hodgdon Manual No. 25, on page 153, lists loads for four of their powders in four bullet weights from 60 to 117 grains, with velocities taken in a twenty-inch rifle barrel. The third edition of the Hornady Handbook, pages 126-127, covers loads for the 60-grain and 117-grain bullets, with six powders for the 60 and five for the 117; respective maximum velocities are 2900 and 2300 fps and velocities were also from a twenty-inch barrel. Hornady listings are for H4198, IMR-3031, BL-C2, IMR-4064, H4895 and IMR-4320 under the 60-grain, with BL-C2 omitted from the 117-grain table.

Before you come down with the pulsating hots to find a .25-35 Contender barrel, I'd like to quote a brief passage from the introduction for the .25-35 in the Hornady Handbook: "By modern standards the .25-35 Winchester, a rimmed bottlenecked cartridge first introduced in 1895, is only marginally useful on deer, appropriate for smaller game only at moderate ranges, and quite unspectacular on varmints."

I trust there is no need for me to point out that the ballistics under discussion were from barrels twice as long as those for the Contender.

6mm T/CU

This is the newest of the Thompson/Center Ugalde (T/CU) cartridges, of which the 7mm T/CU came first, followed by the 6.5mm T/CU. Thus, we now have remarkably efficient wildcats based upon the .223 Remington case, taking bullets of .243-, .264- and .284-inch diameters. To the present, there have been no T/CU cartridges bearing caliber designation in inches, although a comparable cartridge taking the caliber .270 (.277-inch diameter) bullet seems rather promising in speculation.

As yet, no manuals or handbooks carry load data for the 6mm T/CU, although I have pre-publication listings from both Hornady and Sierra and propose to reproduce both of them here.

A loaded round of 6mm T/CU with a fired case that nearly separated on its second loading. Problem here was some overly brittle brass. Careful positioning of the sizing die and canny choice of brass avoids all this.

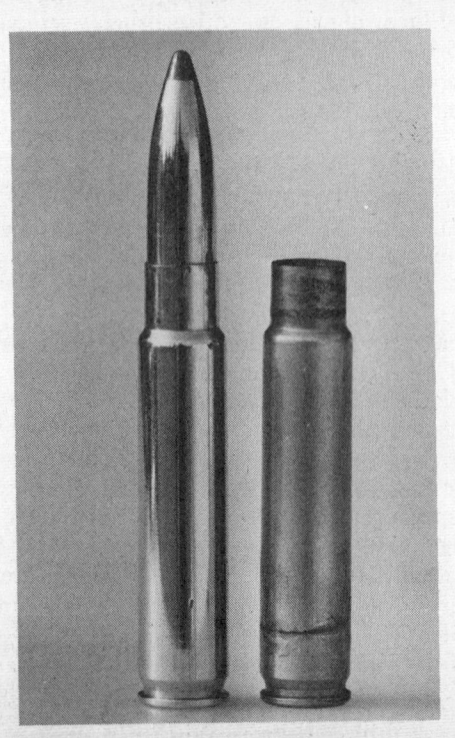

The sheets of load data from Hornady carry this notice: *Notice of Disclaimer: This data was developed in the Hornady Mfg. Co. test facilities under controlled conditions. As components and conditions may vary, the reloader should approach maximum loads with extreme caution. Hornady Mfg. Co. disclaims all responsibility for mishaps of any nature which might occur from use of this data.*

Hornady
87-grain S.P.

POWDER   VELOCITY (fps)/ENERGY (fpe)
(Charge weights in grains)

	2000/ 773	2100/ 852	2200/ 935	2300/ 1022	2400/ 1113	2450/ 1160
Reloder-7	18.5	19.5	20.5	21.5	----	----
H322	20.9	21.9	22.9	23.9	----	----
MR2460*	21.6	22.7	23.8	24.8	25.9	26.5
H335	----	22.9	23.9	24.9	26.0	26.5
IMR-4895	----	23.1	24.2	25.2	26.3	26.8
Win 748	23.6	24.6	25.6	27.5	----	----

(*An Accurate Arms powder)

100-grain S.P.

	1900/ 802	2000/ 888	2100/ 979	2200/ 1075	2250/ 1124	2300/ 1175
Reloder-7	18.4	19.6	20.7	----	----	----
H322	20.3	21.4	22.4	23.5	----	----
MR2460	19.9	21.4	22.9	24.5	----	----
H335	20.7	21.9	23.2	24.5	----	----
IMR-4895	----	22.6	23.7	24.9	25.5	26.0
Win 748	22.5	23.8	25.1	26.4	----	----

The data from Sierra carries this notice: *The Leisure Group and Sierra Bullets, a division thereof, cannot and do not accept any liability, either expressed or implied, for results or damage or injury arising from or alleged to have arisen from the use of the data in this manual.*

The first Sierra list is for their 60-grain hollow point bullet, at a cartridge overall length of 2.350 inches.

	2500/ 833	2550/ 867	2600/ 901	2650/ 936	2700/ 971
Norma 200	22.5	23.0	23.5	24.0	24.5
MP-5744*	21.0	21.6	22.2	----	----
IMR-4198	22.0	22.5	23.0	23.5	24.1
Reloder-7	22.0	22.8	23.6	----	----
H322	25.3	25.5	25.7	25.9	----
2230*	----	25.8	26.4	27.0	27.6
Win 748	28.0	28.4	28.8	29.2	29.5

SIERRA BULLETS, 70 to 75 grains

	2200/	2250/	2300/	2350/	2400/	2450/	2500/
(for 70-gr.)	752	787	822	859	896	933	972
IMR-4198	19.9	20.5	21.1	21.7	22.3	22.9	23.2
Reloder-7	20.4	20.8	21.2	21.6	22.0	22.5	----
IMR-3031	24.5	24.7	24.9	----	----	----	----
H322	22.3	22.8	23.3	23.8	24.3	25.0	----
2230*	24.0	24.3	24.6	24.9	25.3	25.7	26.1
Win 748	----	----	25.5	26.0	26.5	----	----
IMR-4895	25.8	26.2	26.6	27.0	----	----	----

SIERRA Bullets, 80 to 85 grains

	2100/	2150/	2200/	2250/	2300/	2350/
(80-gr.)	784	821	860	900	940	981
IMR-4198	18.9	19.5	20.1	20.7	21.2	----
REloder-7	19.0	19.6	20.2	----	----	----
H322	21.1	21.7	22.3	22.9	23.6	----
2230*	----	----	22.5	23.2	23.9	24.6
2460*	----	----	23.0	23.8	24.6	25.5
Win 748	----	23.1	23.8	24.6	----	----
IMR-4895	25.0	25.4	25.8	26.1	----	----

SIERRA BULLETS, 85 to 90 grains

	2100/	2150/	2200/	2250/	2300/
(85-gr.)	833	873	914	956	999
IMR-4198	18.6	19.5	20.1	21.2	----
Reloder-7	18.6	19.1	19.6	20.1	----
H322	20.3	21.0	21.6	----	----
2230*	21.7	22.2	22.9	23.1	----
2460*	----	----	22.0	23.0	24.0
Win 748	23.2	23.5	23.8	24.0	----
IMR-4895	24.9	25.3	25.7	26.1	26.4

SIERRA BULLETS, 100 grains

	2000/	2050/	2100/	2150/	2200/
	888	933	979	1027	1075
IMR-3031	21.4	22.0	22.6	23.2	----
H322	20.4	20.9	21.5	----	----
2230*	21.1	21.6	22.1	22.7	----
2460*	21.6	22.1	22.6	23.1	----
Win 748	----	23.1	23.5	23.9	24.4
IMR-4895	23.2	23.8	24.4	25.0	25.4
IMR-4064	23.0	23.6	24.2	24.7	----
IMR-4320	24.2	24.9	25.6	----	----

(*Accurate Arms powders)

Author's Note: As the Hornady Handbook and the Sierra Manual are updated, data for the 6mm T/CU undoubtedly will be included therein. Should the final listing in either book differ in any detail from data quoted here, the final version in the handbook or manual should be followed.

Data from both sources were taken in a fourteen-inch T/C Contender barrel, with a rifling pitch of one turn in ten inches (1:10). The data from Hornady used cases formed from Hornady/Frontier .223 Remington brass and Federal 205 primers. The Sierra data used cases formed from Federal .223 Remington brass and Federal 205M (the M is for match, not magnum) primers.

The Sierra data carried an additional note: *If military brass is used, reduce all charges by one (1.0) grain. The last listed charge for each powder is a maximum load. NEVER start at a maximum load. Loads less than minimum are not recommended.*

At the time of writing, T/C offers 6mm T/CU barrels only in the Super-14 length.

6.5mm T/CU

It is my sneaking Irish hunch that this one is quite possibly the best of the T/CU threesome, the others being the 6mm and 7mm. I mentioned that to Warren Center and he confided that he felt exactly the same way about it.

From the viewpoint of the hunter, the 6.5mm or .264-inch-diameter bullets have an enviable reputation as putter-downers of game well beyond reasonable expectations. Discussion of the 6.5mm JDJ later offers more comment on that.

At the moment, I'm in the chagrinful position of being under-researched on the cartridge under discussion. Some while ago, I got a ten-inch barrel in 6.5mm T/CU and got in no more than a token nest-egg of research on it. Then, in final boarding phases of getting copy down on the manuscript you're reading, a 6.5mm T/CU Super-14 barrel came waltzing along to take its place in the battery.

As yet, the 6.5mm T/CU is not available as a twenty-one-inch carbine barrel. I have a strong suspicion it might be a superb performer in that length, based upon the fact that the 7mm T/CU does quite well in the carbine barrel.

Even in the ten-inch barrel, the 6.5mm T/CU is quite pleasant to fire, by reason of moderate recoil with bullets of the various available weights.

To the date of writing, load data for the 6.5mm T/CU is not too widely available. It is covered on pages 334 and 335 of the No. 25 Hodgdon Manual for bullet weights from 87 to 160 grains. I have some data worked up by the Hornady lab and propose to quote it here. Presumably, this will be incorporated in a future revision of the Hornady Handbook and, when and if it is, if any discrepancies are noted between the data quoted here and that appearing in the Hornady Handbook, the data in the book should be used.

NOTICE OF DISCLAIMER: This data was developed in the Hornady Mfg. Co. test facilities under controlled conditions. As components and conditions may vary, the reloader should approach maximum loads with extreme caution. Hornady Mfg. Co. disclaims all responsibility for mishaps of any nature which might occur from use of this data.

Having no control over components, equipment and techniques used by others, DBI Books and the writer cannot and do not assume any liability, implied or expressed, for damages or injuries arising from or alleged to have arisen from the use of data appearing anywhere in this book, including the following:

6.5mm THOMPSON/CENTER UGALDE (T/CU OR TCU)

Barrel: 10-inch factory standard; Primers: Remington 7½; Cases: Frontier, re-formed. Factory barrel has .256 to .257-inch bore diameter, .264 to .265-inch groove diameter and rifling twist of 1:8 inches, LH.

100-grain HORNADY SPIRE POINT BULLET

POWDER	VELOCITY/ENERGY				
	(charges in grains)				
	1800/	1900/	2000/	2100/	2200/
	720	802	888	979	1075
IMR-3031	20.6	21.6	22.6	23.6	24.6
H322	24.4	25.2	26.0	26.9	----
IMR-4895	24.0	25.0	26.0	27.0	28.0
Win 748	27.0	28.4	29.8	----	----

129-grain HORNADY SPIRE POINT BULLET

	1700/ 828	1800/ 928	1900/ 1034	1950/ 1089	2000/ 1146	2050/ 1204
Reloder-7	19.7	20.7	21.7	22.2	----	----
H322	21.8	22.8	23.8	24.3	24.8	----
IMR-3-31	22.9	24.0	25.0	25.6	26.0	----
IMR-4895	23.3	24.5	25.7	26.3	26.7	----
Win 748	24.6	26.0	27.4	28.1	28.8	29.5
BL-C2	----	25.9	27.5	28.3	29.0	----

140-grain HORNADY SPIRE POINT BULLET

	1700/ 899	1800/ 1007	1850/ 1064	1900/ 1123	1950/ 1182
H322	21.4	22.4	23.0	----	----
IMR-3031	22.0	23.2	23.8	24.5	25.1
IMR-4895	22.7	24.2	24.9	25.6	----
BL-C2	24.2	25.7	26.4	27.0	----
Win 748	24.6	26.0	26.6	27.3	28.0

7mm T/CU

Gunsmith Wes Ugalde designed the 7mm T/CU — Thompson/Center Ugalde — by opening the neck of the parent .223 Remington case to accept bullets of .284-inch diameter. For the record, 7mm converts to .275-inch and .284-inch is equal to about 7.21mm, but it makes a little better sense if you know that the bore diameter of the 7mm T/CU barrel is .277 to .2785-inch, while the groove diameter is from .284 to .2855-inch. Quite a few calibers are designated in accordance with the bore diameter — distance from one land to the opposite land — rather that the groove diameter that is close to the bullet diameter. The rifling pitch is 1:9, LH.

The 7mm T/CU came along in 1980 as a standard Contender chambering, made in both ten-inch and Super-14 lengths. Unless you need the ten-inch to qualify for a competitive event, the longer barrel has a lot of good things going for it.

In a ten-inch barrel, the 7mm T/CU generates quite a bit of recoil and muzzle blast. In the more energetic loadings, it is not a whole lot of fun to fire. The Super-14 handles the recoil better and coaxes more velocity out of the cartridge in the bargain. If permitted by rules of the given game, a scope sight is immensely helpful in optimizing results with the 7mm T/CU Contender.

I have a special fondness for the 7mm T/CU, because my Super-14 barrel in that flavor was the touchstone that brought me my first Contender group under one minute of angle (MOA). At one hundred yards, 1 MOA is a trifle over 1.04 inches and I put five shots into a spread between centers of .914-inch.

My magic recipe combined Remington 7½ primers, 32.0 grains of Hodgdon BL-C2 powder and a Sierra No.1900 120-grain spitzer bullet at 2.590 inches loa. It put four holes into a center-spread of .530-inch, but the fifth blew it to .914-inch. Even so, that's .870-MOA and comfortably under the mark. For the best four shots, make that .506-MOA. Some amount of credit should be awarded to the 4X Redfield scope that steered the hits so close together. The group came out of the Super-14.

Back in 1982, when I enjoyed my sweet moment of triumph over the fickle muse of ballistics — Expelode, I think she's called — my chronograph of the time must have hiccuped, because the account mentions nothing as to velocity and energy.

Returning to the fray in 1986, this time with a twenty-one-inch carbine barrel in 7mm T/CU to back up the tenner and Super-14, I re-created the winning load of '82 and tried it out of all three. Results ran:

10-inch	1850/912
14	2140/1221
21	2286/1393

A further checkout of a second load — 26.0 grains of H322 behind the 100-grain Hornady No. 2800 JHP — produced these results:

10	1862/770
14	2088/968
21	2228/1103

Two more charge-boosts, out of the carbine barrel, went:

| 26.5 gr | 2308/1183 |
| 27.0 | 2348/1224 |

As in so many other examples of Contender cartridges, getting the most out of the 7mm T/CU involves working out an ultra-precise and finicky adjustment of the full-length resizing die, so as to headspace the chambered round on the shoulder. The 7mm has the least shoulder of the present T/CU series, which doesn't make it a bit easier. I think it shows promise as a caliber for the carbine conversions and look forward to doing more research in that specialized area.

Given a free choice in handgun barrel lengths, the Super-14 is clearly the way to go. There is a good amount of published load data available for the 7mm T/CU and more should be on the way shortly.

.32 HARRINGTON & RICHARDSON MAGNUM

This relatively recent addition to the family of handgun

Below, left, a .25-35 WCF next to a 7mm International Rimmed cartridge. Either can be made from .30-30 WCF.

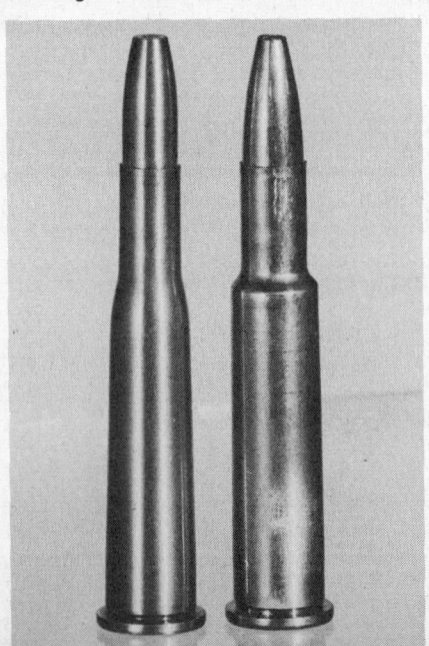

cartridges is intended to offer the shooter a reloadable centerfire caliber with ballistics substantially better than those of the .32 S&W Long, ranging upward to about the level of standard .38 Special loads.

The concept has considerable merit in view of the general desirability of light and handy guns for plinking, informal target practice, taking small game and similar applications. The generic term for small ordnance of that approximate category is "camp gun."

Originally designed and introduced by Harrington & Richardson (H&R), who offered some small-frame, double-action revolvers chambered for it, the cartridge was adopted by other makers such as Charter Arms, Sturm, Ruger & Company, Dan Wesson Arms and perhaps others, by this time. Thompson/Center introduced ten-inch Contender bull barrels for the .32 H&R magnum at about the same time they added the .32-20 WCF. The groove diameter for both Contender barrels is .308-inch, whereas the groove diameter for most other guns in .32 H&R magnum is more like .312-inch.

The working pressure of most factory loads for the .32 H&R magnum is reported to be in the neighborhood of 20,000 c.u.p. — well below the levels of most other magnums. "Magnum" has developed into a sort of buzz-word, with semantic connotations probably neither intended nor envisioned by the people who first applied the term to cartridges long ago.

The example of the .32 S&W Long cartridge is illustrative in this instance. Given a gun of adequate action strength, it is possible to develop loads that muster some thoroughly respectable ballistics. Sadly, however, there have been a great many rather weak revolvers — several of them of the break-action design — chambered in .32 S&W Long and any load data published for the cartridge takes that firmly into consideration.

When reloaded for the Contender and fired exclusively in that gun, it is possible to reload the .32 H&R magnum case to impressive performance levels. As in so many other instances, however, there is the obvious peril and pitfall that such loads might stray and be fired in one of the revolvers incapable of coping with pressures much beyond the 20,000 c.u.p. or so of the factory load.

Hornady markets one of the best bullets currently available for reloading the .32 H&R magnum: their No. 3205, an 85-grain JHP design, .312-inch in diameter. Although .004-inch larger than the groove diameter of the Contender barrels in .32 H&R magnum and .32-20 WCF, the Hornady bullets perform well in both Contender chamberings.

Down in McEwen, Tennessee, at Accurate Arms Company, their head ballistician, Marty Liggins, has run pressure tests on the .32 H&R magnum cartridge with the Hornady No. 3205 bullet and he quotes data and performance as follows:

4.5 grains	Accurate Arms No. 5	854/138
5.0		950/170 MAXIMUM
5.8	Accurate Arms No. 7	864/141
6.5		950/170 MAXIMUM

The four loads quoted should be safe when made up to specifications and fired in any well made revolver in good condition. It goes without saying that the Contender can handle 20,000 copper units in a cartridge with that head diameter and then some. Despite that, I do not feel inclined to suggest, let alone recommend, loads exceeding Liggins' maximums for the .32 H&R magnum in any gun.

7mm INTERNATIONAL RIMMED
7-30 WATERS
7mm REMINGTON B.R.

The first is a wildcat, based upon the .30-30 WCF case, developed under the auspices of Elgin Gates, head of the International Handgun Metallic Silhouette Associaton (IHMSA), Box 1609, Idaho Falls, ID 83401. As you might suppose, the .30-30 case is necked down to accept bullets of 7mm (.284-inch) diameter. Gates can furnish both Contender barrels and loading dies for the 7mm IR, as well as load data for it.

For some reason, Thompson/Center has refrained from offering the 7mm IR as a standard Contender chambering. Rather, they introduced the 7-30 Waters cartridge initially as a caliber option for the Contender carbine conversions, then added pistol barrels in both ten-inch and Super-14.

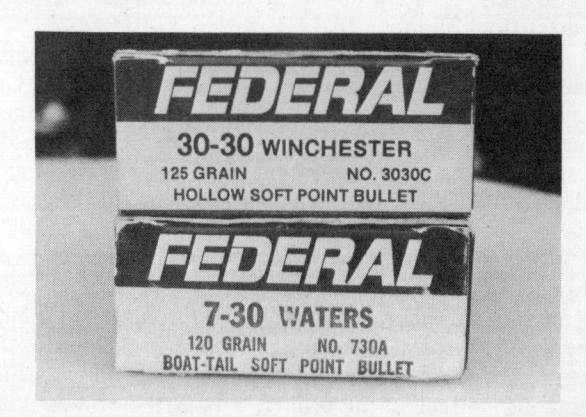

Never a wildcat, the 7-30 Waters, designed by noted ballistician Ken Waters, started its career as a factory load from Federal. Below, for comparison, the 7mm International Rimmed, .30-30 WCF and 7-30 Waters.

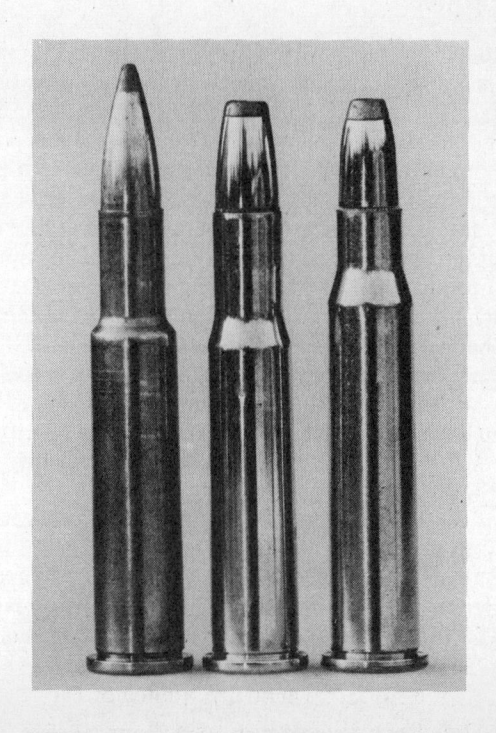

GUN DIGEST BOOK OF HANDGUN RELOADING

The 7-30 Waters has a shorter neck than the 7mm IR, with a corresponding moderate gain in case capacity. As a result, either cartridge offers some gain over rounds such as the 7mm T/CU or the 7mm Remington Bench Rest.

T/C personnel, by the way, express considerable doubt that they will ever chamber Contender barrels for the 7mm Rem BR round, for which Remington offers their Model XP-100 bolt-action single-shot pistol. The general consensus is that some of the loads quoted for the 7mm Rem BR — originally worked up for the XP-100 — develop pressures overly vigorous for the basic Contender action.

.30 M-1 CARBINE

A standard offering for the ten-inch octagonal barrels of the Contender in its early days, the .30 M-1 barrels were dropped from the line for several years, then reinstated in the No. 14 catalog published early in 1987. The '87 version of the .30 M-1 barrels are standard in ten-inch bull configuration, with a Super-14 version available from T/C's Custom Shop.

Load data for the .30 M-1 is to be found in any manual or handbook and data quoted for use in the eighteen-inch carbine barrel will be somewhat faster than can be obtained, even in the Super-14 barrel. Data quoted for use in the Ruger Blackhawk single-action revolvers can be topped substantially by the ten-inch Contender and the Super-14 will go somewhat faster.

When reloading old military .30 M-1 cases, you can expect to encounter a unique problem that I've never noted in any other caliber. About one to three cases out a hundred will have some amount of fatigue in the cup of the spent primer. As a result, the decapping pin will punch out the exposed face of the primer, leaving the primer skirts tightly held in the pocket. I know of no way to salvage such cases and they are a real trial to your patience and sunny disposition.

Accurate Arms No. 9 powder is an excellent performer in the .30 M-1 case, as are 2400, 296 and either of the 4227 powders.

.32-20 WINCHESTER CENTER FIRE (WCF)

This is an old cartridge, stemming from the days of black powder. The second portion of its designation indicates that it once was loaded with 20.0 grains of black powder. Up through the mid-Thirties, it was regarded as a powerhouse in handguns, until the .357 magnum appeared in 1935. Interest in the cartridge waned considerably for the next many years, but it hung on somehow. At one time, factories produced a handgun load and a somewhat more powerful rifle load in .32-20 WCF. The rifle load seems to have been dropped from production.

Both Winchester and Remington catalog two .32-20 loads, with 100-grain bullets in a choice of lead or JSP. In the ten-inch Contender, all four loads average pretty close to 968/208, with the JSP loads grouping slightly better than the lead ones.

The .32-20 was added to the Contender roster rather recently, as a result of a new target game called NRA Hunter Silhouette. The rules called for straight-sided cases and they decided to sanction the .32-20 despite its remarkably vestigial bottleneck. Contender barrels for this cartridge are available only in the ten-inch length, with groove diameters of .308- to .3095-inch and a pitch of 1:10 inches.

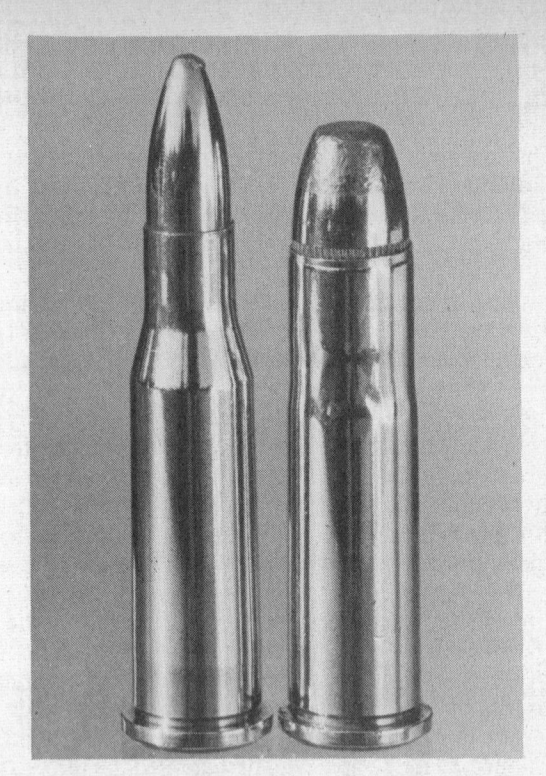

The .32-20 WCF, right, dates from black powder days. The .218 Bee, left, was created by necking down the .32-20 WCF and giving it a slightly greater case length.

Other barrels for the .32-20 usually have had larger groove diameters, varying somewhat. Typical bullet diameters for the cartridge have ranged from about .311 to .313-inch and loading die sets reflect that expectation.

Most sizing dies do not resize the case neck down far enough to get a good grip on bullets of .308-inch diameter, the size best suited to the Contender's groove diameter. In a single-shot action such as the Contender, you do not require a terribly tight bullet grip and, as long as the bullet remains in place during handling, there is really no problem.

Thompson/Center's booklet of load data, *Making Your Contender Perform!,* lists a load with 15.5 grains of Winchester 296 behind the 110-Speer JHP Varminter bullet at 1463/523. Duplicating this load in my own ten-inch barrel, I get a rather more impressive 1825/814: a gain of 291 fpe and hardly a trifling margin. The Speer Varminter, by the way, has a diameter of .308-inch.

Fitted with a 4X Redfield scope on Weaver block and rings, my ten-inch .32-20 Contender barrel has shown me a great deal more accuracy than I'd ever have suspected the cartridge could offer. Five-shot groups, from twenty-five yards, off sandbags on the bench, sometimes got down to less than half an inch in spread between centers. The figure of 1825/814 was the highest energy reading I posted on the load just mentioned, which grouped to 1.806 inches. That is more energy than you can readily coax out of the .357 magnum in typical revolvers and a tighter group in the bargain.

It was noted that current .32-20 factory loads average hardly over 968/208 in the unvented, ten-inch Contender barrel. The most potent load is Winchester's #X32202, with the 100-grain JSP bullet; it goes 995/220. When we

CONTENDER DATA

CALIBER .32-20 WCF   PRIMER Rem. #7½   MAXIMUM CASE LENGTH 1.315 inches

BULLET	POWDER	CHARGE, grains	LOA, inches	VELOCITY/ENERGY 10-inch	25-YARD GROUP inches
71-grain Hornady #3200	H4227	14.4	1.530	1684/447	1.067
"	2400	14.4	1.530	2030/650	1.644
"	296	16.0	1.495	1994/627	1.444
"	"	18.2	1.580	2145/726	0.947
85-grain Hornady #3205	H4227	14.4	1.509	1672/528	0.792
"	2400	11.0	1.555	1545/451	0.434
110-grain Speer #1835	H4227	14.4	1.650	1569/601	0.857
"	296	15.5	1.640	1825/814	1.806
110-grain Speer #1845	H4227	14.4	1.708	1568/601	0.957
110-grain Speer #1855	"	14.4	1.840	1569/601	0.900
110-grain Sierra #2110	IMR-4227	14.4	1.800	1527/540	1.061
"	H4227	14.4	1.920	1543/582	0.771
"	"	14.4	1.845	1561/595	0.566
"	296	15.5	1.730	1793/785	0.858
118-grain Lyman #311316, cast	Blue Dot	6.8	1.505	1235/400	0.676

bring up performance on the order of the load from the T/C booklet, doing 1825/814, we are looking at nearly four times the energy of the factory load.

It should be obvious that the factory load is loaded-down to be reasonably safe in typical .32-20 revolvers. If you put one of those hot Contender loads in an old, light-frame revolver of the same caliber, it is pretty certain to blow the revolver into small, high-velocity pieces.

The 1980 edition of Lyman's *Cast Bullet Handbook* lists .32-20 data to a maximum of 1390/493, warning not to use the loads in rifles designed for black powder or in handguns. That suggests that about 500 fpe should be regarded as the ceiling for use in relatively modern .32-20 rifles and it follows that any load that develops 814 fpe should not be used in contemporary rifles of the same caliber.

What I'm saying is that, if you make up such loads, it is your responsibility to make absolutely certain that they do not end up being fired in a rifle or handgun, other than the Thompson/Center Contender. On that firm stipulation, I will report additional Contender loads in .32-20, noting that none of the listed loads showed any indication of excessive pressure in my Contender. It would, however, be a good idea to approach the hotter loads with caution, from about ten percent below the listed charge, increasing as pressures may seem to warrant.

.30 HERRETT

As early as 1970, Steve Herrett was working up experimental wildcat cartridges for use in the T/C Contender. By 1972, the first barrels were made in .30 Herrett and several head of big game were taken with the cartridge.

The .30 Herrett is based upon the .30-30 WCF parent case, shortened to a case length of 1.605 inches and fire-formed. It will handle bullet weights from 110 to 165 grains and load data for the cartridge is abundantly supplied by all the contemporary manuals and handbooks.

Barrels are available in both ten-inch and Super-14. If you can live with the extra bulk, the Super-14 is the preferable way to go.

Some while after the introduction of the .30 Herrett, T/C made Contender barrels available in the standard .30-30 WCF chambering and — particularly in the Super-14 barrel — the .30-30 is a vastly better choice than the .30 Herrett. You do not have to go through the drill of producing the cases and the .30-30 will out-perform the .30 Herrett by significant margins.

The .30 Herrett will not deliver peak accuracy, unless you go to considerable pains to headspace the reloaded rounds on the shoulder of the cartridge, rather than upon

In working up to the .30 Herrett, Steve Herrett tried out a highly necked wildcat based upon the .44 magnum case and somewhat resembling the .17 Bumblebee in the center, here. Herrett's .17-.44 appears near the lower left corner of the lead illustration for this section, was a devil to make.

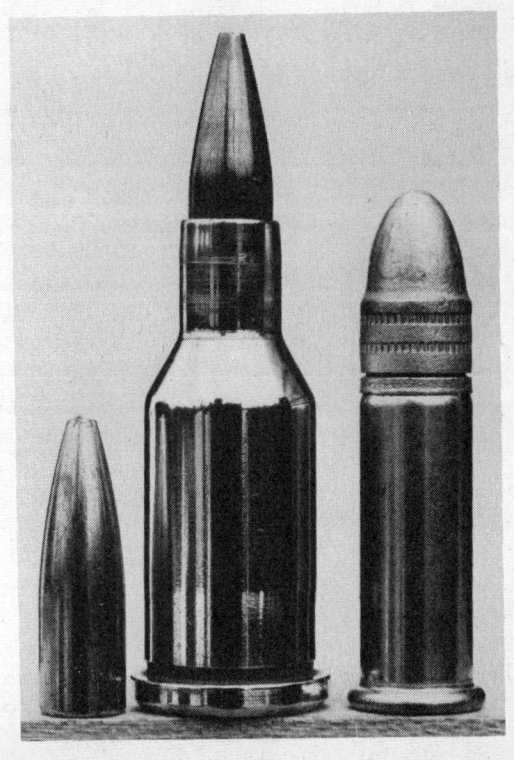

GUN DIGEST BOOK OF HANDGUN RELOADING

the rim. To accomplish this involves a highly critical adjustment of the full-length resizing die of the loading die set.

If the die is turned down in the press just a thin trifle too far, the resulting loads will be short on headspace and will not group all that well. If the die is turned up just a little higher, the resulting reloads either will not chamber or, if chambered, will not fire.

You are looking at a distance of perhaps .004-inch between the two extremes. If you propose to work extensively with the .30 Herrett Contender, I suggest you take portable reloading equipment afield and fiddle with the adjustment of the locking collar of the loading die until you get it just right, then lock it in place for keeps.

.30-30 WINCHESTER CENTER FIRE (WCF)

This cartridge was not added to the T/C catalog until fairly recent times. It needs the fourteen-inch barrel for best performance and that best performance is pretty good: substantially better than that of the .30 Herrett.

Winchester 748 powder usually works best in the .30-30 Contender, with Hercules Reloder-7 as second choice. Unlike lever-action rifles with tubular magazines, there is no objection to using pointed or spitzer-type bullets in the .30-30 when the reloads are to be fired in the Contender, thus widening the choice of bullets considerably. Given a scope sight and a solid firing point, groups as small as one inch at one hundred yards are not an impossible dream.

Large rifle primers must be used when reloading the .30-30 for use in the Contender. Some suggested loads are:

110-grain Hornady #3010
    41.0 grains 748   2230/1215
    29.0 IMR-4198   2310/1304
125-grain Sierra #2120
    41.0 grains 748   2193/1335
    32.0 IMR-3031   2210/1356
130-grain Hornady #3020
    40.0 grains 748   2153/1338
    31.0 IMR-3031   2160/1347
150-grain Hornady #3031
    38.0 grains 748   2090/1455
    28.0 Reloder-7   1916/1223
    30.0 IMR-4064   1900/1203

Generally speaking, bullets heavier than 150 grains do not perform well in the .30-30 Contender.

9mm LUGER (PARABELLUM)

Like the .30 M-1 Carbine, this is a Contender chambering introduced and dropped long ago, then restored to the listing in the No. 14 T/C catalog issued in early 1987.

Load data for the 9mmP — as I customarily abbreviate it — is to be found in lavish abundance in virtually any manual or handbook; all such listings can be used in the Contender to excellent effect.

As most such data is derived from barrels of four- to five-inch length, you will obtain significantly better ballistics in

The 9mm Luger, left, recently was reinstated in the T/C catalog. It's shown with a .45 ACP, once offered as a Contender barrel but probably not apt to return

At left, the 9mm Luger appears next to the somewhat longer .38 Colt Super or .38 Auto +P, as maker terms it. Above, the 9mm Luger next to a 10mm Auto round. Latter can be used in custom barrels from Mike Rock.

the ten-inch Contender barrel and the accuracy is generally apt to be quite gratifying. Your rate of fire will be sharply diminished in comparison to typical autoloading pistols, but there is the consolation that recovery of the fired cases is immeasurably more convenient.

The intriguing accuracy potential of the 9mmP in Contender barrels makes installation of a scope sight worth considering.

.38 COLT SUPER

This is a chambering once offered in the heyday of the ten-inch octagonal barrels, then dropped. It does not appear overly likely that this chambering is apt to make it back into the catalogs, because, unlike the .30 M-1 and 9mmP, the cartridge and guns for it have never won a high degree of popularity with the shooting public.

Standard load data for the .38 Colt Super, found in a broad variety of sources, can be used for reloading this cartridge for your Contender and some increase in ballistics over the listings — which are taken from five-inch barrels — can be expected.

.357 HOT SHOT

The .357 Hot Shot, like the .44 Hot Shot, was brought out at the time the .45 Colt/.410-bore barrels were discontinued for the first time, with the .357 version following the .44 by quite some time.

At present, availability of either the loaded rounds of .357 Hot Shot or the .357 Hot Shot capsules is more than somewhat iffy. The .44 was much the more popular of the two and, as noted elsewhere, actually performed quite well. The current T/C catalog continues to list the .44 loads and components. A box of twenty .44 Hot Shot cartridges is $19.95 and a box of fifty .44 Hot Shot capsules is $19.50, with a choice of #6 or #7½ shot.

It should be emphasized that the Hot Shot capsules and loads, of either size, are for use *only* in the Contender Hot Shot barrels with the detachable choke installed. The ribs of the choke are necessary to shred the tough plastic covering of the capsule and release the shot pellets. If fired in a Hot Shot barrel without the choke, the Hot Shot capsules go forth unbroken, keyholing madly, with little if any pretense to accuracy.

If you have a .357 Hot Shot barrel and some of the .357 Hot Shot capsules, the recommended powder charge is 9.0 grains of Bullseye, rated to give the half-ounce charge of shot a velocity of 1202 fps.

Reloading procedure is much the same as for standard bullets. The case is resized, reprimed and the mouth lightly belled to accept the base of the capsule, which is seated and given a moderate roll-crimp.

.357 MAGNUM

Introduced on Monday, the eighth of April, 1935, the .357 magnum has become one of the most popular of all handgun cartridges. In essence, it is the much older .38 Special cartridge, with its case lengthened by .135-inch to prevent chambering in revolvers of the older, weaker caliber. Guns for the .357 magnum are engineered to handle peak pressures to a maximum of 46,000 copper units, as compared to 18,900 c.u.p. for the .38 Special or 22,400 c.u.p. for the +P .38 Special. The cylinders of .357 mag-

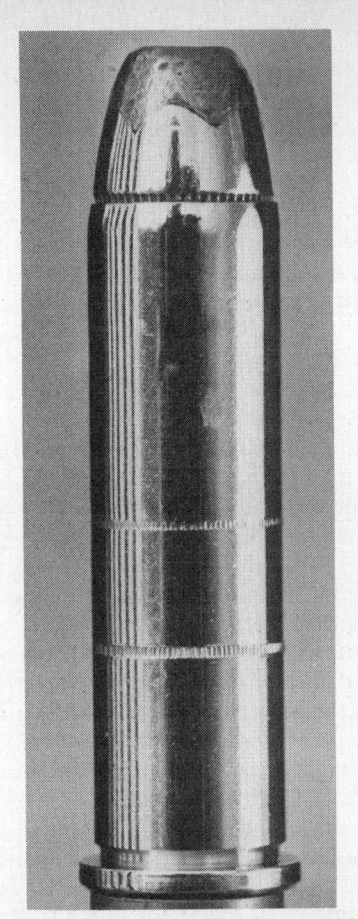

Introduced in April of 1935, the .357 magnum is one of the best performers in the Contender, particularly out of the Super-14 barrel, capable of unusual power.

num revolvers were not lengthened correspondingly, requiring a cartridge overall length for the newer cartridge that is about the same as the .38 Special.

The .357 magnum is an outstandingly fine performer in the T/C Contender, particularly in the Super-14 barrels, where it is not difficult to obtain impressive velocities and remarkable accuracy, particularly if a scope sight is installed. Unlike revolvers, the Contender imposes no arbitrary limit as to overall cartridge length, provided the full-diameter portion of the exposed bullet does not interfere with the commencement of the rifling, or *leade*.

Nominally, the .357 magnum cartridge is loaded with bullets of .357-inch diameter in the jacketed versions, or with cast bullets sized to .358-inch. The T/C specifications for .357 magnum Contender barrels show a bore diameter of .346- to .348-inch; a groove diameter of .356- to .358-inch and a rifling pitch of 1:14, LH.

Purely in line of experimental research, I have used nominal 9mm bullets, with a diameter of .355-inch, in reloads for the .357 magnum Contender and, in general, the results have been highly encouraging. Such maverick reloads also work quite well in the Colt Python, with its .354-inch groove diameter.

The lightest jacketed .357-inch bullets generally available weigh 110 grains. Bullets in 9mm range down to 88 grains, in the example of Speer's #4000 JHP design. I loaded the Speer #4000 ahead of 18.0 grains of Hercules

Blue Dot powder and clocked it out of the Super-14 Contender at a five-shot average of 2639/1361 and, with a 4X Redfield scope on top, it delivered a center-to-center spread of 1.020 inches at twenty-five meters.

Moving on up to Speer's 100-grain #3983 JHP — likewise intended for the 9mm — I used a charge of 23.7 grains of Winchester 296 powder and got an average of 2321/1196 in the Super-14 barrel. The twenty-five-meter group had a center spread of .260-inch for five shots, which works out to about .908 minute of angle (MOA).

Compared to typical revolvers, the Super-14 Contender extracts quite a bit of additional velocity and energy from the .357 magnum cartridge. For example, the Federal #357B factory load, with a 125-grain JHP bullet, goes 1436/572 in a four-inch Model 19 S&W, 1544/662 in a six-inch Colt Python, 1575/688 in an eight-inch Dan Wesson Arms — and 2072/1191 in the Super-14! Indeed, the Super-14 raises the .357 magnum to energy deliveries that compare well with typical .44 magnum loads in revolvers of average barrel length.

My favorite cast bullet for the .357 magnum is Lyman's mould #358156, a Thompson-type semi-wadcutter with gas check. Lyman rates that bullet at 155 grains and says that 14.0 grains of Hercules 2400 powder drive it to 1299/581 in a vented four-inch test barrel; pressure, 41,900 c.u.p. In the alloy at hand, my bullets weighed 158 grains and 13.9 grains of 2400 drove them out of the Super-14 barrel at an average of 1677/987, grouping 1.816 inches at twenty-five meters, with four of the holes in a spread of only .550-inch.

Using the 125-grain Hornady #3573 JSP bullet and 18.5 grains of 296 powder, the Super-14 averaged 1874/975 and grouped to .739-inch at twenty-five meters. The 140-grain Hornady #3574 JHP went 1645/841 and grouped 1.087 inches on a charge of 17.2 grains of Du Pont IMR-4227 powder.

By way of checking for random flukes, I made up two five-round lots of test ammo with the 88-grain Speer #4000 (9mm) JHP ahead of 23.7 grains of 296 powder. Fired at two different aiming pasters, off a sandbag rest on the bench at twenty-five meters, one lot averaged 2342/1071 and grouped to .794-inch. The other averaged 2342/1072 and grouped to .708-inch. The 1-fpe discrepancy can be explained by the tendency of the chronograph to round-off decimals on the figures for velocity. Apart from that, I thought the results were surprisingly consistent.

A lot of research has been lavished upon heavier bullets in the .357 magnum Contender by handgun metallic silhouette fans, who need weightier slugs to cope with the 200-meter ram targets. The big three bulletmakers — Hornady, Sierra and Speer — all offer bullets in the weight class of 160 to 200 grains, carefully designed to apply the maximum amount of decisive shove-over to the reluctant rams.

Such bullets can be used in the Contender without obvious problems, because the front of the bullet can be loaded to extend ahead of the case mouth somewhat farther than if the reload were to be fired in a revolver.

The #25 Hodgdon Manual, on page 367, lists maximum loads for the following bullet weights, in the .357 magnum Contender ten-inch barrel:

160-grain	15.5 gr. H4227	1550/854
	16.0 H110	1618/930
170-grain	14.0 H4227	1280/619
	15.0 H110	1394/734
180-grain	13.5 H4227	1261/636
	14.0 H110	1284/659

As always, maximum loads such as these should be approached with caution and avoided if indications of excessive pressure occur. Ballistics from the fourteen-inch barrel would be somewhat higher than those shown.

Winchester's *Ball Powder Loading Data* has a listing for 200-grain lead bullets in the .357 magnum: 12.4 grains of 296 for 1335/792 at a pressure of 35,000 c.u.p.; barrel length is not specified.

.357-.44 BAIN & DAVIS

In essence, this is the .44 magnum case necked down to accept a bullet of about .357-inch diameter. There was a much earlier version, with a rather sharp shoulder, which designer Keith Davis modified to the present more gradual taper.

Thompson/Center offered Contender barrels in this caliber for a time, but that was in the days of the octagonal barrel, with the forend held in place by a spring clip. The B&D round, along with the .44 magnum, were among the cartridges whose recoil was apt to be vigorous enough to rip the forend free of the barrel by sheer inertia and send it clattering about the shooter's feet. That was distracting and a trifle annoying. By the time the bull barrels came along, about 1974, the .357-.44 B&D was fairly well defunct as a catalog offering and, if any bull barrels were ever chambered for the B&D round, I never heard of it.

Another single-shot pistol, the Merrill Sportsman, made by Rock Pistol Manufacturing, of Brea, California, was offered in .357-.44 B&D, among several other calibers.

The Hodgdon Manual, for the past several editions, down through #25, has listed a table of data for the .357-.44 B&D; it appears on page 369 in #25. Data was worked up in a ten-inch Contender barrel and it takes bullet weights up to 158 grains, to a maximum velocity of 1604 fps and 903 fpe. Hodgdon lists data for two of their powders: H4227 and H110. The data in the Hodgdon Manual

Here's the barrel marking on one of the early octagonal barrels in .357-.44 Bain & Davis, long since discontinued.

should be regarded as maximum for use in the Contender and the octagonal barrels chambered for that cartridge.

In the Rock/Merrill pistol, it is possible to use heavier bullets and to drive them faster. Using the 180-grain Speer JSP bullet over 20.0 grains of Winchester 296 powder, I got 1760/1273 and a group spanning 1.838 inches at fifty yards, out of the Rock/Merrill.

Jim Rock, an avid silhouette competitor, favors the 200-grain Sierra JSP bullet over either 20.0 grains of 296 or 22.0 grains of Winchester 680, using Federal .44 magnum cases as the parent brass and CCI-350 magnum large pistol primers. It should be emphasized that these loads are *not suitable for use in the Contender.*

.357 REMINGTON MAXIMUM

This is an interesting newcomer to the cartridge board and it finds a notably happy home in the Contender. It has been marketed in a single-action Ruger and in the heavy-frame double-action Dan Wesson Arms revolver. I've fired it in both makes of revolvers — out of the 7.5- and ten-inch length in the Ruger — and I can report that the cartridge works best of all in the Contender. That's not to badmouth the Ruger or the DWA; only to note that the R-Max — as I'm prone to abbreviate the name — and the Contender seem almost supernaturally compatible with each other.

The cartridge, by the way, came into being at the suggestion and instigation of IHMSA honcho Elgin Gates, who favored calling it the .357 Supermag.

I have both ten-inch and Super-14 Contender barrels for

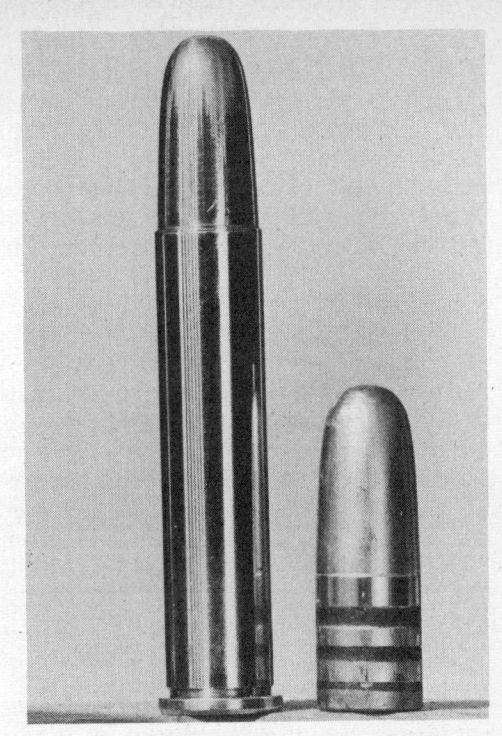

Here, one of the Lyman 358318 gas-checked bullets is shown next to a round loaded with the same slug. Sad to say, Lyman no longer makes this mould, though it works remarkably well in the .357 Remington Maximum round.

A .357 Remington Maximum, left, with 245-grain Lyman No. 358318 cast bullet; center, the same in Remington factory loading and, right, the .357 Remington magnum.

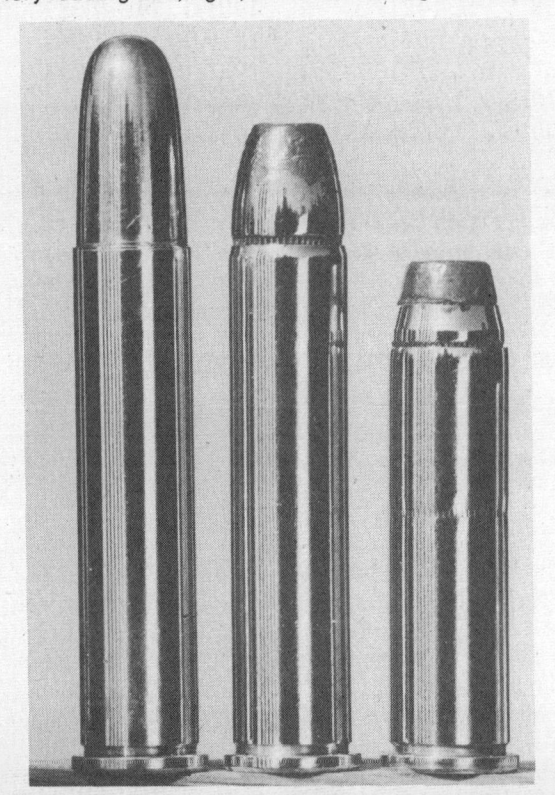

the R-Max. The ten-inch shoots awfully well and the Super-14 shoots even better: a state of affairs that has led me to do the bulk of my research out of the latter. Both of the barrels carry scope sights, because I'm interested in exploring the outer limits of cartridge capability, without hamstringing the project down to the level of my own ability to cope with iron sights: a state of affairs that no more than rarely surges to the lower levels of mediocrity. That is an explanation, not an excuse.

The rifling specifications for the R-Max Contender barrel are exactly the same as for their .357 magnum barrel: .346-.348-inch bore diameter; .356-.358-inch groove diameter; rifling pitch one turn in fourteen inches (1:14), left-hand twist. I crank that data in at the start, by way of noting that, to date, I've never been able to assemble and fire a load for the R-Max that did not stabilize in a satisfactory manner, giving accuracy better than the mill-run average.

Lyman used to list two moulds for rifles that used bullets of .358-inch diameter. They were the #358315 and #358318, at respective weights of 204 and 245 grains, when cast in Lyman #2 alloy. Both were round-nose designs with gas checks and both are exceptionally good performers in the R-Max Contender barrels. It's doubtful if either of them would fit comfortably into the more confined spaces of loads for the available revolvers, but they shoot nothing but great in the Contender.

The sad news is that Lyman has dropped the #358318 design from their catalog, as they've dropped so many admirable others. You can still get the #358315 mould and, if you think you're going to need one some day, it might be well to get one before they drop that one, too.

My customary bullet-casting mix runs about one pound

of linotype metal to six pounds of reclaimed wheel weights and may vary slightly from that hopeful specification, depending upon the remeltable salvage that I've recycled into the pot since that last time it was fired up. I've been casting bullets since the days when Harry and Bess Truman were in the White House, but I've never managed to put my hands upon even so much as one small pig of Lyman #2 alloy. I note that to explain the minor vagaries in end bullet weight.

Which is by way of explaining that, the last time I did a concerted effort on the R-Max, my Lyman #358318 bullets were scaling 252 grains instead of the textbook 245. At the time, I had a hunch that the 4198-type powders might do well in the R-Max, but that hopeful theory did not work out in practice. Ballistics with Du Pont IMR-4198 were well below those obtainable with other powders and it would have been much the same with H4198, I'm pretty certain.

I used a charge of 18.5 grains of IMR-4198 and the Lyman #358315, at 209 grains, went 1337/830; 1301/947 for the 252-grain Lyman #358318. The target was at fifty meters and the #358315 grouped 1.995 inches; 1.092 inches for the #358318. The latter did much the tightest group of seven bullets tested on that charge.

On 20.9 grains of Norma N-200 powder, the #358318 went 1417/1174 and did a 1.595-inch group.

Increasing the charge weight of Norma N-200 to 24.0 grains for some of the lighter bullets, I got nearly all of the best groups for that session, although the energy figures were exceeded by some of the other bullet/powder combinations.

On the mentioned 24.0 grains of N-200, results were:

140 Sierra #8325 JHC	1640/836	1.022 in.
140 Hornady #3574 JHP	1613/809	1.257
158 Speer #4211 JHP	1610/910	0.928
160 Hornady #3572 JTC	1585/893	0.747

The ballistics peaked up quite briskly when I switched over to Winchester 680 powder, at a charge weight of 26.5 grains, with these results:

125 Speer #4013 JHP	2249/1404	1.741 in.
140 Hornady #3574 JHP	2079/1344	1.020
140 Speer #4203 JHP	2170/1464	1.804
140 Sierra #8325 JHC	2137/1420	1.994
140 Hornady #3574 JHP	2222/1535	1.212

You noticed that the same bullet was listed twice in that group? The difference is that the first load was in the Remington case and the second in the Federal case. The latter has less internal capacity, resulting in nearly two hundred added fpe; well, all right, 191 more. I included that to point up my earlier comment on the difference between the three makes of R-Max cases currently available.

Accurate Arms markets a powder called MP5744 that works quite well in the R-Max The following loads displayed no obvious signs of high pressure in my Super-14 Contender, but *I do not recommend them for use in revolvers* and suggest a cautious approach in other Contenders; on 23.3 grains of MP5744:

Bullet at left is Lyman's No. 358315 at about 200 grains, still in their catalog, last time I looked. Bullet at right is the redoubtable Lyman N. 358318, as discussed.

140 Hornady #3574 JHP	2222/1535	1.542 in.
150 Speer #4207 FMJ	1875/1171	0.995
158 Speer #4211 JHP	1905/1274	2.382
180 Speer #2435 JSP	1820/1324	1.748
207 cast, SAECO #351	1937/1725	6.823
170 cast, Lyman #358429	2049/1585	7.448
209 cast, Lyman #358315	1926/1722	3.581

All of the loads with 23.3 grains of MP5744 powder were put up in Federal cases. All of the loads in the entire test series used the Remington #7½ small rifle primer. To put the foregoing data into perspective, the Remington #357MX1 factory load gave this performance:

158 Remington JHP   2012/1421   1.282 in.

The newest Winchester data book lists:

158-grain JSP 23.4 grains 680	1780/1112	34,400 c.u.p.
180-grain FMJ 19.0 grains 296	1670/1115	46,900
180-grain FMJ 19.7 grains 680	1550/960	38,300

The loads listed previously, with 26.5 grains of 680 powder, with bullets up to 140 grains in weight, can be judged against the official Winchester listing. As noted, they showed no obvious signs of excessive pressure in my Super-14 barrel, but I don't recommend their use in .357 Remington Maximum revolvers and would urge a cautious approach for use in other Contenders of that caliber.

I remain unconvinced that such a thing as the "inherent

accuracy," often referred to in the firearms press, actually exists in the Real World. Assuming it does, for the sake of discussion, I sincerely believe that the .357 Remington Maximum cartridge, in the T/C Contender Super-14 barrel, shows more of that good stuff than hardly any other twosome I can bring to mind, right offhand. It is entirely true that the Super-14 in .35 Remington can out-punch the R-Max, but the .35 Rem kicks harder — naturally enough — and it seldom groups quite as well in my own experience.

.357 HERRETT

Refer to the discussion of the .30 Herrett for several comments that apply to both of the Herrett Contender wildcats. The .357 Herrett came along some while after the .30 Herrett, both cases being based upon the .30-30 WCF parent brass.

The .357 Herrett, like the .30 Herrett, gives its best performance if the full-length resizing die is adjusted with exquisite precision to headspace the chambered round on the point of contact between the cartridge shoulder and mating surface of the chamber. Unlike the .30 Herrett, the .357 has a pretty scanty shoulder and, for that reason, it takes some thoroughly precise manipulation of the full-length resizing die to get everything coming up roses as to accuracy.

A well established cartridge, data listings for the .357 Herrett can be found in many sources of load data. The data booklet from Accurate Arms does not list the .357 Herrett and a phone call to their head ballistician, Marty Liggins, turned up the fact that he had listed the .357 Herrett at one time, but pulled out the listing to make room for some other entry. He looked up and was able to supply the data, suggesting 29.0 grains of Accurate Arms 5744

From left, .30 Herrett, .357 Herrett and .30-30 WCF. The .357 Herrett headspaces on its shoulder and does not have all that terribly much shoulder for the purpose.

powder for the 125-grain bullet at 2110 fps and 1236 fpe. For the 158-grain bullet, he quoted 28.0 grains of 5744 for 1920/1294.

In much the same manner that T/C's introduction of the Contender barrels in .30-30 WCF eclipsed the .30 Herrett, their addition of the .35 Remington Contender barrels to the catalog sort of trumped the .357 Herrett's ace. The .35 Remington is not entirely bereft of problems, viewed as a Contender cartridge, but it is capable of doing better work than the .357 Herrett, due to its greater powder capacity.

The .357 Herrett is trimmed to a length of 1.765 inches, as compared to 1.605 inches for the .30 Herrett. Other suitable powders for this cartridge include such numbers as 2400, H4227, IMR-4227, H110, 296 and 680.

.35 REMINGTON

This cartridge probably delivers more energy than any other one in the standard catalog offering from T/C. As with all bottlenecked cartridges, optimum accuracy requires that the cases be resized with great care, so as to headspace the chambered round by contact between the shoulder and the mating area of the chamber. The problem in this instance is that the .35 Remington has no more than a slight amount of shoulder with which to work and the business of getting the requisite fit offers more challenge than most reloaders are apt to feel they need.

When firing new factory loads in my .35 Remington Contender, I have sometimes noted a bright ring at the mouth of the fired case, suggesting that the force of the firing pin blow drove the case mouth into contact with the commencement or leade of the rifling.

I have found it helpful, when making up reloads, to seat the bullet far enough forward to bring it into light contact with the leade, for the sake of improving the headspace or support in the chamber.

Bullets up to 200 grains can be used in the 35 Remington Contender, but the potential velocity makes heavier bullets not overly practical.

The No. 25 Hodgdon Manual, on page 371, lists loads for three bullet weights, with three powders apiece, for the .35 Remington in the Super-14 Contender barrel and several other loads for it are covered in the booklet available from Thompson/Center.

10mm AUTO

This is a cartridge of comparatively recent origin, initially designed for the Bren Ten pistol and later adapted to the Colt Government Model autoloader.

Mike Rock (101 Ogden Ave., Albany, WI 53502) has produced some number of Contender barrels in 10mm Auto chambering. I have not had the opportunity to work with a 10mm Contender barrel, but I can offer some loads I've worked up for use in the five-inch barrel of the Bren Ten, as a suggested starting point for reloading this cartridge to use in the Contender. Listed velocities and energies are derived in the Bren Ten and one could expect somewhat higher figures from the Contender, be it the ten-inch or Super-14 type.

Jacketed bullets for the 10mm have a diameter of .400-inch and cast bullets usually are sized to .401-inch. Dave Corbin (Box 2659, White City, OR 97503) can furnish swaging dies in .400-inch diameters, as well as a die for drawing down .44 jackets to the proper diameter.

The 170-grain bullets used in developing the following data were Sierra No. 8500 JHPs, originally in .410-inch diameter. They were put through the draw-down die and then given a "bump-up" in the nose-forming die of the .400-inch set. The bump is quite necessary to re-establish the intimate contact between core and jacket. Merely putting a slightly larger bullet through the draw-down die results in a bullet of thoroughly unacceptable accuracy capability.

10mm Auto in 5-inch Bren Ten

BULLET	CHARGE (grains)	FPS/FPE	REMARKS
200-gr. cast	9.3 Blue Dot	1176/614	
	10.3 AA-7	1182/621	
153-gr. JHP	11.8 AA-7	1195/539	
	12.1 AA-7	1369/637	
170-gr. Sierra	9.0 HS-7	965/352	worked action
	9.6 HS-7	1039/408	
	11.0 AA-7	1103/459	
	11.4 AA-7	1145/495	
	11.6 AA-7	1170/517	
	11.8 AA-7	1195/539	
	12.0 AA-7	1195/539	
	12.2 AA-7	1219/561	
	12.4 AA-7	1244/584	
	12.6 AA-7	1259/598	see note 1 worked action
	8.6 Blue Dot	942/335	
	8.8 Blue Dot	1012/387	
	9.0 Blue Dot	1022/394	
	9.5 Blue Dot	1111/466	
	10.0 Blue Dot	1131/483	
200-gr. Norma FMJ	9.0 HS-7	1025/467	
	9.7 AA-7	1035/476	
	10.0 AA-7	1056/495	
	10.5 AA-7	1096/534	
	11.0 AA-7	1143/580	
	8.6 Blue Dot	1049/498	
	9.0 Blue Dot	1049/498	
	9.0 Blue Dot	1049/498	
	8.2 AA-5	1160/598	

NOTE 1: This load was tested by Marty Liggins at the Accurate Arms ballistic lab and pressure averaged 30,000 c.u.p.: a typical working pressure for the 10mm Auto cartridge.

AA-5 and AA-7 are Accurate Arms powders. HS-7 is a Hodgdon powder and Blue Dot is a Hercules powder.

.41 MAGNUM

As in the example of the .44 magnum and several other cartridges, loading data for the .41 magnum is abundantly available from a number of sources and all such loads should function nicely through T/C Contender bull barrels in ten-inch or Super-14 size.

You can expect to get more velocity out of the ten-inch bull barrel than most of the data sources show, with even more coming out of the Super-14 barrel, if you are foresighted and fortunate enough to have one of those at your disposal.

Sad to say, the Super-14 barrels are no longer to be had in .41 magnum, although I'm sure you could still coax one out of the T/C Custom Shop on a special-order basis. The ten-inch barrels for .41 mag remain in the catalog, so far, and it is a barrel I'd suggest grabbing on to, if you don't have one, already.

As noted, any load shown in a reliable manual or handbook can be used in the T/C Contender of either barrel length to excellent effect. It is highly probable that you will obtain more velocity than most data sources suggest.

By way of illustrating the useful effect of bore length, I have made up three different loads in .41 magnum and fired them out of five different guns, ranging in barrel length from 2.75 inches to the Super-14, with velocities and energies noted. The 2.75-incher is a custom-modified and converted Model 28 S&W. The four-inch is an M657 and the six-inch is a '64 vintage Model 57 S&W. The other two are T/CCs of indicated barrel length.

Load 1: 14.2 grains of Du Pont 800-X, more recently termed IMR 800-X, behind a 184-grain cast bullet from a Hensley & Gibbs No. 291 mould at a cartridge overall length of 1.657 inches. Five-shot average ballistics were:

2.75	1255/643
4	1388/787
6	1409/812
10	1647/1108
14	1755/1259

Load 2: 20.2 grains of Accurate Arms No. 9, behind the same 184-grain H&G No. 291 cast bullet with conical point, at 1.631 inches cartridge overall length:

2.75	1240/628
4	1410/813
6	1422/826
10	1651/1114
14	1759/1264

Load 3: 23.6 Winchester 680, behind a 210-grain Hornady No. 4100 JHP bullet at 1.610 inches cartridge overall length:

2.75	1053/517
4	1183/652
6	1183/653
10	1534/1098
14	1640/1255

Comments: The six-inch revolver has seen a lot of duty in the twenty-odd years of its career to date, while the four-inch is new and only marginally broken in, thus far. Bore dimensions of the latter are, presumably, a bit tighter than the older and longer gun. In Load 3, the mystifying figures for the four-inch and six-inch are due to the rounding-off to the nearest whole by the computer that's part of the chronograph. The marked improvement in ballistics out of the Contenders, in comparison to the revolvers, is typical in my experience.

In .44 magnum revolvers of six-inch barrel length, you'll do well to coax 1150 fpe out of manual load data. The ten-inch Contender comes close to that figure and the Super-14 tops it comfortably. In all three of the quoted loads, the Super-14 shows an edge of about 150 fpe over the ten-inch.

In my purely personal opinion, the .41 magnum is one of the most interesting Contender calibers to be had and I'm awfully glad I have my Super-14 in hand!

.410-BORE

This is the smallest diameter of shotshell in commercial production, unless you count specialized loads made for

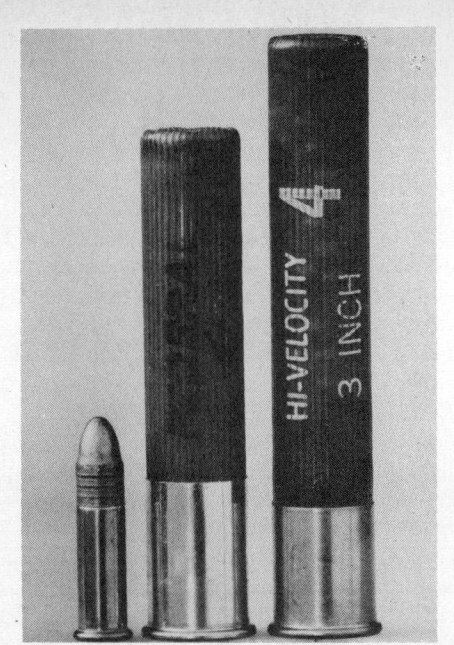

A round of the familiar .22 LR appears quite petite next to the standard, 2½- and 3-inch loads of .410-bore.

use in handguns, rather than shotguns. Marketed briefly, several years ago, the .410-bore Contender barrels came back onto the market in 1985 and met an enthusiastic reception from Contender fans.

Most shotgun shells are designated in gauges, representing the number of round lead balls of bore diameter required to weigh one pound. Thus, the 12-gauge is about .729-inch diameter and twelve such balls weigh one pound. Based upon the same system, the .410-bore would be about a 67-gauge and, on the same system, a ".410-gauge" would have a bore diameter of about 2.248 inches and a ball of that diameter would weigh 2.44 pounds. If you want it to the last decimal, the .410-bore is equal to a 67.587433-gauge.

The Contender barrels are offered as a combination to handle both the .45 Colt and the .410-bore shotshells of either the standard 2.5-inch or magnum 3-inch length. An internal choke is used with the shotshells and the choke *must* be removed when firing .45 Colt ammo! A fold-down rear sight is used for .45 Colt and folded forward for use as a shotgun. The bores are rifled, causing the charge of shot pellets to spin, which would result in a broad, open-centered pattern, if left uncorrected. The internal choke arrests the swirling motion and delivers a remarkably tight and effective pattern, when installed.

Recently, the line of twenty-one-inch barrels for the Contender Carbine conversion was augmented by one for .44 magnum and another for the .410-bore shotshell. It is illegal to possess or use a smoothbore barrel in a pistol, but legal in a shoulder-stocked gun if the barrel is sixteen inches or longer and if the overall length is 26.5 inches or more. Thus, the .410 carbine barrel is a smoothbore, with a ventilated rib and it cannot be used to fire solid-bullet cartridges, such as the .45 Colt.

Standard .410 shotshell load data can and should be used when reloading for the Contender of that bore diameter.

It has recently come to our attention that the .45 LC/ .410-bore Contenders, as well as the American Derringer Corporation derringers in .410-bore chambering are illegal to possess in California, under penalty of a one-year prison sentence. It might be well to check for legality in your own jurisdiction, with that in mind!

.44 MAGNUM

This is an extremely popular and somewhat legendary cartridge that was introduced about 1956 as a round for the large or N-frame Smith & Wesson revolvers. It was created by adding about .125-inch of length to the basic .44 Special cartridge and beefing up the construction of the guns to fire it, so they could handle peak pressures on the order of the SAAMI industry maximum of 43,500 copper units of pressure. Since that time a great many other handguns have been chambered for .44 magnum, including the T/C Contender. It has also been used with some degree of success in various rifles.

As a general observation, in revolvers of typical barrel length, the .44 magnum can be expected to deliver about 1150 foot-pounds of energy and a great many handgunners regard it as the world's most powerful handgun cartridge — a dubious claim, as a casual reference to several entries in the book at hand would indicate.

Loading data for the .44 magnum is lavishly available in just about any source of such information and any loads listed therein should be amply safe for use in the Contender action. When fired in the Contender — particularly in the Super-14 — you can expect to generate substantially higher ballistics than when firing the same load in revolvers. The absence of a gap between the front of the revolver cylinder and the rear of its barrel accounts for a great deal of the

Another obsolete Lyman mould and one that deserves to be is the old No. 427103. It comes out large enough to size to .430-inch, weighs 356 grains and appears here in a .44 magnum case. Sadly, it won't stabilize in .44 guns.

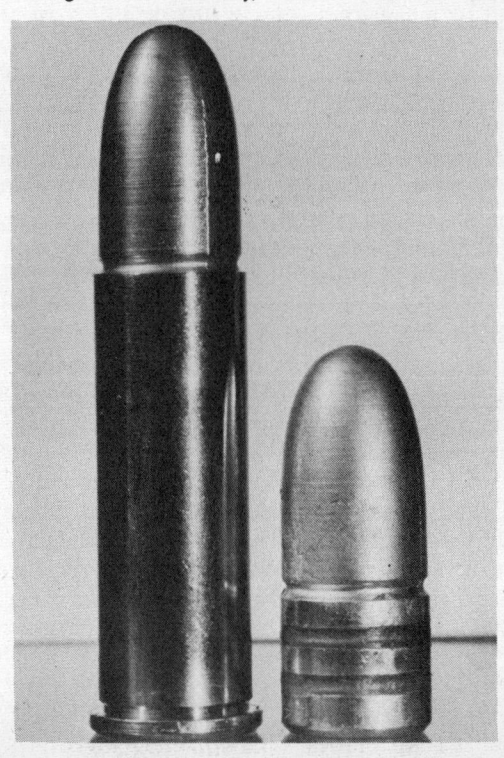

The .44 magnum Contender will handle the .44 Special, here with another cast bullet from an obsolete Lyman mould. It's the No. 429251, a round nose at 255 grains.

added performance, as does the increased length of barrel.

At the time the original .44 magnum Contender barrels were introduced, the bull barrel configuration was still in the future and the old octagonal barrels still retained the forend by means of a snap or ball detent. Now, the .44 magnum cartridge delivers a thoroughly impressive amount of recoil and the rather modest weight of the Contender with octagonal barrel made firing the .44 magnum a most unforgettable experience! As a usual thing, the forend would be ripped from its catch by inertia and shooters of those early .44 magnum Contenders usually wrapped tape around the forend to keep it from coming loose. Even with the bull barrels, Mag-na-port vents at the muzzle will be found to be a most useful improvement.

If the proposed utilization is such as to make installation of a scope sight practical and desirable, the T'SOB scope mounts and rings from SSK Industries probably will be found to be most satisfactory with the .44 magnum and similar high-recoil chamberings. Not all handgun scopes are capable of standing the violent gaff of .44 magnum recoil and a few makes and models are apt to self-destruct after some number of shots have been fired. Thompson/Center's own R-P — for Recoil-Proof — handgun scopes have proven quite dependable for heavy-recoiling cartridges, including the .44 magnum, in my own experience, and other sturdy scopes I've used include those by Redfield, Leupold and — when they were still available — Weaver.

While lightweight bullets can be fired to good effect, the specialized capabilities of the Contender make considera-tion of heavier bullets worthwhile. Most .44 bullets — which are actually .429-inch to .430-inch in diameter — top out at about 240 grains, with Hornady's 265-grain SP about the heaviest one commercially available. For those who cast their own bullets, SSK Industries can supply moulds, custom-made for them by NEI, to cast bullets up to about 370 grains in weight. All of the SSK/NEI bullets have truncated cone noses and display excellent accuracy ahead of a wide variety of suitable powder charges.

It is a curious enigma of ballistics that the .44 magnum in the Contender comes surprisingly close to equaling the performance of the much longer .444 Marlin cartridge, for which SSK can also supply Contender barrels.

Powders that I have used in the .44 magnum Contender to excellent effect include Accurate Arms No. 9, Hercules 2400 and Blue Dot, Hodgdon H110 and H4227, IMR-4227 and Winchester 296 and 680.

Unlike revolvers, the chamber of the .44 magnum Contender will allow a somewhat greater overall cartridge length, affording added flexibility in reloading for it.

For those Contender owners who feel a need for even longer barrels and somewhat hotter ballistics, T/C now supplies twenty-one-inch barrels for the Contender carbine conversions, for use with shoulder stocks.

.45 COLT/LONG COLT

An old cartridge, but still a good one in the proper handgun, the .45 Colt — sometimes termed the .45 Long Colt — is superbly adapted for use in the Contender. While factory loads are held to maximum pressures of 15,900 c.u.p. — and usually well below that point — most of the modern sources of load data include a special section of .45 Colt loads such as the one in the No. 10 Speer Manual headed, ".45 Colt for Ruger or Contender only." Such loads are held to a maximum pressure of 25,000 c.u.p. and add 288 fps to the top listing for the 200-grain bullet, in comparison to the fastest load for that bullet in the section for standard data.

In Contender barrels, the .45 Colt usually is encountered in the combination barrels with removable chokes, likewise handling the .410-bore shotshells. What we might term straight .45 Colt barrels have not been listed in the catalog since the heyday of the octagonal barrels. At the time, I had a six-inch .45 Colt barrel — an available option for a short while in those days — and, as you might suppose, its recoil was impressive, even with factory loads.

At present, it is possible to obtain .45 Colt Contender barrels through the Thompson/Center Custom Shop, in either ten-inch or Super-14 configuration, with the standard rear sight adjustable for both elevation and windage. By removing the rear sight, a scope can be installed. T/C offers an excellent set of mounting rail and rings for the purpose. As a matter of moderate personal frustration, I have a Super-14 barrel in .45 Colt, but it has arrived so recently that I've not yet had a chance to find out what it can do.

The .45/.410 combination barrels have a foldable rear sight, to be raised for firing .45 Colt, but it is not readily adjustable for either elevation or windage and thus makes it difficult to realize the full potential of the .45 Colt cartridge.

Here and there through gun-oriented literature, you may

encounter references to the fact that the .45 Colt case is woefully weak and incapable of coping with pressures of the hotter reloads. I do not subscribe to that particular belief and I'll explain why. Dick Casull chose the .45 Colt case as the starting point for his .454 Casull wildcat cartridge and had some .454 Casull cases made up at a length slightly greater than the 1.285 inches of nominal case length for the .45 Colt. His reasoning was obvious: He wanted to prevent use of .454 Casull loads in other revolvers designed to handle pressures no higher than 15,900 c.u.p. It is the same approach as the creation of the .357 magnum, with a longer case to prevent its higher-pressure loads from being used in the .38 Special.

In specifying the custom .454 Casull cases, the only modification was the slightly greater length. Casull did not beef up the wall thickness or head construction. He then went on to develop some loads in the .454 Casull cases to work at some really mind-boggling pressure levels and the cases stood up quite nicely. To my mind, that suggests that the basic .45 Colt case is reasonably adequate for most reloading purposes.

As was noted, the .45 Colt traces its origin back to the black powder days of the previous century and you still may encounter some .45 Colt cases in the early "balloon head" design. By looking into the mouth of the empty case under good light, you'll be able to see that the area around the primer pocket protrudes up from the head, with a depression surrounding it, rather than the solid head construction of the latterday "web head" cartridge cases.

Reloaders — as I well know, being one myself — tend to glom onto empty cases from any viable source and some of those old balloon head jobs may crop up in your supply of cases. It is a good idea to inspect newly acquired lots of cases for the possible presence of balloon heads and, if any are discovered, to segregate them well away from your normal supply of reloading components. Better yet, crush the mouth of such cases with a pair of pliers, by way of positively preventing them from getting back into the system. Balloon heads may turn up among several other ancient calibers, such as .44 Special, .45 Auto Rim, .38-40, .44-40 and the like.

Getting back to the .454 Casull, it is designed to be fired in — and should *only* be fired in — the revolvers manufactured and chambered for it by Freedom Arms Company, Box 1776, Freedom, WY 83120. Should you encounter listed load data for the .454 Casull and endeavor to try the same loads in your .45 Colt Contender or Ruger, the results are highly apt to be disastrous.

Instead, work with the loads listed in the handbooks and manuals for the .45 Colt and use the ones stipulated for the Contender and Ruger, if you like. Don't go beyond those. The Contender — and particularly the Super-14 Contender — will deliver somewhat higher velocities than the data charts quote, along with some respectable accuracy, depending upon the sighting system used.

.45 WINCHESTER MAGNUM

Some years ago, Wildey — pronounced "Will-dee" — Moore designed a gas-operated autoloading pistol to be known as the Wildey and persuaded Winchester to produce two different cartridges for use in it. They were named the 9mm Winchester magnum and the .45 Winchester magnum; commonly voiced as Win-mag to conserve jaw motion.

The 9mm Win-mag bore a close resemblance to a round of .30 M1 carbine, left a bit longer on the vine — about the same overall length, but a trifle fatter. To the present, the 9mm Win-mag has not gotten much of anywhere. In that respect, it is in much the same boat as its parent vehicle, the Widley pistol. L.A.R. made a few of their *Grizzly* auto pistols to handle it and then encountered difficulty in obtaining the ammunition of that size, they reported. They also made and still make the Grizzly in .45 Win-mag and the increased length of the cartridge results in a gun that resembles the M1911-pattern Colt auto, but longer, front-to-back, through the grip section.

Meanwhile, T/C took a thoughtful look at the .45 Win-mag and decided it has something to offer in the Contender, so they produced barrels of that chambering in both ten-inch and Super-14 lengths. They decided, quite justifiably, that the 9mm Win-mag offered nothing that several of their existing chamberings didn't excel, so to the best of my knowledge, there never was and never will be a 9mm Win-mag barrel for the T/CC.

I obtained one of the early .45 Win-mag Contender barrels in ten-inch bull format and worked up some tentative data for the cartridge, then sent the barrel along to old friend Dave Andrews at the CCI/Speer plant. He, in turn, used it to generate the load data that is on pages 398-400 of the tenth edition of the Speer Manual. I refer you to that work. Other books cover the cartridge, but SM-10 handles it best, I think.

.45 SILHOUETTE

Frank C. Barnes is a well-known wildcatter, cartridge authority and author of *Cartridges of the World,* currently in its fifth edition and published by DBI Books.

On page 160 of COTW-5, Barnes tells how he developed the .45 Silhouette cartridge as a more efficient performer in the Contender and similar handguns, as compared to the .45-70 round. Earlier, with some amount of success, Barnes had created the .308x1.5 Barnes cartridge by shortening the .308 Winchester case to an overall length of 1.5 inches. He used somewhat the same approach again in shortening the .45-70 case from its usual length of 2.1 inches to 1.5 inches and did his reload development in a Siamese Mauser, initially chambered for the .45-70, with its barrel shortened to twenty inches. He lists the following loads:

		20-inch barrel
300-gr. cast	36.0 4198	1612/1731
"	44.0 FFg black	1170/912
300-gr. JHP	35.0 4198	1485/1469
"	38.0 4198	1670/1858
"	40.0 4198*	1810/2183
		12-inch barrel
300-gr. cast	35.0 4198	1355/1223
300-gr. JHP	38.0 4198	1585/1674
"	40.0 4198*	1725/1983

*compressed charge

Barnes notes that RCBS can furnish the trim die and loading for the .45 Silhouette and goes on to suggest a rifling twist of 1:16 or 1:18, with about .25-inch of freebore.
— *Dean A. Grennell*

J.D. Jones Of Hand Cannon Fame — Or Infamy — Utilizes Contender Actions For His Big, Big Bores!

From left, the .17 Remington, .17 MACH IV and .17 K-Hornet. J.D. Jones, via SSK Industries, can furnish Contender barrels for the .17 Remington.

J. D. JONES proprietor of SSK Industries, (Rte. 1, Della Drive, Bloomingdale, OH 43910), is the source for ultra-intrepid chamberings for use in the Contender. He can furnish Contender barrels of his own manufacture that handle the rather horrific cartridges of his own design. Jones' term for the resulting artifact is "hand cannon." Once one has sampled the unique experience of firing one of these lusty rascals, the term does not seem all that far-fetched.

I understand that Jones reserves the hand cannon designation for barrels in calibers that are .375 JDJ and larger, including a few other cartridges in the general category. For one example, the .45-70 Government cartridge gets lumped into the HC class and it was on the scene for several decades before Jones emitted his first lusty squall.

Jones has supplied me with some notes and data on quite a few of the cartridges for which he can supply barrels to use in the Contender. He also performs conversions on

Remington Model XP-100 single-shot pistols and his offering of load data notes includes some numbers that belong in the XP-100, rather than in the Contender. The XP-100 is somewhat stronger and better able to cope with exalted peak pressures than is the Contender.

Rather than sprinkle the Jones load data willy-nilly through the entire compendium, I am going to insert it all in lump, making an effort to start with the smaller cartridges and work my way up through the entire gamut of them. Be advised that the ranking may be off by an entry or two, in some places.

.17 ACKLEY BEE

This is a wildcat, based upon the .218 Winchester Bee parent brass, designed by noted gunsmith, P.O. Ackley. Data taken in a 12.5-inch barrel, using Winchester cases, CCI-400 primers, with the firing done at ambient temperatures ranging from 10 to 45 degrees F. All bullets for caliber .17 cartridges in this series are the Hornady 25-grain JHP, No. 1710.

14.4 N-200	3024/508	"Accurate"
14.1 Re-7	3056/519	
13.4 H4227	3145/549	
14.0 H4198	3124/542	

.17 REMINGTON

This is the only non-wildcat caliber .17 cartridge. Remington continues to produce loaded ammo for it, quoting a muzzle velocity for the 25-grain bullet of 4040 fps and 906 fpe from a rifle barrel of unspecified length.

The loads quoted here are maximum and must be approached with caution, starting with charge weights of the given powder at least ten percent lower than those listed here. As an item of possible interest, velocities are also quoted from a 22.25-inch rifle barrel, as well as from the fourteen-inch Contender barrel.

	14"T/CC	22¼" Rifle	
25.0 IMR-4320	3506/683	4031/902	most accurate, rifle
24.0 IMR-4895	3396/640	3968/874	most acc., TC/C
24.0 N-202	3417/648	4019/897	
23.0 IMR-4064	3194/566	3766/566	Inacc., both guns
23.0 W-748	3307/607	3748/780	excess. flash, T/CC
Rem. factory load	3323/613	3847/822	accurate in rifle

.22 REMINGTON JET (RCFM)

Velocities taken from a ten-inch Contender barrel, using R-P brass, CCI-550 (small pistol magnum) primers, with 45-grain Sierra .224-inch diameter bullet at cartridge overall length of 1.720-inches.

14.0 2400	2547/648
14.5 2400	2616/684
13.0 H110	2492/621
13.5 H110	2502/626
14.0 H110	2492/621
14.0 IMR-4227	2425/588
15.0 IMR-4227	2563/657

.22 PPC

Velocities taken in a fourteen-inch SSK barrel, using Sako cases, CCI-BR4 match primers and Hornady 53-grain Match bullet, at ambient temperatures of 72-74 F. A Leupold M8-2X scope was mounted on the barrel and 100-yard groups, from a sandbag bench rest ran from .5-inch to .75-inch in center spread.

20.5 IMR-4198	2707/863
21.0 IMR-4198	2760/897

.223 REMINGTON

Velocities taken in a T/C Super-14 barrel, using military brass and CCI-400 primers, with 50-grain Hornady SX bullets, at ambient temperatures from 48 to 60 F. Barrel carried a T/C 3X R-P scope and the groups ran about 1½ inches in spread at 100 yards.

21.5 IMR-4198	2871/915	48 deg. F
21.5 IMR-4198	2954/969	60 deg. F

.225 WINCHESTER

Nominally a rifle cartridge, the quoted data was developed in a fourteen-inch SSK barrel, using Winchester brass, CCI-BR2 primers and 50-grain Hornady SX bullets. Ambient temperatures ranged from 36 to 52 degrees F and the charge weights of the maximum loads listed should be reduced judiciously, if there is a possibility the loads will be fired at higher temperatures. Ambient temperature at the time of firing can — and usually does — have a significant effect upon peak pressures, as well as velocities.

29.5 IMR-3031	2787/863
30.0 IMR-3031	2903/936
30.5 IMR-3031	2954/969
31.0 IMR-3031	3063/1042
31.5 IMR-3031	3125/1084
32.0 IMR-3031	3178/1122
31.5 IMR-4064	2928/952
32.0 IMR-4064	2982/987
32.5 IMR-4064	3067/1045
33.0 IMR-4064	3172/1117

It is noted that the load with 31.0 grains of 3031 was used to take prairie dogs at distances out to 296 paces.

Do not fire .225 Winchester factory loads in the Contender!

.226 JDJ

This is a wildcat developed by Jones, based upon .225 Winchester parent brass. Test firing was done at ambient temperatures ranging from 35 to 40 degrees F and, as has been noted in other entries, charge weights of the given powder should be reduced if the loads are apt to be fired at higher temperatures. Brass is by Winchester — no other manufacturer makes this case — and primers are as noted.

38.5 H4831	2831/1121	63-grain Sierra bullet Fed. 215 pr.
35.0 H414	2732/995	60-grain Hornady Fed. 215
32.0 BL-C2	2637/849	55-grain Hornady SX Fed. 215
36.0 H205	2650/983	63-grain Sierra Win. 250
32.0 N202	2622/916	60-grain Nosler Win. 200*
34.0 N202	2824/886	50-grain Hornady SX Win. 200*
33.0 N202	2969/1175	60-grain Hornady SP Fed. 215

*Fire-forming loads

6mm JDJ

The 6mm JDJ is a wildcat developed by Jones and it is based upon the .225 Winchester parent brass, with the neck opened to accept bullets of .243-inch diameter, fire-formed to a fairly sharp shoulder. The 6mm is the only one of the .225-based JDJ wildcats that does not have a forty-degree shoulder. This is due to a misunderstanding with the man who ground the reamer for it.

The supplied data does not specify the length of the test barrel, but in high probability it was either 12.5 or 14 inches.

CHARGE (grains)	BULLET (grains)	FPS/FPE	PRIMER
27.0 IMR-4198	60 Sierra	2920/1136	CCI-200
32.0 IMR-3031	60 Sierra	2889/1112	CCI-200
32.5 IMR-3031	60 Sierra	3048/1238	CCI-200
34.5 IMR-4895	60 Sierra	2955/1164	CCI-200
34.5 IMR-4064	60 Sierra	2961/1168	CCI-200
35.0 IMR-4064	60 Sierra	3052/1241	CCI-200
36.5 H380	60 Sierra	2821/1061	Rem 9½M
26.0 IMR-4198	75 Sierra	2684/1200	CCI-200
26.5 IMR-4198	75 Sierra	2780/1287	CCI-200
30.0 IMR-3031	75 Sierra	2643/1164	CCI-200
30.5 IMR-3031	75 Sierra	2750/1260	CCI-200
32.5 IMR-4895	75 Sierra	2701/1215	CCI-200
33.0 IMR-4895	75 Sierra	2753/1262	CCI-200
31.0 H4895	75 Sierra	2729.1241	Fed 210
33.0 IMR-4064	75 Sierra	2744/1254	CCI-200
33.5 IMR-4064	75 Sierra	2795/1301	CCI-200
33.0 IMR-4320	75 Sierra	2763/1274	CCI-200
37.0 IMR-4350	75 Sierra	2726/1238	Rem 9½M
35.5 H380	75 Sierra	2697/1212	Rem 9½M
30.5 H335	75 Sierra	2696/1211	Rem 9½M
29.5 H322	75 Sierra	2724/1236	Rem 9½M
36.0 H414	75 Sierra	2659/1178	Rem 9½M
36.5 H414	75 Sierra	2694/1209	Rem 9½M

.257 JDJ

This is another of Jones' wildcats based upon the .225 Winchester case. The upper attainable ballistics with typical .257 bullet weights are:

75-grain	2825/1329
100-grain	2625/1530
120-grain	2400/1535

Without necessarily going to the maximum loads, here are some typical .257 JDJ loads.

75-grain Hornady No. 2520JHP
33.0 H4895	2430/984
29.0 H322	2209/813
30.0 H322	2308/887
31.0 H322	2504/1044
36.0 W748	2565/1096
37.0 W748	2646/1166
38.0 W478	2721/1233

87-grain Speer No.1241
29.0 H322	2195/931
30.0 H322	2278/1003
31.0 H4895	2299/1021
32.0 H4895	2426/1137

87-grain Sierra No. 1610
35.0 W748	2405/1118
36.0 W748	2514/1221
37.0 W748	2591/1297

100-grain Speer No. 1405
29.0 H4895	2155/1031
30.0 H4895	2259/1133
30.0 IMR-4320	2081/962
31.0 IMR-4320	2209/1084
35.0 IMR-4350	2209/1084
36.0 IMR-4350	2303/1178

117-grain Sierra No. 1630
29.0 IMR-4320	2056/1098
30.0 IMR-4320	2126/1175
33.0 IMR-4350	2058/1101
34.0 IMR-4350	2141/1191
35.0 IMR-4350	2194/1251
36.0 IMR-4350	2293/1366

Remington No.9½ primers and fire-formed Winchester cases were used in working up the listed data.

6.5mm JDJ

Speaking personally, this is my favorite among the several Jones wildcats based upon the .225 Winchester case. I have one of the SSK barrels, with a T'SOB scope mount. I find the cartridge quite pleasant to fire, from the standpoint of recoil, although the report is rather sharp and ear protection is pretty mandatory.

With a 4X Redfield pistol scope topside, my best group to date spans just .750-inch between centers for five shots at one hundred yards. The load was 35.0 grains of IMR-4320 behind the 85-grain Sierra JHP, No. 1700. SSK dope sheets list this load for 2644/1320 out of a fourteen-inch barrel and the five-shot average in my 12.5-inch barrel was 2453/1136. Yes, that seems like quite a bit of discrepancy for 1.5 inches of missing bore but, as noted, the load is gratifyingly accurate — better than any number of rifles I've owned or still own.

Another promising load for my 6.5mm JDJ is 32.5 grains of IMR-4320 behind the 120-grain Speer No. 1435 JSP bullet. That one's good for 2185/1272 in my barrel and the best five-shot group at one hundred yards to date measures .983-inch between centers.

I worked out the trajectory table for that 120-grain load and, if it is zeroed for 175 yards, the bullet path will be a maximum of 2.56 inches above the line of sight at 100 and 2.06 inches low at 200. In other words, it stays within 2.6 inches from line of sight clear out to two hundred yards. In my book, that's a flat-shooting handgun, with sub-MOA capabilities to boot!

J.D. Jones has taken a lot of game with the 6.5mm JDJ, including a number of African species, out to some pretty extended distances. He says that, as a class, the 6.5mm (.264-inch diameter) bullets display markedly superior game-downing ability. He feels this is due, at least in part, to the fact that several rifle cartridges for bullets of that diameter muster no more than moderate velocities. For that reason, the bulletmakers tend to design for decisive expansion at velocities down in the lower 2000-fps region.

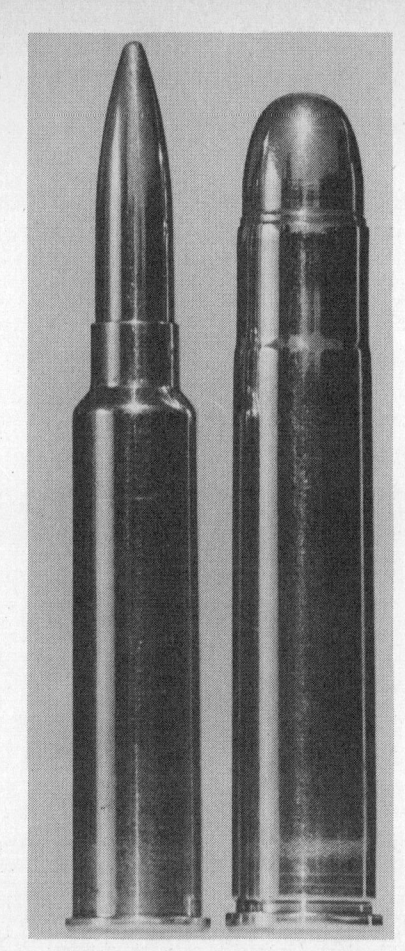

The 6.5mm JDJ, at left, is based upon the .225 Win. case, while the .411 JDJ, at right, is based upon the .444 Marlin case, taking bullets of .410-inch diameter.

J.D. Jones comments: "The bullets work right. It also does well as a silhouette cartridge. It's a deadly antelope cartridge at three hundred yards in good hands. The 100-grain bullets are favored by quite a few shooters. I personally use the 34.0 grains of IMR-4320 under the 120-grain bullet for everything. The 120-grain Speer at 2400+ fps has taken over one hundred deer and antelope for me without a failure or a recovered bullet."

.270 JDJ

J.D. Jones comments that this wildcat on the .225 Winchester case, "...does decently with 130-grain bullets at 2400 fps and with 110s at 2500. You have to be careful in your choice of bullets, however; some are too tough and others too fragile. It seems exceptionally accurate with just about anything it's fed. Does well at metallic silhouettes and the 150-grainers are positive on the rams."

100-grain Speer
32.5 Reloder-7   2739/1666
34.0 Reloder-7   2794/1734
37.0 H4895   2637/1544
110-grain
34.5 IMR-4320   2382/1386
36.0 IMR-4320   2522/1554
35.5 IMR-4064   2378/1382
130-grain
35.0 H4895   2385/1642
35.5 IMR-4064   2340/1581
140-grain
35.0 H4895   2375/1754

7mm JDJ

J.D. Jones reports that he chose the .225 Winchester case as the parent brass for his series of wildcats in the smaller bores because of its substantically greater strength, as compared to similar cases such as the .30-30 WCF, a popular starting point for many other wildcatters.

As with others in the line, the 7mm JDJ has a shoulder angle of forty degrees, with the neck opened to accept bullets of .284-inch diameter — the nominal size of the 7mm.

139-grain Hornady Bullet (Charge weight in grains)
32.0 IMR-4895   2128/1398
32.5 IMR-4895   2175/1460
31.0 IMR-3031   2157/1436
31.5 IMR-3031   2200/1494*
31.0 IMR-4320   2045/1291
31.5 IMR-4320   2091/1350
32.0 IMR-4320   2103/1365
32.5 IMR-4320   2189/1479
33.0 IMR-4320   2211/1509

*exceptionally accurate

120-grain Sierra
33.0 IMR-4895   2162/1246
34.0 IMR-4895   2226/1321
35.0 IMR-4895   2305/1416
36.0 IMR-4895   2390/1522

.300 SAVAGE

Jones checked out some loads in a ten-inch bull barrel for the Contender, reporting 2125/1504 for the 150-grain Federal factory load; 1928/1486 for the 180-grain Federal factory load; 2106/1281 for 35.0 grains fro 4198 — Hodgdon or Du Pont, not specified — behind a 130-grain Hornady and 1837/1244 for 35.0 grains of 4198, as above, behind the 150-grain Hornady FMJ.

I include this, because it's the first reported instance I've seen of anyone using the .300 Savage cartridge in a Contender — or in any other handgun, for that matter.

.309 JDJ

This is one of the barrels available from SSK Industries. The cartridge is based upon the .444 Marlin case, necked to accept bullets of .308-inch diameter. Quoted data is from a fourteen-inch barrel. Large rifle primers were used and large pistol primers should not be used. The suggested fire-forming load is 40.0 grains of IMR-4064 behind the 150-grain bullet.

150-grain Hornady
52.0 IMR-4350   2355/1848
165-grain Sierra
50.0 IMR-4350   2197/1769
51.0 IMR-4350   2265/1880

GUN DIGEST BOOK OF HANDGUN RELOADING

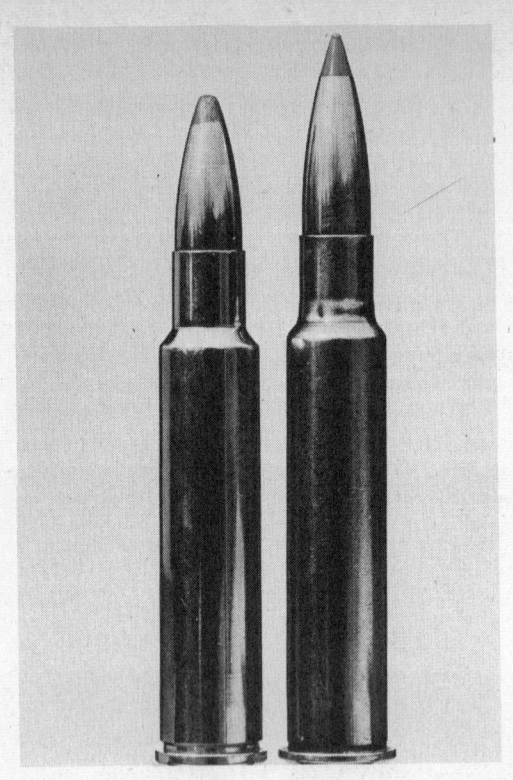

Photo at left shows the .309 JDJ (left) with the .30-40 Ackley Improved and photo at right shows the same cartridges with the familiar .30-30 WCF at left for scale comparison. The .309 is based upon the .444 Marlin and the .30-40 Ackley Improved is formed from the .30-40 Krag, the U.S. service round of 1898.

It is best to avoid use of IMR-4320 with this cartridge, as it tends to exhibit advanced indications of high pressure on a small increase in charge weight.

.358 JDJ

This is another example of the JDJ wildcats on the .444 Marlin case, the .358 JDJ is made by simply running the .444 case up into the regular full-length resizing die of the loading set. The operation requires no trimming, reaming or fire-forming: Just size the case, load it and you're ready for bear, or such other target as you may select. In addition to offering new barrels in .358 JDJ, SSK Industries can rechamber existing Contender barrels in such calibers as .357 magnum, .357 Herrett or .35 Remington to take the .358 JDJ.

Jones prefers the 250-grain Speer spitzer at about 1950 fps, but regards the .375 JDJ as superior to the .358 for African game, due mainly to the tougher construction of the .375 bullets.

Suggested loads for the .358 JDJ are:

250-grain Speer

48.0 IMR-4064	1959/2131
46.0 H4895	1939/2088
44.0 H322	1920/2047

.375 SUPER MAG

This is the basic .375 Winchester case, trimmed back to a length of 1.605 inches. SSK Industries can furnish Contender barrels in .375 Super Mag chambering and Joe Wright, head of the Thompson/Center Association, ordered such a barrel from SSK.

Wright reports that he has tended to avoid the use of Ball-Type powders in the .375 SM Contender. In many instances, it is not possible to load them to a density of ninety percent of case capacity or better. Ball powders, with air space behind the bullet, are prone to erratic ignition, hence Wright's reasoning on the matter. He favored either H4227 or IMR-4227 with the 220-grain Hornady jacketed bullets and Accurate Arms MP-5744 with cast bullets weighing from 250 to 260 grains.

Wright notes that maximum velocities attainable in the .375 SM for the Hornady 220-grain bullet are on the order of 1700 fps, as compared to about 2100 fps obtainable with the same bullet in the .35 JDJ cartridge, the latter being based upon the .444 Marlin case.

Using a Super-14 barrel with 1:12 twist, .375 SM cases from IHMSA — manufactured for them by Winchester — and Federal No. 155 magnum large pistol primers, with the 220-grain No. 3705 jacketed flat tip bullet at an overall cartridge length of 2.145 inches, Wright reports the following velocities for the progressively increasing powder charges, with weights quoted in grains.:

20.0 H4227	1618
20.5 H4227	1671
21.0 H4227	1701
21.5 H4227	1725 MAXIMUM
22.0 MP-5744	1575
22.5 MP-5744	1575
23.0 MP-5744	1653
23.5 MP-5744	1682 MAXIMUM

With the 235-grain Speer No. 2471 spitzer soft point, at

an overall length of 2.355 inches, other components as before, Wright reports:

19.5 H4227	1581
20.0 H4227	1604
20.5 H4227	1613
21.0 H4227	1641 MAXIMUM
21.5 MP-5744	1518
22.0 MP-5744	1555
22.5 MP-5744	1586
23.0 MP-5744	1597 MAXIMUM

.375 JDJ

This is one of the best, most effective of the .444 Marlin-based JDJ wildcats, in part due to the remarkably satisfactory bullets available in the .375-inch diameter. "Prairie dog accuracy plus elephant killing penetration, without objectionable recoil," is how Jones describes it.

The 220-grain Hornady spire point at 2150 fps is the best deer bullet and it will take them out to three hundred yards, according to its users. The 220-grain Hornady spire point, at about 1960 fps, is the best all-around big game bullet, giving exceptional accuracy, penetration and down-range performance. The 300-grain Hornady spire point is the best short-range stopper and the Hornady 300-grain solid almost always penetrates elephants' heads, from any angle. The 46.5 grains of H4895 behind any 270- to 300-grain bullet is a superior load, Jones notes.

Suggested .375 JDJ load data:

220-gr. Hornady SP	40.0 IMR-4198	2147/2252
270-gr. Hornady SP	44.0 IMR-3031	1893/1751
	46.5 H4895	1960/1877
300-gr. Hornady	46.5 H4895	1931/2485

The .375 JDJ, formed from the .444 Marlin case, is probably the most successful JDJ round to date.

.411 JDJ

I've had the pleasure of working with a 12.5-inch SSK barrel for this highly intriguing cartridge. It is the .444 Marlin case, necked to accept bullets of .410-inch diameter: In other words, typical bullets as used in the .41 magnum. The .411 JDJ is to the .41 magnum as the .357 Remington Maximum is to the .357 magnum — somewhat more so, if anything!

The primary problem with the .411 JDJ is the scarcity of ready-made bullets in weights greater than 210 to 220 grains. Years ago, Lyman listed a mould — their No. 410426 — that dropped big, round-nosed slugs at a typical weight of about 245 grains, depending upon composition of the casting alloy. I've had a lot of good success with that bullet, but as with so many other favorite Lyman mould designs, it's been dropped from their catalog.

Having one of Dave Corbin's CSP-1 swaging presses with dies for .410-inch diameter bullets, as well as several other sizes, I've been able to draw down some of his longer .44 jackets and go on to swage some thoroughly decent and satisfactory bullets for effective use in the .411 JDJ, at weights to up around 310 grains or so.

One of my most successful .411 JDJ recipes goes together like so: Starting with some Lyman No. 358315 cast bullets, fitted with gas checks and lube/sized, I seated them in some jackets, originally .70-inch .44 size. The improvised cores were a tight fit in the drawn-down jackets, but I used a homemade punch to convert my drill press into an improvised arbor press and rammed them home with that.

I then put them through the regular nose-forming die of the Corbin set in the Model CSP-1 Corbin bullet swaging press and they came out weighing a nice, round 250 grains, on the nose. The cores, of typical pot-luck casting alloy, were considerably harder than if made of pure lead and it took a bit of muscle to work the handle of the swaging press, but the CSP-1 is hell for strong and stood up to the gaff, nicely.

I put five of those bullets ahead of 47.0 grains of Hodgdon's H322 powder and they averaged 1788 fps and 1775 fpe. At a distance of fifty yards, firing off the bench, four of them cut a neat cloverleaf pattern and even the inevitable flyer only boosted the group spread between centers to .828-inch.

When a load such as that cuts loose, you have no slightest doubt that you just fired a handgun, but thanks in no small measure to the presence of four Mag-na-port vents at the muzzle, the recoil is endurable. Fun, no; endurable, yes. My private term for that bullet is the .41 Hardcore.

Jones has worked up a considerable body of data for use of the 210-grain Sierra bullet in the .411 JDJ, mainly in a fourteen-inch, but with a few listings out of a ten-incher. For example, 42.5 grains of 4198 — again, the maker is

It's rare to find factory .41 bullets heavier than 220 grains and the .411 JDJ needs more weight. This is the 310-grain JSP bullet I made up on my Corbin CSP-1 bullet press, as described in the text above the photo.

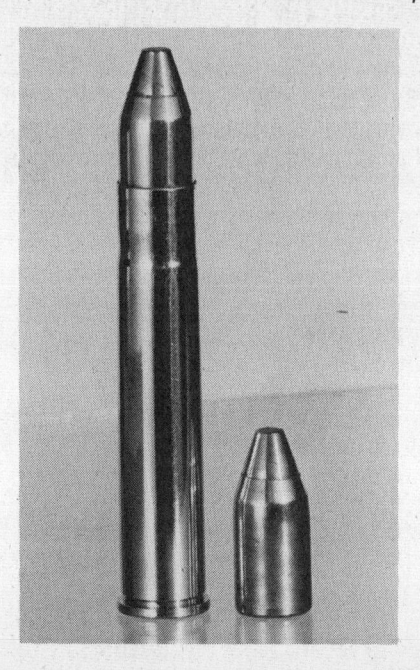

From left, the .411 JDJ, .444 Marlin and .45-70; Jones can supply Contender barrels for all three.

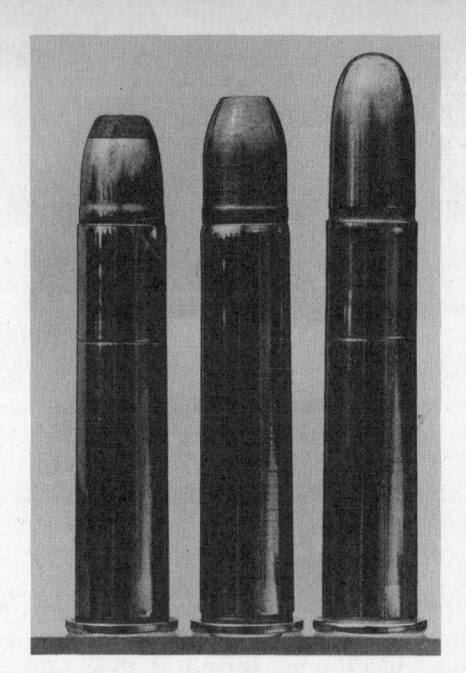

Three different .45-70 loads and the one at right carries the 500-grain Hornady FMJ; a real stomper!

not specified — goes 1669 fps in the ten-inch, 1827 in the fourteen. Boost that to 45.0 grains and the respective velocities are 1688 and 1878. With 42.0 grains of IMR-4227 and the same bullet, the ten-inch gets 2108 and the fourteen-inch 2218 fps. That last load should be regarded as maximum and approached with suitable caution.

.444 MARLIN

Introduced in early 1964 for Marlin's lever-action rifle, produced as a factory load only by Remington, the .444 Marlin is to the .44 magnum as the .357 Remington Maximum is to the .357 magnum, more or less. Contender barrels are available from SSK Industries in lengths to fourteen inches and SSK's T'SOB scope mount makes a useful addition to the barrel. With a two-power scope on the fourteen-inch test barrel, the Remington factory loads proved capable of surprising accuracy.

The Remington load is tailored to the Marlin rifle and has a cartridge overall length (COL) of about 2.525 inches or so. A greater COL may not feed and eject from the rifle. In the SSK Contender barrel, it is possible to let the bullet extend somewhat farther from the case mouth, considerably extending the versatility of the cartridge. The SSK test barrel has a rifling pitch of 1:20 RH twist and thus is capable of stabilizing somewhat heavier bullets than some of the typical .44 magnum barrels, which may have pitches as slow as 1:38.

Hornady makes a 265-grain JSP bullet, their Index #4300, which is about the heaviest .44 bullet commonly available. Load data in the Hornady Handbook takes their 265-grain to a top of 2200/2849 in the twenty-four-inch barrel of the Marlin Model 336, whose barrel is rifled with a 1:38 pitch. The top listed powders are IMR-4198 (45.3 grains); IMR-3031 (52.8 grains) and H322 (53.8 grains).

With the 265-grain Hornady bullet in the fourteen-inch SSK Contender barrel, 54.0 grains of H322 delivers 1892/2107 and excellent accuracy. Even in the fastest twist of the SSK Contender barrel, best accuracy is obtained with bullets in the weight range of 240 to 265 grains. The recoil of the warmer loads is eminently noticeable, as you might suppose. Having the barrel Mag-na-ported is quite helpful in reducing recoil and muzzle-flip.

.45-70 GOVERNMENT

The 12½-inch barrel used in developing the quoted data is from SSK Industries. Large rifle primers were used and large pistol primers should not be used.

300-grain Sierra #8900
55.0 IMR-3031	1720/1971
54.0 H322	1741/2020
45.0 IMR-4198	1726/1985
48.0 Reloder-7	1691/1905
59.0 IMR-4064	1742/2022

350-grain Hornady #4502
52.0 IMR-3031	1686/2210
50.0 H322	1618/2035
43.0 IMR-4198	1633/2073
47.0 Reloder-7	1611/2018
58.0 IMR-4064	1725/2313

400-grain Speer #2479
51.0 IMR-3031	1631/2363
49.0 H322	1600/2274
42.0 IMR-4198	1532/2085
46.0 Reloder-7	1506/2015
54.0 IMR-4064	1624/2343

500-grain Hornady #4504
46.0 IMR-3031	1417/2230
48.0 H322	1392/2152

For this barrel, stay below Lyman Handbook listing for 1886 Marlin rifles. Do not use in Contender barrels other than those from SSK Industries. *Not for use in Springfield "Trapdoor" rifles! — Dean A. Grennell.*

There Are No Huntable Tigers These Days, But Lee Jurras' Howdah Gun Is Good For Knocking Down Mountains!

A HOWDAH WAS — and perhaps still is — a sort of wicker cockpit capable of being attached to the topside of an Indian elephant. No African elephant would ever stand still for such foolishness. The maharajahs, nizams and similar upper-class gents of an earlier India were much given to punting about on their howdah-topped pachyderms, perhaps with a loaf of bread, a jug of wine and a bevy of nautch dancers to prevent boredom.

Occasional problems arose: In touring the rain forests and jungles of India's sunny clime, there was a certain risk of encountering a hungry tiger; one of the most fanatically carnivorous of all creatures. If that happened, the idyllic bliss of the moment was apt to be rudely disrupted by a momentary apparition of striped pelt and flashing fangs.

The situation led to the development of what came to be termed a *howdah pistol*. In the thin split second apt to be available, one might just barely be able to grab a pistol and

This is my Howdah, in .460 Jurras, with the original blond walnut woodwork. When and if I fire it, which is none too often, I usually substitute Pachmayr rubber for the walnut. Note lion's head receiver logo.

trigger one effective shot with it. The howdah pistols were, for the greater part, single-shots, quite broad in bore diameter and powered by black powder. Their recoil was noteworthy, but with a hungry tiger airborne in your direction, recoil is just about the last thing you're apt to worry about.

The heyday of the howdah pistol fairly well came and went with the Nineteenth Century. In the early latter half of the Twentieth Century, Lee Jurras happened to comtemplate the concept and it kindled a gleam in his eye. Jurras already had made notable waves in the ammomaking field, having founded the Super Vel firm.

The Contender was well embedded in the state of the handgun art by that time and Jurras chose to use the basic chassis for his latterday, high-tech version of the howdah pistol.

Jurras selected the .500 Nitro Express case as the parent brass for his family of cartridges, as presently manufactured by Brass Extrusion Laboratories, Ltd., 800 W. Maple Ln., Bensenville, IL 60106. In its original form, the .500 NE is 3.200 inches in length, produced to serve as the starting point for a series of elephant-blaster rifle cartridges. A generous amount of brass is trimmed from the necks in converting them to Howdah cases.

Jurras, by the way, christened the actual pistol, The Howdah, marking it accordingly. A few disgruntled purists complained that it was a fine howdah-do.

There are five basic Jurras Howdah calibers and four of the cartridges have a case length of 1.750 inches. The exception is the .500 Jurras, which has a case length of only 1.65 inches. Relevant case dimensions are given under the heading for each caliber.

The cartridges are designated: .375 Jurras, .416 Jurras, .460 Jurras, .475 Jurras and .500 Jurras. We will cover them here as a family, rather than spotting them in among all the other cartridges covered here.

.375 JURRAS

The shoulder is 1.260 inches from the base and diameter at that point is .560-inch. Shoulder angle is thirty degrees and neck length is .350-inch. Bullet diameter is .375-inch. CCI-200 large rifle primers were used in developing the load data. Ballistics are quoted in the usual fps/fpe format.

CHARGE	BULLET	BALLISTICS	COMMENT
28.0 SR-4759	235 Speer	1605/1345	Mild
26.5 SR-4759	270 Hornady	1450/1261	Mild
25.0 SR-4759	300 Sierra	1325/1170	Mild
35.0 H4198	235 Speer	1670/1456	Mild
36.5 H4198	235 Speer	1745/1589	Good working load. Consider 37.0 gr. MAX working load.
34.0 H4198	270 Hornady	1640/1613	Good working load.
35.5 H4198	270 Hornady	1670/1672	Consider 36.0 gr. MAX.
37.0 IMR-3031	300 Sierra	1430/1363	Mild
37.0 H322	300 Sierra	1525/1550	Mild, some unburned powder.

The comment, "mild," refers to chamber pressure, *not* to recoil. It is suggested that starting loads be at 1.0 grain less than listed, increasing if indications warrant. No extraction difficulties or flattened primers were encountered with the listed loads. The 285-grain Speer Grand Slam bullets can be used with the data quoted for the 270-grain Hornady bullet.

.416 JURRAS

Base to shoulder, 1.300 inches; diameter at shoulder, .560-inch; shoulder angle, thirty-five degrees; neck length, .350-inch. CCI-200 large rifle primers were used.

29.0 SR-4759	300 Barnes	1410/1325	Mild. Good working load.
39.0 H4198	300 Barnes	1425/1353	Consider 40.0 gr. MAX.
42.0 IMR-3031	300 Barnes	1360/1232	Some unburned powder.
43.0 H322	300 Barnes	1345/1205	Mild. Powder too slow

Refer to .375 Jurras comments.

Here are all five of the Jurras Howdah cartridges, in all their looming majesty. From left, the .375 Jurras; .416 Jurras; .460 Jurras; .475 Jurras and .500 Jurras. The .460 takes bullets of .458-inch diameter.

Brass Extrusion Labs supplies the .500 Nitro Express case, right, that serves as parent brass for all of the Howdahs. I've never figured what to do with all that excess brass.

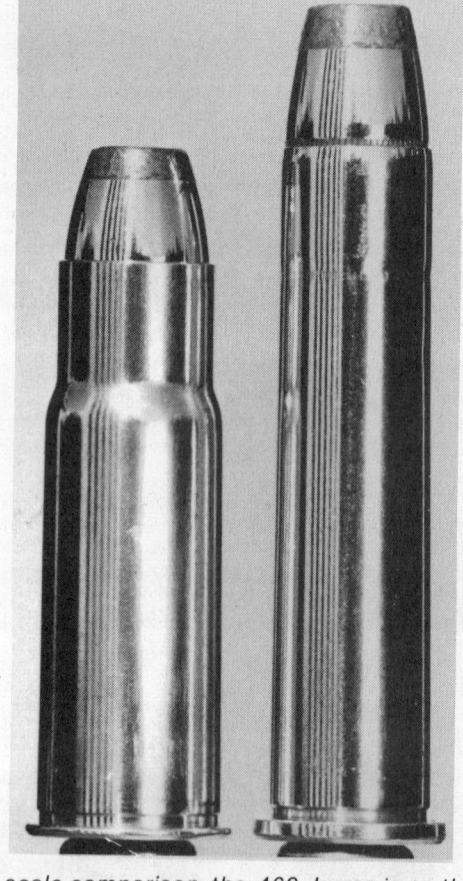

For scale comparison, the .460 Jurras is on the left, next to a .45-70, which has somewhat less capacity.

.460 JURRAS

Base to shoulder, 1.300 inches; diameter at shoulder, .560-inch; shoulder angle, twenty-five degrees; neck length, .350-inch. CCI-200 large rifle primers used. Bullet diameter, .458-inch.

30.0 SR-4759	350 Hornady	1290/1294	Mild.
29.0 SR-4759	400 Speer	1205/1290	Mild.
27.0 SR-4759	500 Hornady	1055/1236	MAX load, this powder.
39.0 H4198	350 Hornady	1270/1254	Mild. Consider 41.0 gr. MAX.
38.0 H4198	400 Speer	1210/1301	Consider 39.0 gr. MAX.
40.0 IMR-3031	400 Speer	1090/1056	Consider 42.0 gr. MAX.
36.0 H4198	500 Hornady	1195/1586	Consider 37.0 gr. MAX.

Refer to .375 Jurras comments. The heavier listed loads provide more than ample recoil and about all the fun a shooter is apt to want on a chronographing session.

.475 JURRAS

Bass to shoulder, 1.250 inches; diameter at shoulder, .560-inch; shoulder angle, twenty degrees; neck length, .400-inch, CCI-200 large rifles primers used. Bullet diameter, .475-inch.

27.0 SR-4759	500 Barnes	1030/1178	Good load.
28.0 SR-4759	500 Barnes	1115/1381	MAX working load.
35.0 H4198	500 Barnes	1080/1295	Good load. Consider 30.0 gr. MAX

Refer to .375 Jurras comments.

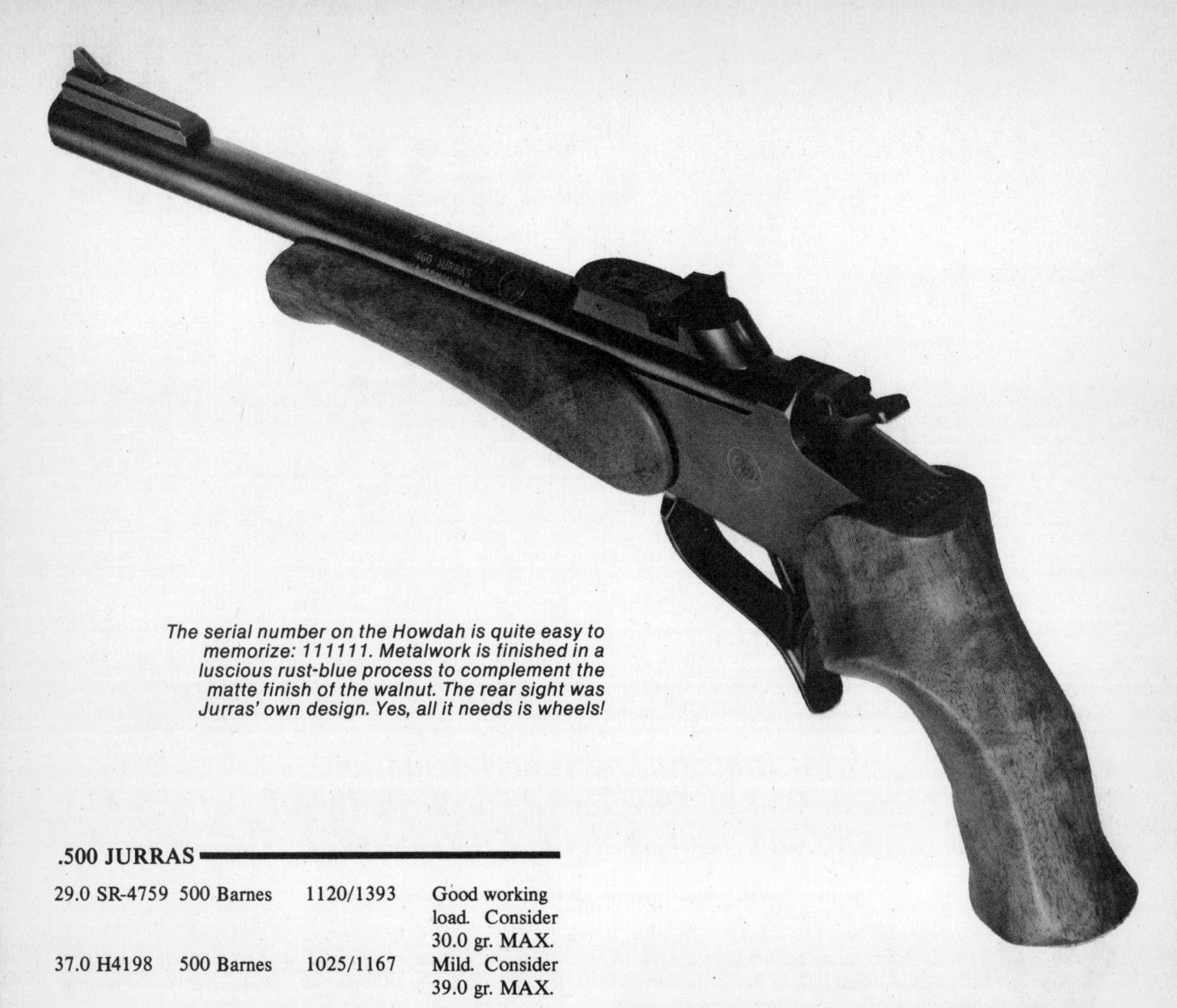

The serial number on the Howdah is quite easy to memorize: 111111. Metalwork is finished in a luscious rust-blue process to complement the matte finish of the walnut. The rear sight was Jurras' own design. Yes, all it needs is wheels!

.500 JURRAS

29.0	SR-4759	500 Barnes	1120/1393	Good working load. Consider 30.0 gr. MAX.
37.0	H4198	500 Barnes	1025/1167	Mild. Consider 39.0 gr. MAX.

Refer to .375 Jurras comments.

I have a Jurras Howdah in .460 Jurras and it's the only one of the five cartridges with which I've worked. Even so, I've not worked with it all that extensively. The paper figures as to muzzle energy do not prepare you for the personal interface of firing such loads. Upon occasion, I have launched well over a foot-ton — 2000+ fpe — at one pull of the trigger and sometimes it wasn't much worse than a bad cold. The thing is, bullet weight is a — pun unavoidable — heavy factor in equations for felt recoil and there is absolutely no bypassing the hard fact that 500 grains is about 1-1/7 ounces, all in one massive lump.

Earlier the afternoon of the present day, I was lofting 180-grain bullets at an average velocity of 2017 fps for a corresponding muzzle energy of 1625 fpe — a lot more than any energy figure quoted for the Jurras Howdah cartridges — and the recoil was no big thing, at all. If you were about to ask, I was using a Contender with Super-14 barrel in .44 magnum, to be specific. I would much rather shoot five hundred rounds of that load than one with the .460 Jurras, with a 500-grain Hornady at 1195/1586. Please accept my sincere word that there is no comparison in terms of felt recoil.

I remind you that the original intent of the howdah pistol was to deliver one humungous whap to a hostile critter at halitosis distance and, as noted, you tended to not even notice the recoil under such circumstances. When you're trying to steer a weighty projectile across the screens of your chronograph, you most certainly do notice the recoil, with every micro-erg, or whatever, of attention that you can bring to bear.

Taken as a class, the Jurras Howdahs are not much fun to use for perforating tin cans, but if I ever found myself at ground zero, with a starving tiger's trajectory coinciding with my locus, I think I'd be awfully grateful if a loaded Jurras Howdah was handy. It is a highly specialized design concept. And it surely does buck back in your hand.

— *Dean A. Grennell*

HANDLOADING THE OLDTIMERS

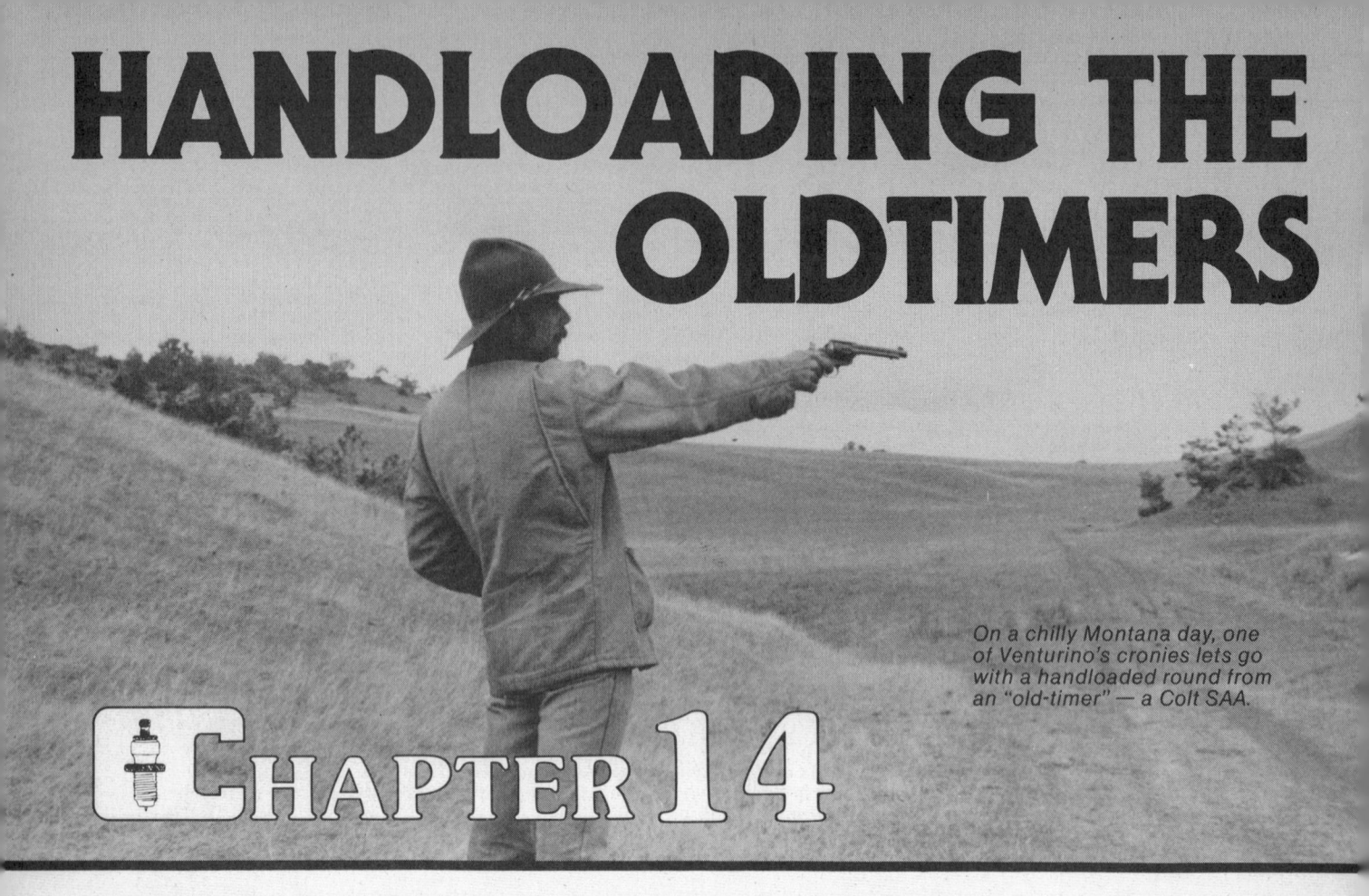

On a chilly Montana day, one of Venturino's cronies lets go with a handloaded round from an "old-timer" — a Colt SAA.

CHAPTER 14

With Thought And Experimentation, Loads Can Be Developed For Virtually All Of The Obsolete Relics Of Yesteryear

WITH THE abundance of modern handguns on today's market, it is not necessary to utilize a used revolver from a bygone era. Generally speaking, they are less powerful, less accurate and less sophisticated than current handguns. Still, there is an aura surrounding them. Perhaps they remind their owners of days when such handguns were not intended for sport or recreation, but for deadly serious business. They may be family heirlooms and their present owners wish to experience shooting the way grandpa or great-grandpa did.

Regardless of the rationale, there is no reason old and obsolete handguns of almost any caliber cannot be put to use safely today. There is however one consideration: that those guns be given a clean bill of health by a gunsmith who knows the finer points of antique weapons. The steel in some of these handguns exceeds one hundred years in age and the care they were given in their heyday has a great deal to do with whether they should be fired. For instance, if a six-gun's chamber was badly fouled with black powder and left uncleaned, it could have developed pits so deep as to render it unsafe for shooting. Also, some guns will have minute cracks from stresses in the steels that only a good

gunsmith will perceive. Of course, lock-up and positive functioning are also important factors that require expert considerations.

However, once your obsolete six-gun is given a clean bill of health, shooting it can be fun and certainly educational. Tools for virtually every caliber ever made still are obtainable today or can be special ordered.

To discuss handloading the obsolete calibers let's first divide them into two general categories. First are those guns for which factory ammo still is in production; the second category concerns those for which brass must be formed from some other caliber.

Chief among the obsolete calibers for which ammo still is made are the .32-20, .38-40 and .44-40. Handguns for these calibers can be early Smith & Wesson double-actions, some Smith & Wesson single-actions, Colt double-actions, the ubiquitous Colt Single Action Army model and a variety of lesser known and seldom encountered variations. Newly produced .44-40 handguns are relatively common. Colt still produces the SAA in that caliber as do several of the replica importers. In fact, the Santa Ana, California, firm of EMF has European-made copies of the Peacemaker in .32-20, .38-40 and .44-40 calibers. In 1986, Smith &

Above: It's true that much of the handgunning in the Old West was done with the ever-popular Colts, but lots of other makers were represented. Here's a sampling: from the top, Smith & Wesson 2nd Model #3, Colt Single Action, Colt SA Bisley and the Remington 1890. Right: The three cartridges are widely regarded as obsolescent — .32-20, .38-40 and .44-40 — but handloads make 'em percolate.

Wesson even made a special run of .44-40 double-action revolvers and rumor has it that Ruger is considering both the .32-20 and .44-40 as limited production chamberings.

Let us say that you have grandpa's old six-gun in .32-20, .38-40 or .44-40 caliber. These guns are likely to be labeled .32 WCF, .38 WCF, or .44 WCF — the WCF meaning Winchester Center Fire — instead of having the double digit label. To start handloading, one must acquire cases, but these are easy to get: merely shoot up some factory ammo. The next most important ingredient is the proper bullet. With the .32 or .44 that is relatively easy. Bullet moulds made for the new .32 H&R magnum are proper for the .32-20, and the lighter weight .44 Special or magnum mould designs will work fine in .44-40. Also some factory-made projectiles are available. Hornady and Sierra once made jacketed hollow points of .312 inch, while Hornady still makes a swaged lead semiwadcutter of .314 inch. All work fine in .32-20 six-guns. Among .44 jacketed bullets there are 200-grain JHPs from Hornady and Speer, plus 180- and 210-grain JHCs from Sierra. Within limitations, all of these can be used in reloading the .44-40.

The .38-40 is in a different boat. When jacketed bullets were produced for the .38-40, they measured .400 inch. Cast bullets are intended to be .400 to .403 inch. That puts the .38-40 in a class of its own, for no other round uses bullets of that diameter. There may be an exception in the new 10mm Auto ammunition just hitting the market. Supposedly bullets for either can be interchanged, which does not help .38-40 shooters much, since as yet no one has seen fit to make commercial bullets for the 10mms. Basically then, .38-40 shooters must cast their slugs.

In using factory-made bullets for these obsolete calibers, there are limiting factors. One is that jacketed bullets are harder on the mild steels of older barrels and can wear them excessively. Generally, 1900 is considered the cut-off time. Guns manufactured after that will be more suitable for jacketed bullets; guns made before that should be

Left: The .38 WCF marking on the SAA barrel has confused plenty of people. It means that the revolver uses .38-40 fodder. It's even more confusing to find that the round is not a true .38, but more of a .40.

Author Venturino contends that best results in handloading the old-timers will come with the use of modern cast bullets. An array of Mike's cast bullet loads in (left to right) .32-20, .38-40 and .44-40 posed with bullets.

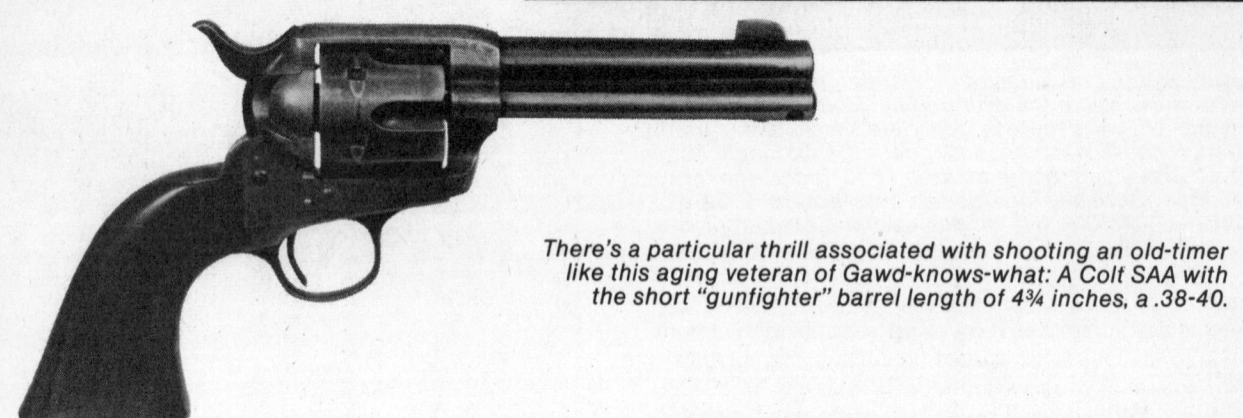

There's a particular thrill associated with shooting an old-timer like this aging veteran of Gawd-knows-what: A Colt SAA with the short "gunfighter" barrel length of 4¾ inches, a .38-40.

used only with lead alloy bullets.

Other factors that must be considered — and not with just jacketed bullets — are bore and chamber mouth diameters. This is one area wherein many experience difficulties in loading for antique six-guns. Holding to specifications in Nineteenth Century manufacturing was not as easy as today. Also those specifications seemed to change during production. Consequently one finds old Colt Single Actions with dimensions that vary greatly from published specifications.

For instance, I once had a Colt SAA .32-20 with a .314-inch bore. However, bullets pulled from .32-20 factory ammunition by both Winchester and Remington measure only .310 inch. Accurate shooting with that combination was a sometimes thing.

Another good example comes with the .44-40. Early .44-40 Colt single-actions had chambers so tight they required .425-inch bullets for cartridges to chamber freely. But then the barrels were about .427 inch in the grooves. I once was reloading for two .44-40 Colt Frontiers at the same time. One dated from 1883, while the other was from 1895. Reloading data from all recognized sources gave .427 inch as nominal bullet diameter. When cast bullets were sized to .427 inch and seated in .44-40 cases, the rounds would not fully enter either Colt's chambers. A sizing die of .425 inch was obtained and the ammo chambered perfectly. However, those undersize bullets did not deliver adequate accuracy. Incidentally, factory ammo in this caliber from both Winchester and Remington carries .425-inch bullets and it is not noted for pinpoint accuracy.

Conversely, modern six-guns for .44-40 have their specifications straightened out somewhat. Every modern Colt Frontier I've encountered will chamber cartridges loaded with .430-inch bullets, if the thinner walled Winchester

Venturino has developed safe and accurate data for many of the old-timers. From the left: .32-20, .38-40, .41 Long Colt, .44 American and his favorite .44-40.

These are old guns, made many years ago, and not in an era of sophisticated metallurgy. They will work, but they ought to be gone over by a good gunsmith before they are fired. And keep the handloads light!

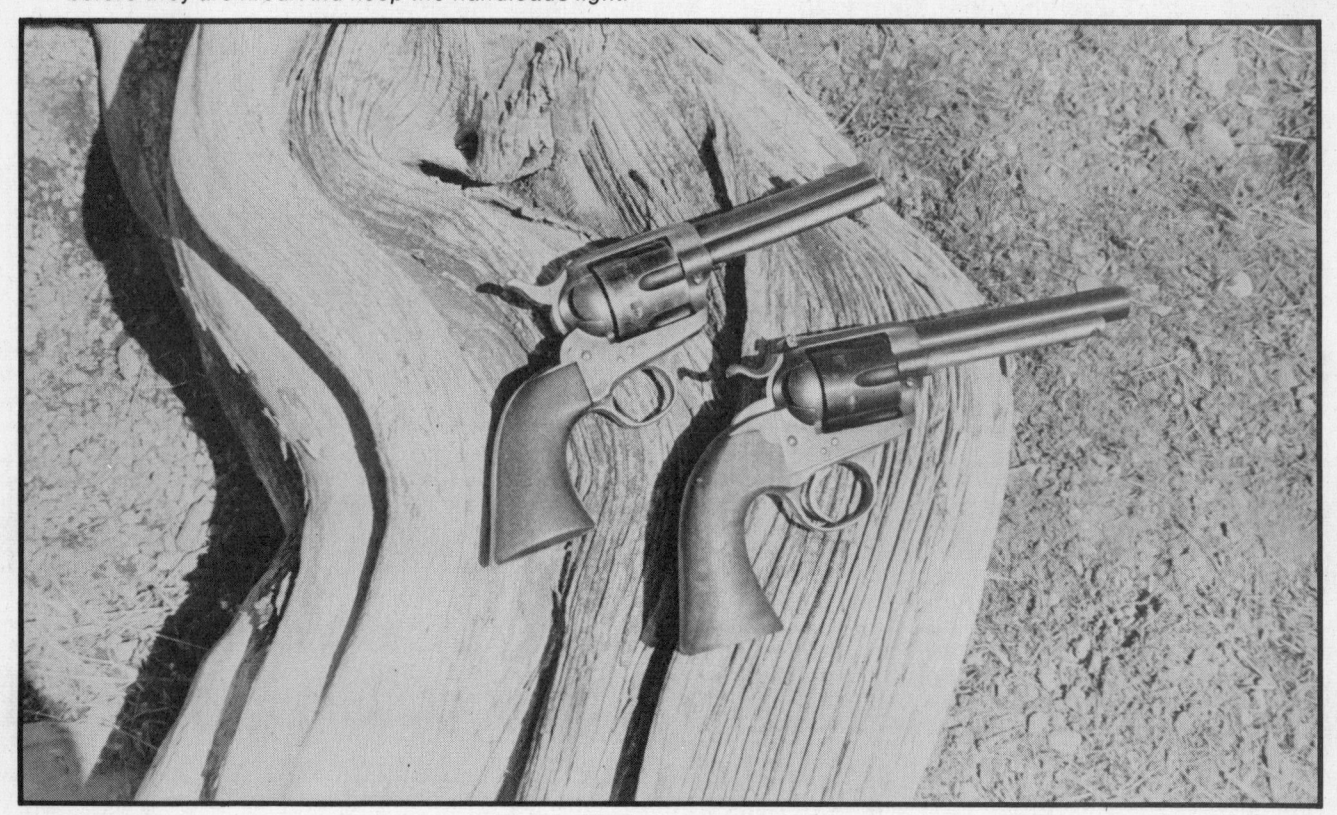

cases are used. And they give fine accuracy.

Specifically, I prefer cast bullets which resemble those used in vintage factory loads in weight. Such moulds are produced by NEI, Lyman, RCBS, Saeco, Rapine Associates and others. Traditional factory ammo bullet weights were: .32/100 grains, .38/180 grains and .44/200 grains. Any cast bullet hitting those by plus or minus 10 grains is suitable. The cast bullet sizing diameters I have settled on are .312 inch, .403 inch and .428 inch respectively. However, I would like to repeat that each gun should be slugged in both the bore and chamber mouths to determine what size is best for it.

Once cases and bullets are in hand, putting them together is no problem. Loading dies are available from all the major manufacturers such as Lyman, RCBS, Redding, Bonanza, Hornady and others. The handy carbide die sets available for most other handgun calibers cannot be used with these three due to their slight bottleneck configuration. However, I find that no great problem. These cases are thin and size easily. For case lube I used the water-base white stuff by Lee. Used when dry, it still lubes well and, if applied sparingly, it need not be wiped off after resizing.

The type of powder used is important, however. Again, this is something in which your gunsmith might have a say. If the six-gun is of old enough vintage, he might advise you to reload only with black powder. Again, if it is of post-1900 vintage, mild smokeless loads should do no harm. I have loaded all three of the above calibers with black powder and found the shooting fun, but clean-up a bit on the messy side. It is interesting to watch the fouling build up. With some .44-40s, trying to rotate the cylinder after about thirty rounds becomes difficult.

In the realm of smokeless powders the best choices are charges of relatively fast-burning powders, ranging be-

GUN DIGEST BOOK OF HANDGUN RELOADING 235

Right: The curious .41 Long Colt cartridge had some peculiarities. It wasn't a .41, but rather a .40; it used a unique heeled bullet with hollow base. Guns haven't been made for it since 1930s...

...but Winchester bowed to popular demand and a hefty order from one particular dealer, producing a single batch of the ammunition back in the 1970s. And from that ammo comes modern, reloadable brass.

tween Bullseye and Unique in quickness, that will duplicate the approximate velocities for black powder loads. In the .38-40 and .44-40, I use 6.5 to 7.0 grains of W231 or 7.0 grains of Unique. The .32-20 does fine with 4.5 to 5.0 grains of Unique. One thing to keep in mind is that these old guns were not intended to be magnums and trying to soup them up to powerful levels is not recommended.

An antique caliber that lies between those for which current factory ammunition is available and those for which cases must be formed is the .41 Long Colt. From my experience, I must say that the .41 LC is an interesting caliber. Factory ammo has not been a standard item for some years, but in the mid-Seventies Winchester made a special run of about a million rounds. It is commonly encountered at gun shows and C. Sharps Arms (P.O. Box 885, Big Timber, MT, 59011) keeps some in stock. Therefore, modern ready-made cases are available with a little searching, as are reloading dies. Both RCBS and Redding list the .41 Long Colt among their offerings. The real problem is in getting the proper bullets.

Originally, the .41 LC carried a "heel" bullet of about .406 inch. Heel-type bullets, for those unfamiliar with them, are styled just as those found in modern .22 rimfire ammunition. They have an undersized shank fitting inside the case mouth, while the rest of the bullet is the same size as the outside of the case mouth. In my opinion, they are a pain. They will work with black powder, but with smokeless it is difficult to get complete combustion of the powder, even with Bullseye.

After the switch to smokeless loads, ammunition companies changed the .41 Long Colt from a heel-type bullet to one with a hollow base, just like a Minie ball. The theory is that powder gases will expand the bullet's skirts, causing it to grip the rifling and, in practice, it does work. You may not get pinpoint accuracy, but 2.50- to 3.00-inch twenty-five-yard groups are possible. That is surprising, considering that started bullet diameter is .386 inch and groove diameter of most .41 LC barrels is .403 inch. That means the bullet must expand about .017 inch to grip the rifling. Of course, this calls for an extremely soft bullet alloy.

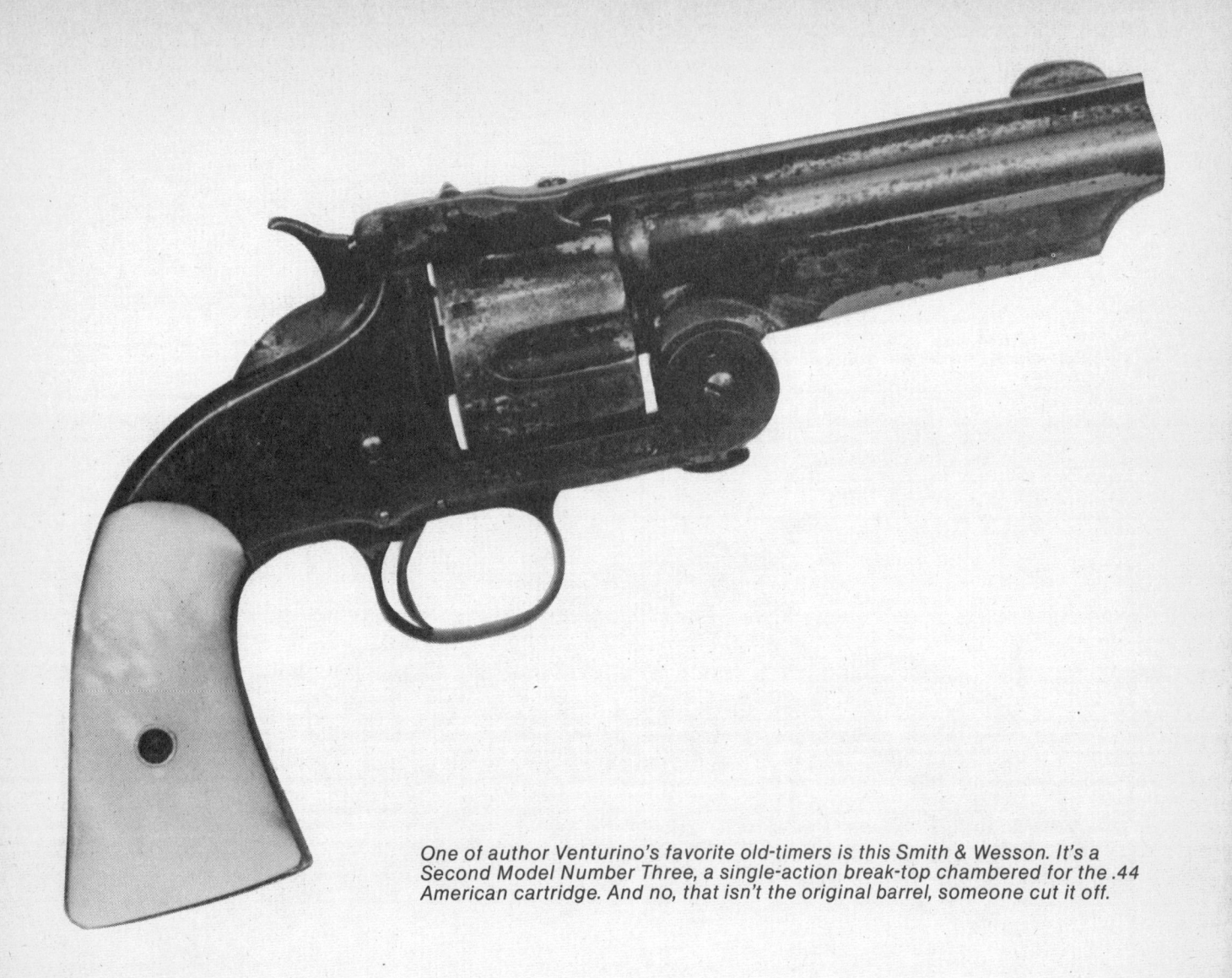

One of author Venturino's favorite old-timers is this Smith & Wesson. It's a Second Model Number Three, a single-action break-top chambered for the .44 American cartridge. And no, that isn't the original barrel, someone cut it off.

When I first began loading the .41 LC cartridge, the heel-bullet was my first try and I gave up in disgust. Then I began searching for a hollow-base mould. I searched and searched and searched. You see, Lyman was the only source for such a mould; its number was 386178. Several years ago, they discontinued it and trying to find one is near impossible. Finally I located one belonging to a doctor in California. He would not sell it, but did loan it to me for a time. Bullets from it loaded perfectly in .41 LC cases and shot quite well.

Today, at least two firms can supply hollow-base moulds. One is Hoch Bullet Moulds (Box 132, Fruita, CO 81521). Their moulds are custom-made on a one-on-one basis. Another company offering a .41 LC hollow-base is Rapine Associates (P.O. Box 234, R.D. #1, East Greenville, PA 18041). Loading the .41 Long Colt today is entirely feasible and as easy with most any other centerfire caliber.

Lastly among obsolete handgun calibers are those for which factory-produced brass is not available. This is an area for more experienced handloaders. In a couple of instances the case-forming chores are simple. Consider the .44 Russian and the .45 S&W Schofield. Their dimensions are compatible with the .44 Special and .45 Colt. The letter cases need only be shortened to .97 inch and 1.10 inch respectively, then loaded with the appropriate bullets. Both the .44 Russian and the .45 S&W Schofield require the same diameter bullets as standard .44 Special and .45 Colt.

With some of the more rare obsolete calibers, the reloading chores get more difficult. Only a year ago, I picked up a Smith & Wesson #3 .44 American revolver in good shooting condition. Not knowing a single fact about the cartridge, I bought the handgun out of curiosity.

Reloading for it turned out to be possible, even fun, but there was a lot of education to be had before it was successful. For one thing, the .44 American takes a .425-inch bullet and all its original factory loads used heel-type bullets. Also the .44 American cartridge case was not merely a variation of the .44 Special case. Its rim diameter was nominally .506 inch as opposed to .514 inch for the

GUN DIGEST BOOK OF HANDGUN RELOADING

Right: After doing it the hard way, Venturino found that RCBS makes the necessary dies to form, load and cast proper bullets for the aging .44 American.

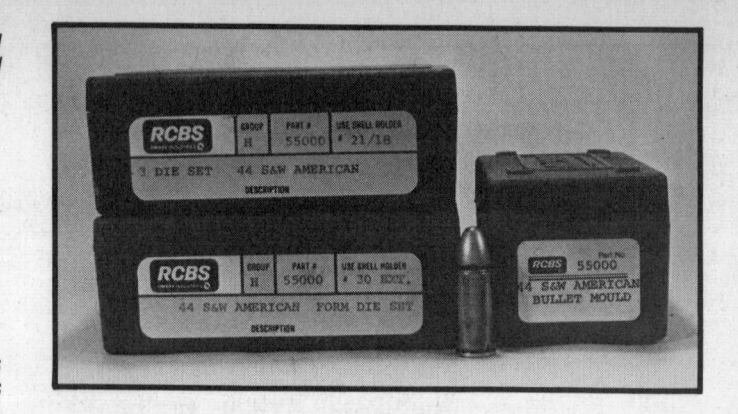

Below: If this one looks suspiciously new, that's because it is. EMF and other firms market replicas of many black powder-era guns. An 1890 Remington.

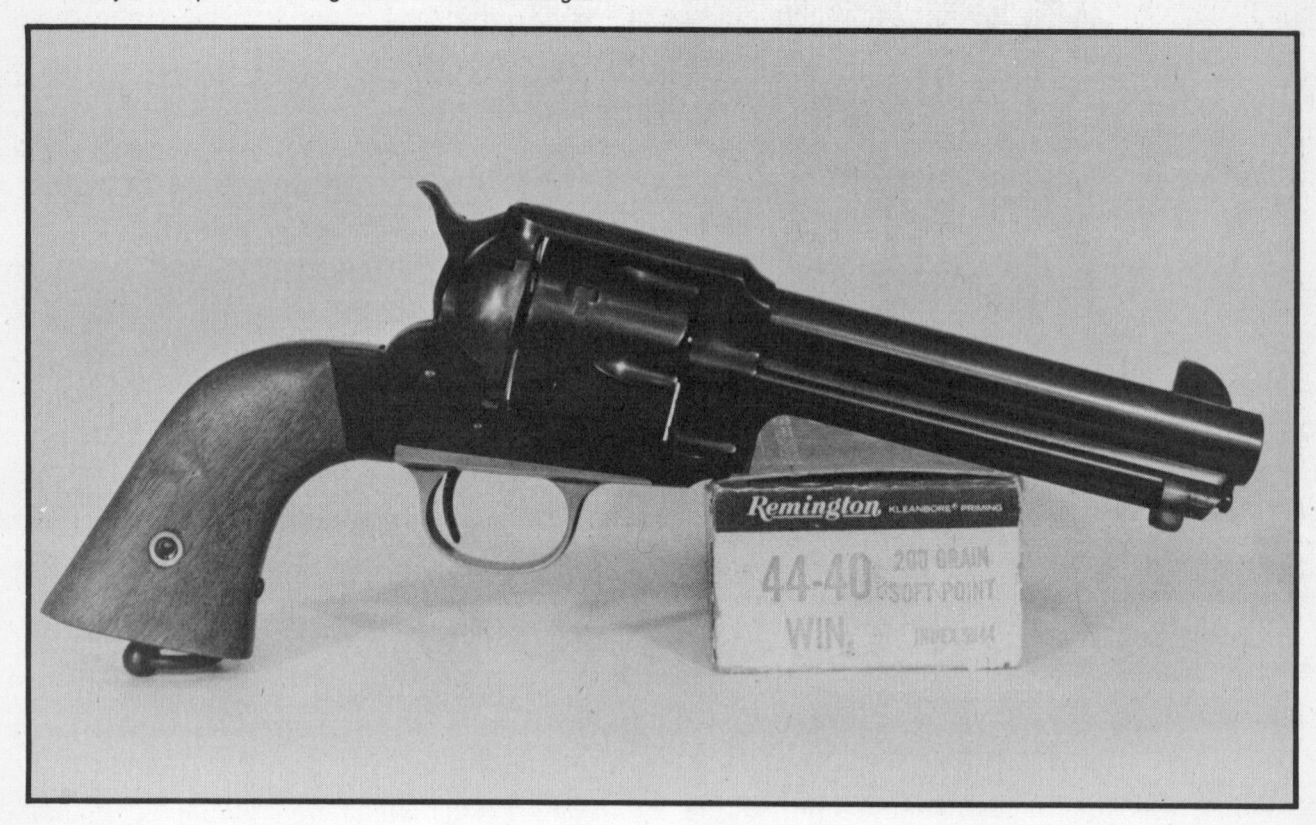

Russian, Special and magnum .44s. Those latter cases will not enter a .44 American chamber unless lathe-turned to paper thinness, then they will split after a firing or two.

But I was able to get my Smith & Wesson #3 .44 American shooting. Instead of trying to squeeze down large cases, I found it for more practical to blow out .41 magnum brass after it was shortened to 1.00 inch in length. Originally, I used several other reloading dies to "cobble" loads together. A standard .44-40 cast bullet was squeezed down to .425 and loaded over 20 grains of FFg black powder. Fifty-foot groups were the size of my palm and I was so happy with the results that I wrote them up for a 1986 issue of a handloading magazine. Shortly after the magazine appeared, I received from RCBS a package containing .44 American reloading dies, case forming dies, printed instructions on how to form the cases from standard .41 magnum brass and, of all things, a bullet mould whose bullets are shaped and weigh just like the original .44 American slugs. A call to RCBS confirmed that they do not even catalog such a mould, but they did have the cherry on hand.

My lesson from the .44 American was twofold. One was that, if you have some basic handloading experience and a good assortment of tools on hand, you can cobble most any cartridge you desire. Secondly, I learned that if you will just bother first to contact the real experts, such as the people at RCBS, they can give you an amazing amount of information right away and often supply you with the tools to load about anything without resorting to cobbling.

A few weeks ago, a fellow dropped by the house whose interests parallel my own. He was toting a Colt factory conversion of a Model 1849 percussion frame. That little five-shooter, in near mint condition, was intended for a short .38 centerfire cartridge. I admired it briefly and handed it to its owner saying, "I'm sure, with some work, ammo could be made from something or other for that." He just looked at me with a gleam in his eye and produced a box of loads. He had shortened .38 Special brass until it would

There's not a reason in the world why you can't shoot such guns as this 1902 Model Colt DA Frontier revolver. This is the large trigger guard version collectors call the "Alaskan" or "Phillipine" Model. It's a .45 Colt.

Authentic! The puff of powder smoke, the hard-kicking Colt Single Action, even the real-life Stetson. I'm damned if you can't almost smell the fresh buffalo chips.

chamber. He had forced a .36 caliber lead ball into the case mouth until it was held securely, then dipped the ball in melted wax. Under the round ball was a case full of black powder. I know that because I helped him shoot all his ammo away that afternoon. It was great fun, and that little pocket pistol made over a century ago was surprisingly accurate.

Most of our historic handguns can be fired and offer great entertainment, if the handloader is dedicated and, importantly, careful. If you have any of grandpa's old Colts, Smith & Wessons, Merwin & Hulberts, Remingtons or who knows what lying about, dust them off and consider shooting them. You will enjoy it. — *Mike Venturino*

CHAPTER 15

SPECIAL SITUATION RELOADING

ONCE IN a while, you can come up against a reloading situation not fully covered by reference works on the subject. Possibly, you're working with a caliber that's comparatively new and covered little, if at all, by the manuals and handbooks on reloading. Or perhaps you envision a need for a type of load you've never worked with before. These things happen, now and then.

I wish I could set down a treatise so mercilessly exhaustive that it would provide full chapter and verse for every conceivable special situation. A certain degree of innate modesty suggests I'm not quite capable of bringing that off, as evidenced by the fact that I tend to sink clear to my ankles every time I try to walk on water. What I propose is to discuss some special situations I've dealt with in the past, hoping the details of the given situation may prove helpful in your own time of need.

At the start, let me say that the development of load data from scratch is a project heavily fraught with the direst kind of peril available. It is by no means the sort of activity I'm inclined to sanction or advocate. At various times, I've had no choice but to do so and I'll freely confide it never fails to scare me mottled chartreuse. With puce polka dots.

In plausible theory, what you do is start out with a charge of the given powder, then work up gradually, keenly alert for indications that the peak pressures are edging toward the back-off point. That's really all there is to it. Sur-r-re it is!

Some Thoughts, Comments And Observations On What To Do — If Anything! — In The Absence Of Explicit Instructions

Opposite page, RCBS recently added a power kit to replace the usual operating crank on their case trimmer for use with an electric drill, thus taking a great deal of effort and drudgery out of trimming cases. Above, the 1950 Target Model Smith & Wesson .44 Special, as discussed in the adjacent text.

Confidentially, settling upon the exact starting charge weight so safe to verge upon the ridiculous is — well, as they used to say while spraying the Royal Iranian palace with insecticide — that's about where the Flit hits the Shah.

It is a well known fact that several powders nominally intended for reloading shotshells can be used to excellent effect in reloading handgun cartridges. Not all shotgun powders are listed in the loading manuals for use with handgun ammo, however. It is, undoubtedly, a natural human trait to speculate upon the possibilities they might offer, when thus utilized.

At least, that was the thought that sauntered idly through my mind on a day long ago. I had a nice old Smith & Wesson revolver of which I was fatuously fond. It was their 1950 Target Model in .44 Special, with a four-inch barrel, that I'd owned for seven or eight years, at the time. I'd bought it second-hand, not too long after Elmer Keith had been working with similar guns, using a generous quantity of Hercules 2400 behind the bullet he'd designed for Lyman. Listed in the Lyman catalog as #429421, it was a semi-wadcutter of about 240 grains.

A friend who worked at the ballistics lab of one of the major bulletmakers has since advised me that they ran Elmer's hot 2400 load through their pressure gun and got a reading in the high 30,000 c.u.p. brackets: well up into typical levels for the .44 magnum and quite a long way above the SAAMI ceiling of 15,900 c.u.p. for the venerable .44 Special round.

Be that as it may, in my enthusiasm, I'd made up and fired quantities of Elmer's pet load through my beloved S&W — a Model 24, by current nomenclature — and it hadn't come down with the pip or shown any alarming symptoms. On one occasion, in line of duty as an auxiliary cop, I'd driven one of the big slugs clear through a Wisconsin whitetail deer, lengthwise, at that. The buck dressed out at 180 pounds; a good-sized one, by local standards.

Recollections such as that contributed toward giving me more courage than I really needed, as it turned out. I hankered to try out this one powder in the .44 Special. For purposes of the present discussion, I'd prefer not to identify the powder and, for the record, I no longer recall just how much of it I decanted into the case, but the level of the charge looked about right, so I went on to seat some man-

GUN DIGEST BOOK OF HANDGUN RELOADING

Left, the four-cavity mould for Hensley & Gibbs' #333, a lightweight wadcutter for use in .357, .38, or 9mm loads. Below are two of the 333s, as cast, two more lube/sized simultaneously for convenience of loading into a single .357 magnum cartridge case.

ner of bullet, deemed suitable for the purpose, into the neck. In those days, I enjoyed the sybaritic luxury of being able to make up a load in my basement workshop, then walk through a door to the garage and fire it onto a sturdy bullet trap. That's what I did and, for good measure, I made up six rounds in all, stuffing one into each of the S&W's chambers. Out in the garage, I clipped a target in front of the trap and commenced firing. It punched in a nice group and the recoil seemed energetic, but manageable.

It wasn't until I'd fired all six and eased the cylinder open, pressing on the tip of the ejector rod that I became aware I had a problem. The poor old Smith had learned of it a bit sooner. To say the cases resisted extraction is to understate the facts. I went back through the door and into the shop, dug up a short piece of hardwood dowel and managed to tap the cases free of the chambers, one at a time.

An awakening suspicion was dolefully confirmed when I reloaded with six rounds of a load known to be quite mild and innocuous. Those cases now resisted extraction, also. What I'd done, in my misguided stupidity, was to jug at least two of the chambers, rather severely.

I sent the gun back to Smith & Wesson and they fitted a replacement cylinder. It was so long ago that the tab was a mere $18. I recall that with painful clarity. I still have that old revolver, now a bit over three decades in service, and it remains fully functional. I've not used it for the development of any further harebrained experimental loads. Prior to that fateful occasion, it was purely a tack-driving terror. With the replacement cylinder, it's still a nice old .44 Special, but it never again grouped quite as well.

I'd hope you may have learned something from the foregoing account and I can confide I certainly did.

MULTI-PROJECTILE LOADS

Every now and again, it may seem an intriguing concept to launch two or more bullets at a single press of the trigger. The idea may even have some amount of practical utility, although I've never convinced myself of that, one way or the other. For loading into conventional, straight-sided handgun cases, such as the .357 magnum, it's obvious you need bullets somewhat shorter in length than are used customarily. One of the best choices is Hensley & Gibbs #333, which can be sized to .356-inch for use in the 9mmP or to .358-inch for .38 Specials, .357 magnums or the .357 Remington Maximum. In typical casting alloys, the #333 comes out at a weight of 66 grains or so.

It would be plausible to assume that, when loading two

Right, Speer makes these empty shot capsules in both .38 and .44 calibers, with suggested load data on the boxes. Apart from regular shot, they can be filled with a variety of payloads. Below, a closer look at the H&G #333 bullets, slightly shorter than their own diameter. In typical alloys, they weigh 66 grains.

of the #333s, you'd merely double the weight to 134 grains, look up the recommended charges for bullets weighing close to that — say 140 grains — and use those charges. Actually, that's fairly close, but think on the matter for a bit: You will be seating the two wadcutters without much if any of the front one protruding from the case mouth. That means the base of the rear bullet will be well down into the powder space of the case, as compared to what it would be with a 140-grain bullet, if seated to its usual depth.

Deeper seating causes some amount of pressure increase, as clearly indicated by firing across a chronograph in comparison to identical loads with the bullet farther forward. The preferable alternative, in this example, would be to look up suggested data for the 148-grain wadcutter in the given caliber and use that, rather than dope for the 140-grain bullets.

Three of the H&G #333s add up to about 201 grains and can be loaded into the .357 case, after a fashion. The pertinent question is: How much of what kind of powder do you use then? Some sources carry data for 200-grain bullets in the .357 magnum but, again, it wouldn't necessarily be correct for this application because of the substantially greater seating depth.

A further complexity is that the base of the rearmost bullet probably will be getting back to the point where the case wall commences to taper inward, with a correspondingly smaller diameter. Quite possibly you'll find it necessary to size the rear bullet as if using it for a 9mmP, say .356-inch or so.

One of the slower-burning powders is indicated in a situation such as this and I'd probably settle upon either H4227 or IMR-4227. Both are notably good-natured powders, when used properly — we'll get back to that in just a bit — and are good choices when you want to launch a weighty cargo of payload.

Lay out three of the bullets on a flat work surface and use a caliper to measure the total length of what I guess we'd have to call the bullet column. Use the caliper to measure down from the case mouth and ink a line to mark the point corresponding to the base of the rear bullet. Use an adjustable powder measure, or progressive changes of rotor with a fixed-rotor measure, until you get a charge of either make of 4227 that does not quite come up to the scribed line on the outside of the case.

You may, if you like, cut off a two-inch length of ⅜-inch hardwood dowel, chuck it in the drill press and use a small piece of medium-coarse sandpaper to dress one end to the

GUN DIGEST BOOK OF HANDGUN RELOADING

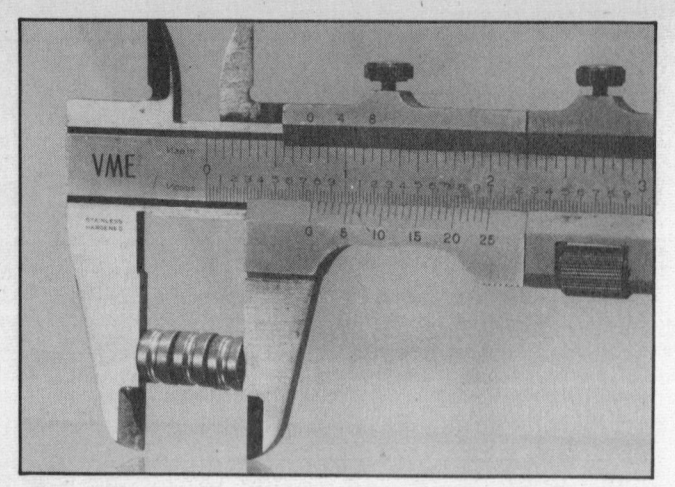

Three H&G #333s measure about .770-inch on the calipers and we tighten the little set screw at the upper center of the photo to hold that setting...

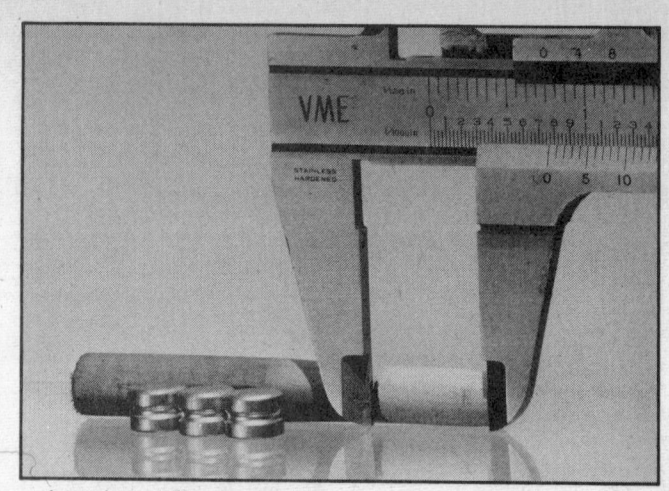

...then draw a line on a piece of 3/8-inch dowel that has been sanded down to fit the .357 case neck. This will serve as a simple gauge for checking powder.

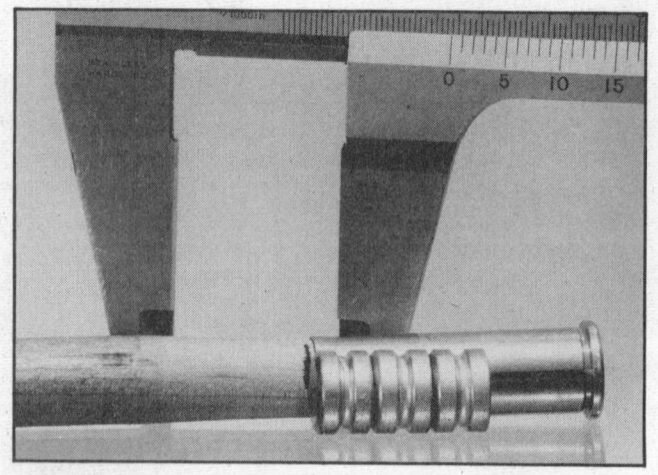

Here, with the dowel inserted in the .357 neck to the line, the three bullets indicate how much room is available for the powder charge, as discussed in text.

point where all or most of the end fits into a .357 case. Then you use that as an improvised depth gauge to check out the tentative charge and see if the base of the bottom bullet will come into contact with the charge or not.

What we're trying to establish here is the amount of 4227 that does not quite fill all of the available space. Later, after trying a load or two, you may wish to increase the weight of the charge perhaps to the point where the charge is moderately compressed. The key point, however, is that you do not do that on your first trial load!

With the greater case length of the .357 Remington Maximum, it is possible to load all the way up to four of the H&G #333s in a single case. My cohort, Wiley Clapp, has expended considerable amounts of time and attention to exploring that particular possibility. Fired with enthusiasm, he even had the cylinder of a Model 581 Smith & Wesson .357 magnum bored out to remove the ledges at the front of each chamber, permitting the use of the .357 R-Max cases in it. The front bullet is seated flush with the case mouth and, thus located, the bullet just barely clears the front of the cylinder.

Clapp calls this his *Quadraximum* load and has a scaled-down cardboard silhouette target tacked to the front door of his office. The area just above the wishbone of the hapless target is well and truly riddled with twenty-four holes: the output of but one cylinderful of Quadraximums!

Clapp targeted the load at several distances, ranging from four to twenty-two yards and the groups are smaller than you'd expect. His first trial load — worked out by steps similar to those just outlined — came to 8.6 grains of IMR-4227 powder behind a quartet of bullets weighing 266 grains at a muzzle velocity of 764 fps for 345 fpe of combined muzzle energy. This was out of a 6½-inch Ruger Blackhawk.

Gradually increasing his charge weights, Clapp worked his way up to 11.9 grains of IMR-4227, which produced an average of 927 fps in his four-inch Model 581 for a muzzle energy of 508 fpe. He reported there still were no indications of excessive pressures in the test gun, but it was at that point he ran out of powder space. In view of all that, I'd hope my reasoning would be apparent for suggesting 4227-type powders as a good starting point for situations of this general nature, where you want to get a heavy payload into motion.

Here's my Remington Model XP-100 in .221 Remington Fire Ball, as it looks these days. The thumb hole stock is from Reinhart Fajen, laminated from layers of birch and walnut. While this is a remarkably strong action, it should not be over-taxed, as discussed here!

Below are photos of two similar but not necessarily identical powders, as offered by IMR Powder Co. (formerly Du Pont) and Hodgdon. Both 4227 and 4198 are really excellent powders, properly used, but don't use 4227 for 4198!

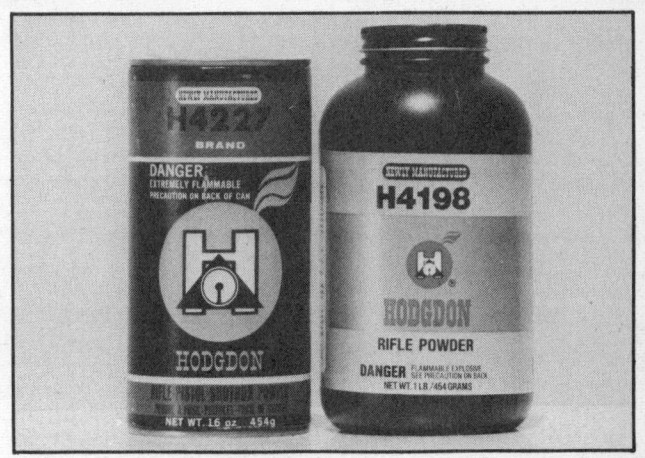

LOAD DATA FROM MEMORY

Earlier, I'd noted that both IMR-4227 and H4227 are notably good-natured powders, when used properly, and said we'd get back to that.

The time was the mid-Sixties and, a short while earlier, I'd completed the research and writeup on Remington's then-new Model XP-100 pistols in .221 Remington Fire Ball. In the course of it, I'd happened upon one particular load that performed like no-tomorrow: pinpoint accuracy, plus some pretty zippy ballistics. A real whizzerballoo!

Thus it came to pass that I planned to get together with some friends for a spot of pleasant, recreational pistoleering on the coming weekend. Wanting to show them the potential of this particular load in the XP-100, I proceeded to load up twenty or so rounds of it. I recalled the recipe for the magic load quite clearly: no need to refer to my notes, I thought.

I was wrong.

I triggered off the first round of the given load and the gun went *foosh-ka-boom*, with a report considerably more ear-splitting than I'd expected. I went to open the bolt and found it well and truly frozen shut. With the aid of a wooden mallet, I managed to coax the action open and drew the bolt back.

The empty case was immovably held in the bolt face. Under what amount of ungodly peak pressure I can't even begin to estimate, the tough brass of the case head had flowed like warm saltwater taffy, back into the extractor and adjacent construction. I had to chuck a dentist's burr into my Dremel Moto-Tool and use that, with great care, to get the last scraps of brass out of the bolt face.

I ended up sending the entire pistol back to Remington for a checkup and such reworking as they felt necessary. Since that time, the snows of twenty-odd winters have fallen and melted away in the springtime sunshine.

I've never had even one little bit of further trouble with that XP-100. I've shot it a lot and it continues to boggle belief with its uncanny grouping capability.

Quite obviously, that first shot called a halt to further work with the Fire Ball on the schutzenfest at hand. When I got back to the loading bench, I finally did what I should have done in the first place: I looked up the original load data...

I'd gotten the charge weight exactly right, down to the

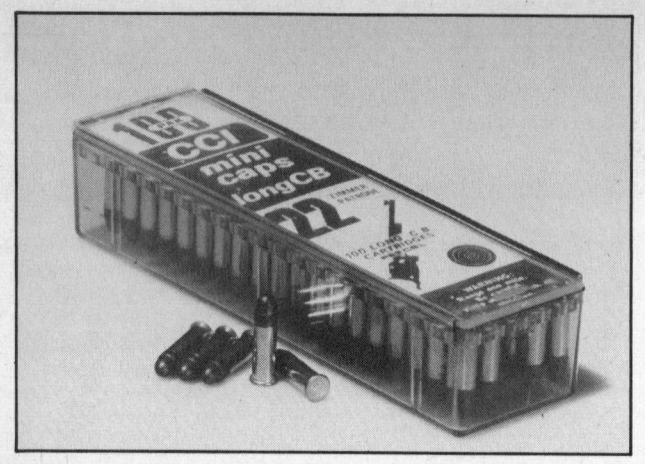

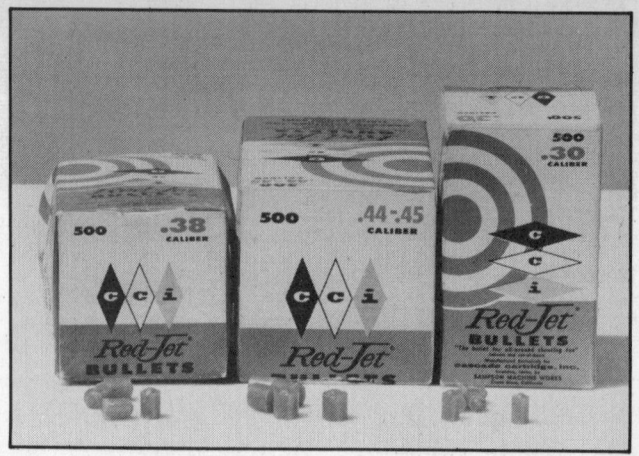

CCI's Long CB Mini-Caps, left, are subtitled Zimmer Patronen, a German term that translates roughly to "room cartridges." As noted, they offer a mild report. CCI also made the Red-Jet wax practice bullets, now discontinued.

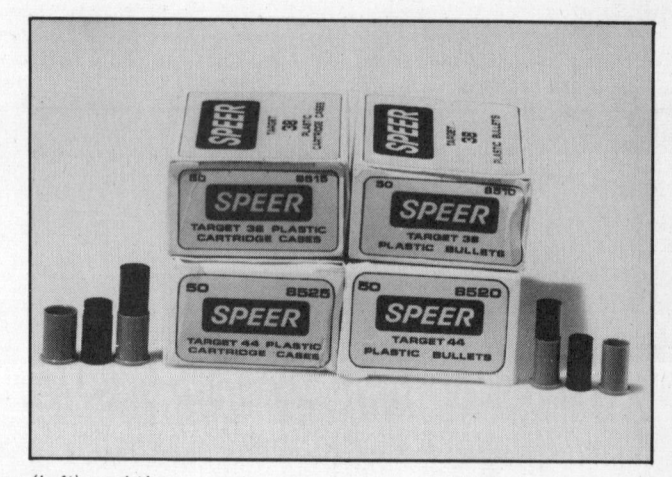

Speer continues to make plastic practice bullets for .45 cases (left) and the combinations of plastic cases and bullets for both .38 and .44 (right). As is noted in the nearby text, such low-power loads need to be handled with caution! All of these centerfire systems are powered by primers alone.

last tenth-grain. What I'd not done, however, was to recall the proper powder. The super-duper load had used 4198 powder, not 4227! Since that time, I've made every earnest effort to avoid quoting or using load data from memory. I respectfully suggest you do the same.

SQUIB LOADS AND SUPER-SQUIBS

Now and then, a situation may arise in which you desire less powder than the customary level; perhaps even more urgently, less noise when the round goes off. CCI's long CB Cap load is fairly close to the lower end of the ballistic scale and, as an added attraction, you can buy the load over the counter at well stocked gun stores.

There are further possibilities, however. The Thompson/Center Contender, particularly with their barrel for the .22 Hornet, is a prime example. You can full-length resize and prime one or more Hornet cases and go on to stuff a caliber .22 airgun pellet in the case mouth. The important thing is to seat the pellet backwards, with skirts to the fore. If you seat it in the normal manner, even the modest force of the primer is apt to blow the nose off the pellet, leaving the skirts lodged in the bore.

The primer will drive the reversed airgun pellet out of the ten-inch Contender barrel at moderate airgun velocities, with enough force to put a dent in soft wood. The sound of the pellet striking the wood will be about as loud as the actual report from the gun. Short-range accuracy is passable, although the point of impact may not coincide closely with the setting of the sights, as used for regular Hornet ammo.

It is possible to add just the merest squinch of a fast-burning powder such as Hercules Bullseye for the sake of boosting the velocity usefully and the report will remain surprising modest.

At various times, assorted makers have offered bullets made of plastic, wax, rubber and perhaps other ingredients I've not heard of. Speer continues to offer a combination of plastic case and fitted plastic bullet, in calibers .38 and .44, usuable in guns such as .38 Special, .357 magnum, .44 Special and .44 magnum. The plastic cases accept standard primers to supply the entire power for firing the plastic bullet.

With any of these squib loads, it is a grievous mistake to

For low-velocity applications in the T/C Contender, a primed .22 Hornet case can be loaded with a .22 air gun pellet, reversed as here, and used indoors.

Two guns in 10mm Auto: Upper is the Delta Elite, by Colt and lower is the discontinued Bren Ten, from Dornaus & Dixon. American Derringer Corporation also makes an over/under derringer for the 10mm Auto.

underestimate their power and treat them with casual contempt. When the Speer plastic bullet system first appeared back in the early Sixties, they sent me some to write up for my publishing outlet of that time. I dreamed up a cute little bullet trap, made by cutting an opening in the front of a corrugated cardboard box, over which I affixed a paper target.

Figuring the plastic slugs just possibly might penetrate the back of the box, I rigged a baffle by draping one of the heavy bathroom towels over a length of dowel and hanging that behind the target opening.

I fired half a dozen rounds at the target, got a pretty good group and happened to check on how the towel was holding up. Not very well, as it turned out. It was so thoroughly riddled that we had to retire it from active duty, accompanied by some rather acrid dialog from my better half. Some of the slugs got through the back layer of cardboard in the box and I had a hard time finding them.

In the early Sixties, CCI marketed some wax practice bullets under the brand name of *Red Jet*. They were offered in several calibers, including .30 size and I still have part of a box of those around the shop somewhere.

If you really want the fine particulars on wax bullets, the man to ask is Bill Jordan, who used them extensively for years while doing exhibitions for the National Rifle Association. If you never saw one of those shows he put on, you have my sincere sympathy.

He would start out slow and easy, asking a young lady from the audience to come up and hold out an empty aluminum pie pan. Jordan's hand would flick, as might a snake's tongue, there'd be the snap of a primer and a caliber .38 hole would appear, quite close to the center of the pan. As his victim gazed at the hole, bulging a bit about the eyes, Jordan would assure her, in that rich, hominy-grits drawl, "Ma'am, you showed real courage — po' judgment, but *real courage!*"

From there, he'd work his way through progressive stages of difficulty until he ended up posing a saccharine tablet, about the size of a BB for a kid's air rifle, homing in upon it with the camera on a closed circuit TV system, so the audience could see what was going on, and doing that whip-crack thing with his wrist that demolished the tiny target with yet another hurtling blob of paraffin wax.

I've never quizzed friend Jordan on the details of his wax loads and you're right in regarding that as regrettable. Be that as it may, it is possible to put in some monstrously impressive pistolry with the things, although it's awfully helpful if you happen to be Bill Jordan; so few of us are...

Lee Precision makes this die set, with carbide sizer, for the 10mm Auto and furnishes it complete with load data and dipper-type powder measure and shell holder.

Colt's Delta Elite replaces the usual rampant Colt stock medallion with a circled red triangle emblem.

THE LOCK-BACK TECHNIQUE

A situation can come to pass in which you want or need to establish a series of reload combinations that will *just barely* work the action of an autoloading pistol, stopping further increases in the powder charge pretty close to that delicate point.

As an example in point, we have the 10mm Auto cartridge, initially developed for the Bren Ten pistol. In the earlier Eighties, there was a vast lot of interest in gun and cartridge combo, perhaps amplified by the scarcity of the guns and even more so by the shortage of magazines to fit them.

I've received phone reports, in mournful tones, from Bren Ten owners who finally acquired the gun of their dreams, bought a box of factory ammo for it and made their blithe way to the nearest shooting facility, only to have the gun disassemble itself in the course of setting off the first magazine full of cartridges. No injuries, no structural damage, but damned disquieting, for all that.

It certainly gives one pause to think, when assigned to develop some load data for the gun and its cartridge. Given some badly needed and warmly appreciated help from old friend Marty Liggins, back at the Accurate Arms ballistic lab, we worked up a series of loads for the gun/cartridge that did not exceed the performance of the factory load to any appreciable extent.

I will confess there have been times when I've taken some amount of urchin glee in working up loads that go well beyond the performance levels of factory loads. It is fairly easy to do with cartridges such as the .38 Colt Super or even the latterday .357 magnum. But this was one instance where, if I could just make the autoloading pistol function, that was all I wanted and I would be glad to stop further increases right at that point.

Look at the situation thoughtfully for a moment. Walk around it and kick the tires. Upon discharging the last round of live ammo, many autoloaders — including the Bren Ten — lock their slides to the rear.

With squib loads, mustering recoil insufficient to work the action, the slide may recoil just far enough to start extracting the empty, then go crashing forward to produce a stoppage in which the empty may get slammed back against the action to ding its mouth and decree its retirement from further reloading.

You need to muster enough recoil to eject the spent case. That can be regarded as the bedrock minimum. It also can be regarded as desirable if the slide is given enough impetus to boot it on back to lock up after the final shot.

So here is how you do it: With the given powder, select a charge weight that is quite obviously too low. That can be iffy, as we've already seen. You are herewith reminded that I never said it would be easy, nor simple — nor, for that matter, iron-clad safe. You venture in harm's way when you develop experimental loads and that means you have to accept and live with any consequences that come along.

In a NGNA situation — new gun/new ammo — I follow certain procedures that have been established over thirty-odd years in the game at hand and, yes, some years were odd, indeed.

First, by looking through the barrel, I verify that one can see daylight coming in from the far end. That has been my m.o. for a gathering number of years. Back in the early Sixties, it paid off quite usefully. A test rifle in an exotic new caliber had been made up and was being passed about from gunwriter to gunwriter.

In time, it came up my turn at bat and the gun came to me

After firing the last round in the magazine, auto pistols, such as this Bren Ten, lock the slide open. As discussed, that can be used as a means of working up loads that are just barely capable of functioning.

Norma's #601001 10mm bullet is a FMJ and the jacket is pure gilding metal, as shown here by the right-hand bullet's complete lack of interest in the block of magnetized iron in the lower right corner of the photo.

from one of the really big names in the business at that time. Feeling a little self-conscious, I made my routine bore-check and beheld no more light than you'd expect to find in the center of a coal briquet. Checking further, I found a jacketed bullet lodged with memorable tenacity in the bore about 1.5 inches from the chamber.

In writing up that particular *wunderkind cartouche,* I sent in copy reporting upon results with 3.5 grains of Unique behind a specified bullet. In those days, I wrote my copy in Wisconsin and blew a fortune on Air-Mail Special-Delivery most months to get it to the publisher in New York in time for the deadline. I never had a chance to see, let alone correct, the typeset galleys. The load came out in the magazine, calling for 5.3 grains of Unique, not my original suggestion of 3.5 grains.

The cartridge in question never quite made it off the launching pad and I continue to regard that as a boon from a benevolent deity.

Tacking back to the original topic, when firing a new auto pistol for the first time, it has come to be my practice to chamber a single round, then remove the magazine. Taking care to wear a sturdy pair of shooting glasses, along with the usual earmuffs, I fire the first shot, watching out of the corner of my eye to see where the brass goes. I retrieve the empty case, examine it thoughtfully and, it if looks okay, I load two in the magazine, chamber one, get a good solid grip on the gun and fire it.

If the second round feeds and chambers without noteworthy incident, I'm inclined to assume the gun functions in a satisfactory manner, unless later events suggest otherwise.

Why the super-cautious methodology? On assorted memorable past occasions, I've encountered autos that left the empty cases with really alarming bulges about the head; even one that blew the head clean off. At other times, I've encountered semi-automatic pistols — nominally so, at least — with delusions that they were submachine guns. If you shoved a fully loaded magazine into the butt and squeezed off your first shot, you got a burst of full-auto fire, seeming as if it would never get that blasted magazine empty. You are more than welcome to adopt these precautionary procedures as your own, with no need to remit royalties.

The lock-back technique — thought I'd never get to it, didn't you? — is a matter of putting one round in the magazine, releasing the slide to chamber it, firing the shot and finding out whether it mustered sufficient recoil to drive the slide rearward far enough to lock the action open. If you've selected your starting charge with sufficiently prudent diffidence, the slide will not lock back and the ongoing drill is just a matter of increasing the charge weight by small increments — 0.2-grain at a time, perhaps — until you reach the point where the recoil impulse is just barely enough to lock the slide to the rear upon firing the single round.

Whether or not you choose to go on exploring the performance of still heavier charges is a question for you to decide. In the example of the Bren Ten, I terminated further increases of the powder charge right about at the point of attaining the lock-back for reasons outlined earlier.

By the by, the reason for removing the magazine before firing the first shot out of new gun with new ammo is to give the high-pressure powder gas a place to vent itself, should the case head rupture. Such a mishap can occur and it has occurred to me on at least one memorable occasion. The ammo in question was not a pie-eyed reload; it was a factory load of excellent credentials. The culprit in that instance was a barrel whose feed ramp had been relieved not wisely but too well. Even with the magazine removed, as it was, the escaping gas mustered more than enough force to crack a really lovely set of Jay Scott stocks in simulated pearl finish. Was I glad didn't have a magazine in place at the time? You can bet your beloved bippy I was!

The Hensley & Gibbs #938 mould turns out a cast bullet with a conical point at typical weights of about 175 grains and often shows highly gratifying accuracy.

Freedom Arms makes this uncommonly sturdy and powerful single-action revolver in calibers such as .44 magnum and the highly energetic .454 Casull.

COPING WITH THE HOT & HAIRY ONES

There are some handgun cartridges that can be worked up to really spooky pressure levels. Examples such as the .454 Casull and the .451 Detonics magnum come to mind, as do quite a few of the JDJ wildcat calibers for use in the T/C Contender or the Remington Model XP-100.

The .454 Casull, despite its caliber designation, works best with bullets of .451-inch diameter if jacketed, or .452-inch if cast. Designer Dick Casull did a lot of his initial development by using the standard .45 Colt, aka .45 Long Colt, case. He also investigated the possibilities of duplex and even triplex powder charges; a procedure I hesitate to mention and do so solely for the sake of adding a friendly Dutch-uncle warning *not* to follow his example. If a rattlesnake has rabies, it is unwise to pat it atop the head in a friendly manner, and working with two or three different powders in the same powder charge comes right along into the same category of unwarranted intrepidity.

In later years, Casull went on to order some custom cases, still basically the same .45 LC in configuration, but with a slightly greater case length, to prevent use of the given loads in conventional .45 LC guns. Still more recently, the .454 Casull case was produced with primer pockets to accept the smaller primer diameter. Small rifle primers are suggested for use in it, rather than the small pistol type, with their thinner primer cups.

Casull has abandoned the multi-propellant powder charges long ere this. He does most of his loading with Winchester 296 powder, using bullets up to 300 grains, perhaps even heavier. The cartridge operates in a big revolver made of stainless steel, with a five-shot unfluted cylinder, as manufactured by Freedom Arms. Other makers have produced revolvers in the same chambering, but their guns tend to fall somewhat short of handling the ungodly pressures generated by the Casull loads, and I would counsel restricting use of the cartridge to the Freedom Arms revolvers exclusively.

If you buy a .454 Casull revolver, Freedom Arms will provide suitable suggested load data. The same gun now is available in .44 magnum and, one presumes, the .44 magnum cartridge can be force-fed to produce some impressive ballistics, when used with it. I have not worked with the Freedom Arms revolver in .44 magnum, so cannot offer any suggested loads for use with it, even if I were inclined to do so and I don't think I am.

In the .454 Casull, I've made up and tested various loads on my own, but have never equalled Casull's performance level. I recall times, earlier in my career, when I didn't seem to know the meaning of fear and was painfully vague about several other words. Candidly, I'm grateful I didn't put hands upon a .454 Casull in those days; not even if it was made by Freedom Arms.

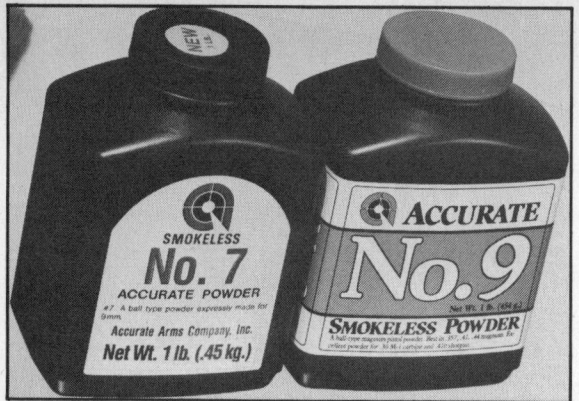

Accurate Arms Nos. 7 and 9 powders are capable of turning up some impressive ballistics but, as with any other powder, require a degree of cautious restraint in dispensing powder charges to avoid unpleasant and/or unfortunate consequences.

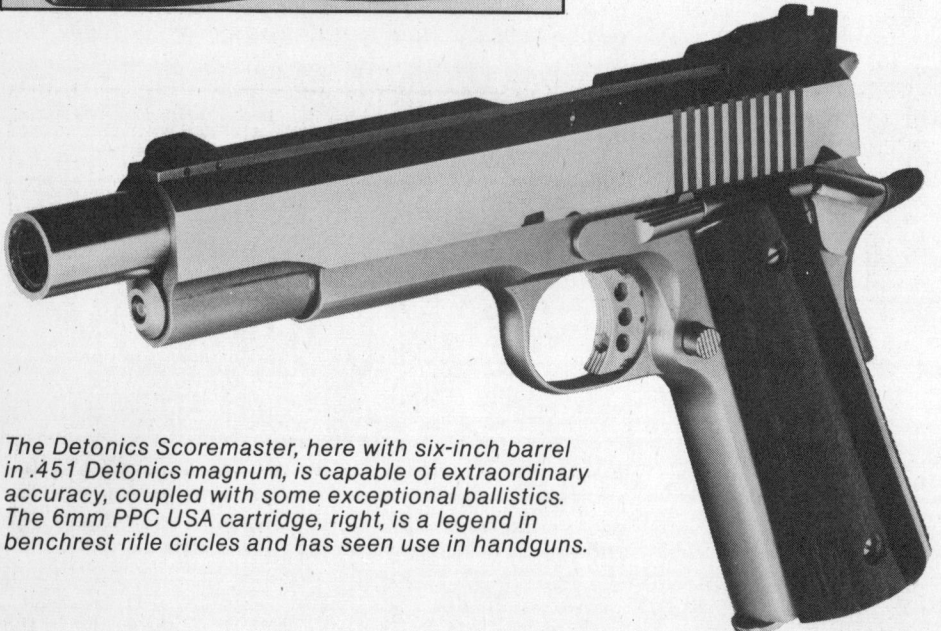

The Detonics Scoremaster, here with six-inch barrel in .451 Detonics magnum, is capable of extraordinary accuracy, coupled with some exceptional ballistics. The 6mm PPC USA cartridge, right, is a legend in benchrest rifle circles and has seen use in handguns.

The .451 Detonics magnum may have about the same construction about the head as the .45 Winchester magnum, I'm not certain. It does have a head that is considerably stronger, more resistant to high pressures than the .45 ACP, a notoriously wimpish case design restricted by SAAMI specifications to peak pressures not higher than 19,900 c.u.p. — and preferably well below that figure. The problem, you see, lies in the portion of the head that hangs out over the feed ramp unsupported by the chamber walls, when the cartridge is locked up and ready to fire. There is nothing between the fury of the burning powder and the mechanism of the pistol but some small amount of brass and that is not a metal noted for its toughness and great strength. Should the case rupture at that point, all hell is well and truly out for noon, as suggested by my experience with the ill-fated Jay Scott stocks mentioned earlier.

The .451 Detonics magnum — or D-mag — with its heavier head construction, is able to work safely at pressures approaching the limits of the 9mmP and .38 Colt Super cartridges, which is to say somewhere up in the 30,000 c.u.p. area. The fact that it's stronger should not be taken to imply that it's indestructible. Not only is it possible but hellishly easy to load more powder into the D-mag case than it can handle with safety.

When using large pistol primers in the D-mag — even at pressures one might term reasonably moderate — the appearance of the fired primer can be downright alarming. Switching to large rifle primers results in a healthier-looking spent primer, but it is well to reduce the charge weight slightly when changing primers, as the rifle primers usually carry a more energetic wafer of priming compound.

Load data, as supplied by Detonics when I got my D-mag Scoremaster in 1985, listed several powders I would have regarded as too fast for the characteristics of the cartridge. These included Hercules Bullseye, which they took to ballistics as high as 1345/743 for 185-grain bullets. The Accurate Arms powders were still fairly new at the time and I tried both AA-7 and AA-9 with the H&G #938 cast bullet at a weight of 175 grains, with fairly good results.

With 18.3 grains of AA-9 behind the 175-grain conical-point H&G #938, firing off the sandbagged benchrest at twenty-five yards, I got a five-shot group spanning hardly as much as one inch of maximum spread between centers, with ballistics of 1348/745. Shifting to AA-7 powder and trying a series of progressively increasing charge weights, I got:

15.0 grains AA-7, 175-grain H&G #938	1410/773
15.7	1420/784
16.6	1520/898
17.0	1557/942

It's pertinent to note that the test session just described took place on a hot day in August, with ambient tem-

GUN DIGEST BOOK OF HANDGUN RELOADING

The .38-45 Clerke cartridge was made by necking the .45 ACP case to accept 9mm bullets, but was tethered to the ACP's low pressure ceiling. By using .451 D-mag cases, it is possible to coax better ballistics from it...

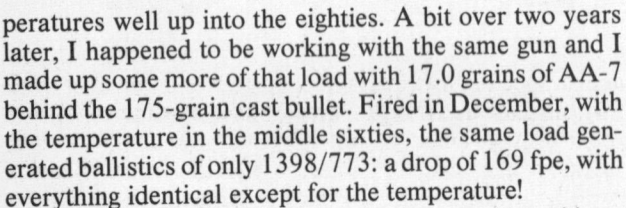

...but control your enthusiasm! This fired D-mag case shows an alarmingly clear outline of the feed ramp: a certain indication that pressure was much too high.

Here are two more D-mag empties, with some bulge over the feed ramp and large pistol primers that show more pressure signs than you'd want to see!

peratures well up into the eighties. A bit over two years later, I happened to be working with the same gun and I made up some more of that load with 17.0 grains of AA-7 behind the 175-grain cast bullet. Fired in December, with the temperature in the middle sixties, the same load generated ballistics of only 1398/773: a drop of 169 fpe, with everything identical except for the temperature!

That brings up what I regard as an important consideration, namely the temperature at the time of firing. What actually counts is the temperature of the powder charge. If you've been firing fairly rapidly over an extended while, getting the barrel and chamber heated, then chamber a round and allow it to remain in the chamber for more than the usual few seconds, you can expect that shot to turn up higher velocities — and higher pressures.

Earlier, we were speaking of the beguiling temptation to try nominal shotgun powders in handgun reloads. Winchester has two shotgun powders for which handgun load data is somewhere between scarce and non-existent: 540 and 571. With the usual cautionary note that the position of a powder in a table of comparative burning speeds means nothing other than the fact that someone thought it belonged about there, I'll note that I've done some amount of experimental work with both 540 and 571. Each has given some good results, without unfortunate consequences. I wish Winchester would quote some load data for both for use in handguns and have spoken to their personnel about the matter, thus far without results.

Using the 185-grain Sierra #8810 match bullet in the .451 D-mag Scoremaster — which has a six-inch barrel — I got the following results with progressively increasing charges of 571 powder:

14.8 grains	1304/698
15.5	1371/772
16.0	1424/833
16.5	1440/851

When fired at 68 degrees Fahrenheit, none of the loads showed any indication of excessive pressures, in my Scoremaster. I would regard 16.5 grains of 571 as maximum for that bullet/cartridge combination, to be approached with suitable caution, from the underside. Obviously, that applies with special emphasis if the temperature is higher than 68F/20C.

The fact that a case is stronger in its head construction, as noted earlier, should not be taken to imply that it is utterly indestructible. If fully supported by exceptionally strong chamber walls, as in the .454 Casull, one sometimes can get away with some ferocious pressures. If on the other hand, the head hangs out over a feed ramp, as it does with the .451 D-mag in an auto pistol, the margin for error is reduced sharply.

I can state without fear of successful contradiction that it is possible to put too much 571 powder into the .451 Detonics magnum case. Increasing the charge weights by only a little bit over the 16.5-grain level, I began to get a thoroughly visible amount of engraving in the case head to

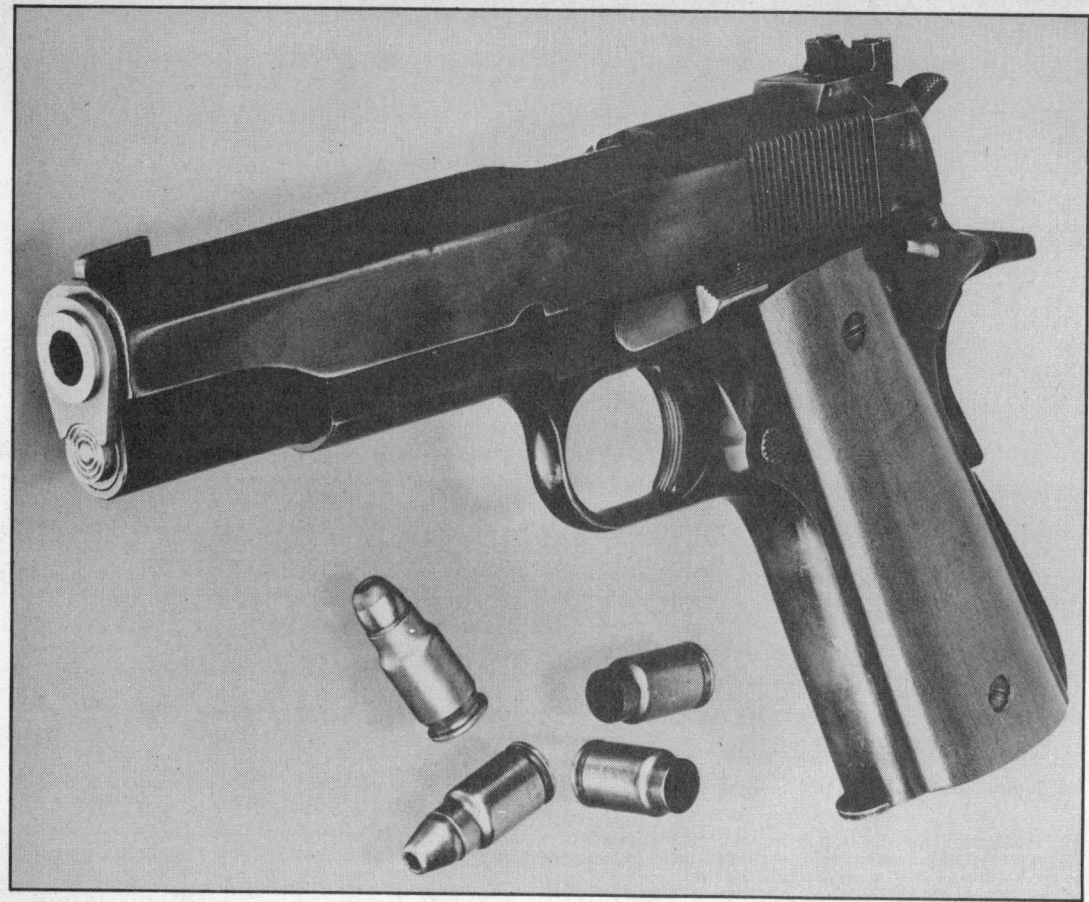

My Bar-Sto .38-45 barrel is set up for this nice old WWII Remington-Rand auto. It has been fitted with target sights and a beaver-tail tang on the grip safety. Using cases made from D-mag brass, it can hit 600+ foot-pounds.

show where the feed ramp began. For obvious reasons, I'm not about to quote the charge weight, but I can confide that it occurs about — or slightly before — the point where you're getting the H&G #938, weighing 178 grains in the casting alloy used at that time, out of the muzzle at 1603/1016.

By way of providing a bit of realistic perspective, a recent research project involved firing thirty rounds of Federal's #44C factory load in .44 magnum out of a Ruger Super Redhawk with a 9.5-inch barrel. The load carries a 220-grain bullet and the thirty-shot average came to 1366/912. Keep that figure of 912 fpe in mind as the discussion continues, please!

There is one particular wildcat cartridge for handguns that has fascinated me far beyond expectable levels for the past many years. I refer to the .38-45 Clerke, developed by John A. "Bo" Clerke. Initially, it involved necking down the .45 ACP case to accept jacketed bullets of .355-inch diameter or .356-inch in cast bullets, despite the nominal caliber designation. The intent was to provide owners of the Model 1911-type pistols in .45 ACP with a spare barrel that could be installed as a subcaliber conversion, using the same magazine and ejector.

I put a lot of work into the .38-45 one year or the other and walked away muttering darkly to myself. The pistol is recoil-operated and you need some amount of recoil to make the action function. Shifting to the lighter bullet cuts down on the recoil (good news), but the lighter recoil fails to work the action (bad news) all too often. If you try to increase the powder charge, for purposes of getting more recoil to work the action, you collide with another hard fact of life: Being based upon the .45 ACP case, the .38-45 is tethered to the same pressure ceiling of 19,900 c.u.p. and any show of undue intrepidity is apt to get you a blown case head and all manner of regrettable grief.

I continued to speculate upon the possibilities of making up the .38-45 cases by using the .451 D-mag as parent brass. With its stronger head construction, it should be possible to obtain substantially better ballistics in a reasonable degree of safety.

The .451-D-mag has a case length of .945- to .947-inch, as compared to the nominal length of .898-inch for the .45 ACP. In point of cold, Real World fact, damned few .45 ACP cases are actually that long and most are some amount shorter, but let's not let me get started on *that* particular harrangue.

Put through the .38-45 set of case-forming dies, the D-mag comes out at a length of about .957-inch. With the .38-45 Bar-Sto barrel removed from the pistol, I used it as a gauge to work out the trim length of the new case by patient trial and error, settling upon .904-inch as the length that put the case neck nicely in touch with its mating ledge in the chamber, for useful support at the time of initial fire-forming.

Confidentially, I've always regarded case trimming as a tedious chore, but RCBS recently added a power conver-

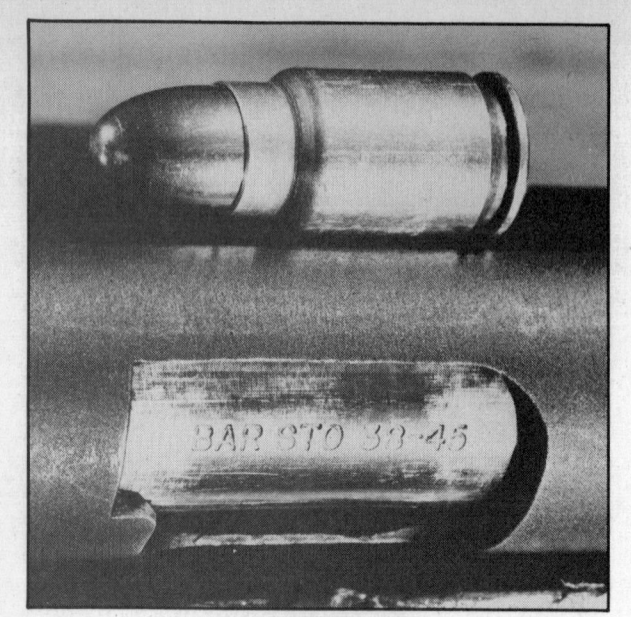

Within fairly recent times, Irv Stone has been putting the brand and caliber on that portion of the barrel visible in the ejection port for clear identification.

When you make a .38-45 Clerke case from D-mag brass, you get what I call the .38-45 Hard Head, with apologies to the wildcat's creator, Bo Clerke.

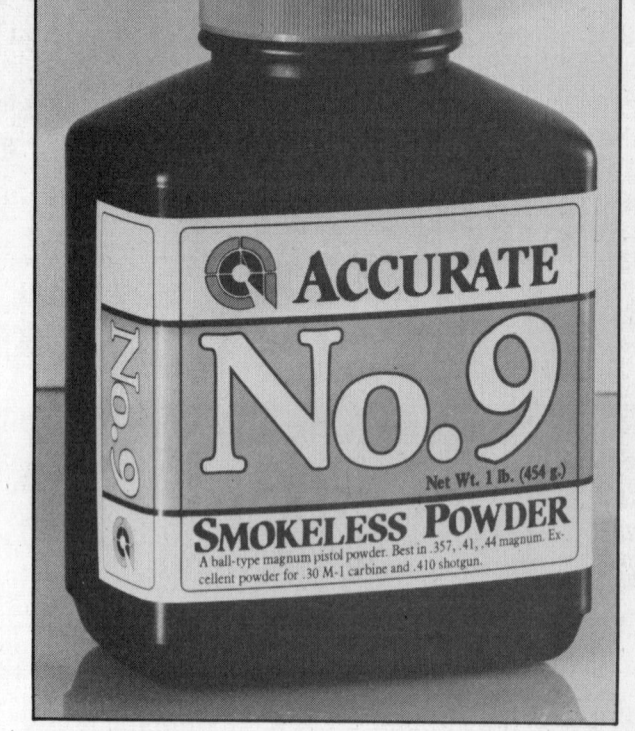

Accurate Arms No. 9 powder has been showing fairly good results in preliminary work with the .38-45 HH, particularly with the lighter bullet weights.

sion kit for their case trimmer. All you do is remove the usual crank and substitute a small washer and hex-socket screw. The kit comes with a matching section of hex stock, relieved to a ball shape at each end. You put that into an electric hand drill and tighten it in place with the chuck wrench. After that, it's just a matter of securing the case in the collet of the trimmer, putting the end of the hex bit into the socket and giving the trigger switch of the drill a twitch as you apply pressure against the case neck. It almost makes case trimming fun!

With the cases formed, trimmed, inside/outside neck-deburred and put through the full-length resizing die of the regular loading set, followed by a pass over a homemade expanding plug in the Lyman M die, I checked and found the empty cases would feed out of the magazine and into the chamber quite nicely. My .38-45 Bar-Sto barrel is set up for use in a WWII vintage Remington-Rand auto, obtained by way of the Director of Civilian Marksmanship (DCM) and gussied up considerably since its long-ago GI days. In addition to that rig, I also have a ten-inch bull barrel for the T/C Contender that is chambered for the .38-45 Clerke. There are not many of those around and Warren Center has the only other one I know of.

The great joy of having the Contender barrel was that it gave me a vehicle in which to conduct the first test-firing of experimental loads, boasting substantially greater action strength than did the auto pistol. If I judged the load to be acceptable in performance, I could go on to try it in the auto.

By way of a special designation for the .38-45 Clerke on the D-mag case, I ran several possibilities across my mental screen, one example being the 9mm-.451 D-mag, but ruled out all such on grounds that, as the Lady Alisande put it in Twain's *A Connecticut Yankee in King Arthur's Court,* they did not seem to fall trippingly from the tongue. What I settled upon, as a name, was the .38-45 Hard Head; enough of a mouthful in its own right, but better than anything else that occurred to me.

I'm pleased to report that the HH performs every bit as well as I'd hoped it would; even better, if anything. I've begun to work with it only in quite recent times and have not plowed anything close to the amount of research into it I hope to, time permitting. Even so, I'll quote the dope on a few of the more promising loads for your possible interest.

The H&G #333 bullet was mentioned earlier, in connection with multi-projectile loads for cartridges such as the .357 magnum. Loaded just one at a time, it often groups uncommonly well, with recoil so feeble as to trigger gales of raucous mirth. I put up a few of these in the HH case ahead of 10.1 grains of Winchester 540 powder, finding I

Winchester 540 worked well with the H&G #333, as discussed in the text here. With the heavier bullets, Winchester 296 has been showing the most promise.

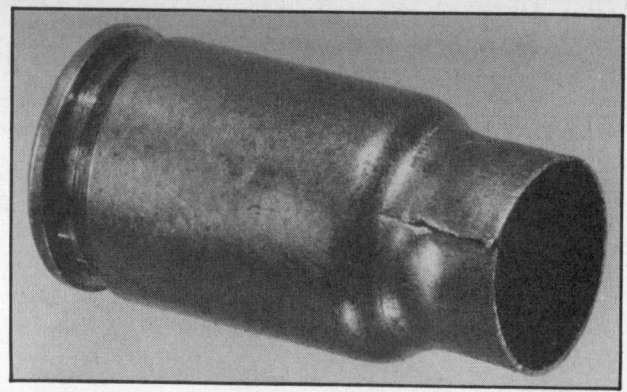

Unless you anneal the neck and shoulder area of the D-mag brass, you tend to get split necks such as this at the time of original case-forming or fire-forming.

Here's one of the .38-45 HH loads with the H&G #333 bullet ahead of 10.1 grains of 540 that converts the old autoloader into a pump-action repeater; see text.

had to seat the front face of the bullet flush with the case mouth. My .38-45 Bar-Sto barrel has little, if any, freebore ahead of the chamber.

The load just described comes out of the Contender averaging 1758/466 and grouping down around one inch at twenty-five yards. The Contender barrel, I should note, has Conetrol scope mounts and a 2X Leupold handgun scope topside.

In the Rem-Rand auto, the same load continued to group extremely well, with its point of impact right at the point of aim. That really was a pleasant surprise. Average ballistics in the auto were 1581/377 and, as you may have been suspecting, it did not work the action. In fact, it just barely twitched the slide and the case remained in the chamber.

Giving the matter some thought, I was delighted it'd worked out that way. Let me explain that you buy the D-mag cases from Richard Niemer, up at Detonics, (13456 S.E. 27th Pl., Bellevue, WA 98005) at $32.75 per hundred, plus shipping costs. You then go on to put a considerable amount of time and effort into turning cases at something like forty cents apiece into their final format. If you lose one, it tends to tarnish your outlook for the rest of the day!

So that first load turned the nominal self-loading pistol into a pump-action repeater which, upon being fired, retained its precious case in place until you got ready to shuck the slide back and catch it.

Some decades back down Memory Lane, a few friends and I formed what we called the BDSA, for Brandon Dump Shooting Association, in dubious honor of the tiny Wisconsin community whose dumping ground was our battlefield. We spent a lot of time in trying to pot the large and voracious dump rats that teemed thickly amid the malodorous refuse. All that good fun — and I certainly thought it was fun! — came to a screeching halt when the Brandon city fathers heard of a marvelous new compound called Warfarin, developed by the Wisconsin Alumni Research Foundation, hence the name. Warfarin proved to be highly irresistible to rats, including those of the Brandon dump, and likewise fatal shortly thereafter. They salted the dump with Warfarin and, within the space of a week or so, deprived us of all those eminently challenging, scurrying gray targets.

What I'm saying is the .38-45 HH load just described would have been the great performer of all time on dump rats. With a dead-flat frontal area barely shy of .100 square inch, moving out at 1581 fps, the effect upon a small varmint should be impressively decisive. However, with the cases valued well up toward a buck apiece, it's obvious you don't want the gun to eject them and leave you pawing

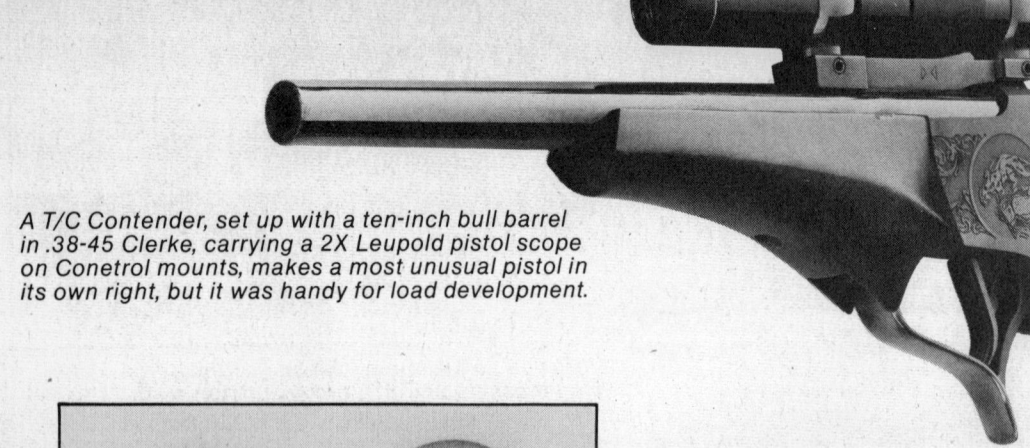

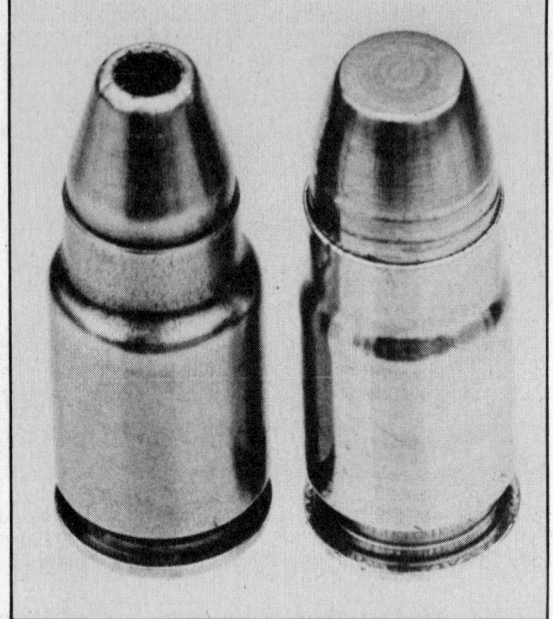

A T/C Contender, set up with a ten-inch bull barrel in .38-45 Clerke, carrying a 2X Leupold pistol scope on Conetrol mounts, makes a most unusual pistol in its own right, but it was handy for load development.

Above, H&G #333, Speer 88 and 200 JHPs; 115 and 124 Hornady; 124 Speer TMJ and 130 Sierra FMJ, all used in working up .38-45 HH loads. Left, a .38-45 HH is to the left of a .41 Avenger, developed by J.D. Jones.

through masses of orange peel and egg shells to retrieve them. Yes, I discovered the great BDSA load of all time, and only about three dozen years too late...

Nearly all of the rest of the loads functioned the action of the Rem-Rand quite reliably and, if you were about to ask, that of the Contender, as well. Here are the average ballistics for some typical loads, out of both barrel lengths:

	5-inch Rem-Rand	10-inch T/CC
10.1 gr 540/115 JHP	1387/491	1530/598
12.9 AA-9/88 JHP	1327/344	1472/423
12.9 AA-9/100 JHP	1336/396	1473/481
12.9 AA-9/115 JHP	1276/415	1462/546
17.0 296/124 TMJ	1499/618	1715/810
17.0 296/124 FMJ/FT	1511/628	1715/809
17.0 296/130 FMJ/RN	1474/627	1699/833
17.0 296/140 JHP (.357-in.)	(see note!)	1697/895

Getting to that note, just mentioned, I'd been firing each load in the Contender, before going on to try it in the auto and the Contender showed moderate resistance to case extraction on that load. Mindful that discretion is much the better part of valor, I refrained from trying it in the Rem-Rand.

It should be noted that the .38-45 Clerke cartridge has never been standardized as to dimensions, permissible peak pressures and a number of other pertinent particulars. I have another barrel for it, whose chamber is shorter in headspace by about .022-inch. The gun it fits has been modified since the original fitting, so that barrel is no longer readily usable. Even if it were, I'd be reluctant to try these loads in it due to the drastically smaller amount of powder space.

I neither recommend the loads just listed nor do I suggest that others try them. Rather, I bring all this up by way of demonstrating the thoughts and methodology of the experimental ballistician in the matter of exploring the possibilities of new and untried loads.

To conduct such research successfully and with a reasonable degree of impunity requires expenditure of a great deal of Factor J, meaning Judgment; *good* judgment. It also is most helpful if you have an Irish guardian angel watching over your welfare at all times. I think of mine as Mr. Mulligan... — *Dean A. Grennell, Life Member, BDSA*